# Ubiquitous Laplacian

An Introduction to Numerical PDEs with Applications in Data Science

## Progress in Data Science

**Series Editor:** Ding-Xuan Zhou *(The University of Sydney, Australia)*

---

*Published*

Vol. 2 *Ubiquitous Laplacian: An Introduction to Numerical PDEs with Applications in Data Science*
by Rongjie Lai and Xiangxiong Zhang

Vol. 1 *Equivariant and Coordinate Independent Convolutional Networks: A Gauge Field Theory of Neural Networks*
by Maurice Weiler, Patrick Forré, Erik Verlinde and Max Welling

Progress in Data Science – Volume 2

# Ubiquitous Laplacian

## An Introduction to Numerical PDEs with Applications in Data Science

Rongjie Lai
Xiangxiong Zhang

Purdue University, USA

World Scientific

NEW JERSEY • LONDON • SINGAPORE • BEIJING • SHANGHAI • HONG KONG • TAIPEI • CHENNAI • TOKYO

*Published by*

World Scientific Publishing Co. Pte. Ltd.
5 Toh Tuck Link, Singapore 596224
*USA office:* 27 Warren Street, Suite 401-402, Hackensack, NJ 07601
*UK office:* 57 Shelton Street, Covent Garden, London WC2H 9HE

Library of Congress Control Number: 2026001699

**British Library Cataloguing-in-Publication Data**
A catalogue record for this book is available from the British Library.

**Progress in Data Science — Vol. 2**
**UBIQUITOUS LAPLACIAN**
**An Introduction to Numerical PDEs with Applications in Data Science**



ISBN 978-981-98-1453-4 (hardcover)
ISBN 978-981-98-1454-1 (ebook for institutions)
ISBN 978-981-98-1455-8 (ebook for individuals)

For any available supplementary material, please visit
https://www.worldscientific.com/worldscibooks/10.1142/14349#t=suppl

Desk Editors: Nambirajan Karuppiah/Lai Fun Kwong

Typeset by Stallion Press
Email: enquiries@stallionpress.com

# Preface

This book is an introductory text intended for graduate students interested in computational mathematics. Some chapters are also suitable for advanced undergraduate students, as well as researchers and practitioners working in related areas. The book introduces numerical methods and selected applications of the Laplacian operator and the Poisson equation. Part I focuses on rectangular meshes and grids, drawing on the second author's lecture notes and projects on numerical PDEs. Part II covers triangulated surfaces and point clouds, derived from the first author's topic courses and projects.

We are grateful to Mr. William R. Francis, Dr. Hao Li, Dr. Chen Liu, and Dr. Xinyu Liu for reading Part I and providing valuable comments. We acknowledge Dr. Xinyu Liu for assistance with writing Section 4.1.5 and Mr. Emmanuel Gil Torres for his contribution to Section 4.4. Our sincere gratitude goes to Mr. Suyi Gao and Mr. Yineng Chen for their efforts in typing the lecture notes that form the basis of Chapters 5 and 6 in Part II. We also wish to thank all the students who participated in these courses or read these lecture notes for their thoughtful discussions, questions, and feedback, which have greatly improved the presentation and clarity of the material. We acknowledge the use of AI tools such as ChatGPT, Gemini, and Grok for assistance in preparing some of the figures.

Last but not least, we are grateful to the National Science Foundation for funding our past research projects that contributed to portions of this book. We are also grateful for the generous support provided by the Book Initiative of the Office of the Provost at Purdue University.

Rongjie Lai and Xiangxiong Zhang

*Department of Mathematics, Purdue University*

# About the Authors

**Rongjie Lai** received his B.S. in Mathematics from the University of Science and Technology of China, his M.S. from the Academy of Mathematics and Systems Science at the Chinese Academy of Sciences, and his Ph.D. in Applied Mathematics from University of California, Los Angeles. He is currently a professor in the Department of Mathematics at Purdue University. His research focuses on computational mathematics, with an emphasis on imaging, data science, and the analysis of manifold-structured data through variational PDEs, computational differential geometry, and machine learning. His work also extends to the mathematical foundations of deep learning. Dr. Lai received an NSF CAREER award in 2018.

**Xiangxiong Zhang** received his B.S. in Mathematics from the University of Science and Technology of China and his Ph.D. in Mathematics from Brown University. He is currently a professor in the Department of Mathematics at Purdue University. His research is focused on applied and computational mathematics with emphasis on design and analysis of computational methods for problems including numerical PDEs and numerical optimization.

# Contents

Chapter 1

# Introduction

The Laplacian operator $\Delta = \partial_{xx} + \partial_{yy} + \partial_{zz}$ naturally emerges in quite a few different types of partial differential equations (PDEs), which are model equations in physical sciences and engineering applications, as well as data science applications.

## 1.1 Linear partial differential equations

Most classical PDEs originate from modeling physical phenomena in science and engineering problems. One thing we should always keep in mind is that these equations are chosen *models*, which are supposed to be valid, suitable or acceptable only under certain assumptions or only within certain context. For beginners learning numerical methods, these equations can be assumed as given and *well-posed*, which roughly means that the equation has a unique and well-behaved solution. Classical equations were mostly derived from *physical principles*, e.g., fluid equations were derived from conservation laws of mass, momentum, etc. Below are a few examples of one dimensional simplified linear PDEs for which a solution $u(x,t)$ with the space variable $x$ and the time variable $t$ satisfying a given initial condition $u(x,0) = u_0(x)$ and the PDE:

(1) *Convection equation:* $u_t = u_x$.
(2) *Diffusion or heat equation:* $u_t = u_{xx}$.
(3) *Dispersion equation:* $u_t = u_{xxx}$.
(4) *Wave equation:* $u_{tt} = u_{xx}$.
(5) *Schrödinger equation:* $\mathtt{i}\, u_t = u_{xx}$ where $\mathtt{i}$ is the imaginary unit.

For the multi-dimensional linear PDEs with a solution $u(x,y,z,t)$, the Laplacian operator emerges in diffusion, wave, and Schrödinger equations:

$u_t = \Delta u$, $u_{tt} = \Delta u$, and $\mathrm{i}u_t = \Delta u$. The diffusion or heat equation with a source term $f(x, y, z)$ can be written as $u_t = \Delta u + f$. The steady state of such a time-dependent problem is defined as a time-independent solution $u(x, y, z)$, and $\frac{\partial}{\partial t}u(x, y, z) \equiv 0$ implies that it satisfies the *Poisson* equation

$$-\Delta u = f, \tag{1.1}$$

which we will focus on in most chapters of this book.

## 1.2 Random walks and the heat equation

For a better understanding of the numerical methods, eventually one must understand the origin of these equations, which often plays a critical role in designing numerical schemes. There are quite a few different ways to derive the heat equation. Here we consider one interesting derivation of the heat equation from random walks in [Evans (2012)], which is based on Einstein's 1905 Brownian motion paper. The one-dimensional random walks are defined on a space-time lattice consisting of $(j\Delta x, n\Delta t)$ where $j$ is an integer and $n$ is a natural number. Initially a particle locates at position $x = 0$ and at time $t = 0$, then it moves to either the left or the right by $\Delta x$ with equal probability at each time $n\Delta t$. If $p(j, n)$ denotes the probability that the particle locates at $(j\Delta x, n\Delta t)$ on the space-time lattice, then it satisfies $p(j, n+1) = \frac{1}{2}p(j-1, n) + \frac{1}{2}p(j+1, n)$ thus

$$p(j, n+1) - p(j, n) = \frac{1}{2}[p(j-1, n) - 2p(j, n) + p(j+1, n)].$$

Under the assumption $\Delta t = \frac{1}{2}\Delta x^2$, it can be rewritten as

$$\frac{p(j, n+1) - p(j, n)}{\Delta t} = \frac{p(j-1, n) - 2p(j, n) + p(j+1, n)}{\Delta x^2}.$$

Suppose there is a smooth function $u(x, t)$ such that $p(j, n) = u(j\Delta x, n\Delta t)$, i.e., the point values of $u(x, t)$ at the lattice give the discrete probability $p(j, n)$. For fixed $j$ and $n$, let $j\Delta x = x$ and $n\Delta t = t$, then we obtain

$$\frac{u(x, t+\Delta t) - u(x, t)}{\Delta t} = \frac{u(x-\Delta x, t) - 2u(x, t) + u(x+\Delta x, t)}{\Delta x^2}. \tag{1.2}$$

By Taylor expansion, the right hand side of (1.2) gives an approximation to the second order derivative

$$\frac{u(x-\Delta x, t) - 2u(x, t) + u(x+\Delta x, t)}{\Delta x^2} = u_{xx}(x, t) + \mathcal{O}(\Delta x^2),$$

which is called *centered finite difference* and will be explained in Chapter 2.

With Taylor expansion in time, (1.2) becomes

$$u_t(x,t) + \mathcal{O}(\Delta t) = u_{xx}(x,t) + \mathcal{O}(\Delta x^2).$$

If taking the limit $\Delta x \to 0$ and $\Delta t \to 0$ with the restriction $\Delta t = \frac{1}{2}\Delta x^2$ satisfied, then the limit gives the heat equation $u_t = u_{xx}$.

It is interesting to see that such a derivation of the heat equation involves the centered finite difference, which is the simplest and the most popular numerical method for approximating the Laplacian operator, as we will learn in this book. The same argument can be extended to the random walk in two dimensions on spatial vertices of an equilateral triangular mesh, see Section 6.3.3.

## 1.3 Numerical schemes

For general PDEs, usually either there are no exact solution formulae or the formulae can be demanding or difficult to compute. One practical goal of numerical methods for PDEs is of course to provide a computationally *tractable* way for generating some accurate approximations of the solution. Whether a computational method is *tractable* may vary, depending on the context of the applications and given computational resources. For example, in many applications, we need to solve (1.1) many times, and some numerical methods can be acceptable for solving (1.1) only once but might be too slow for solving (1.1) many times.

There are many popular numerical methods, which one may not have used but likely have heard of, such as *finite difference*, *finite element*, *finite volume* and *spectral methods*. As shown in Figure 1.1, approximations are obviously quite different in different numerical methods, which is however only a superficial way of understanding numerical schemes for PDEs. As a matter of fact, many of these different numerical methods can sometimes be regarded equivalent methods, especially for solving a one-dimensional problem. In Chapter 3, we will see that the finite difference method and a special implementation of the continuous finite element method are exactly the same for solving (1.1) on a rectangular domain. On the other hand, we will also learn the key fundamental difference between the finite element method and the traditional finite difference method in Chapter 3.

(a) In *Finite Difference* method, point values are used to approximate a function.

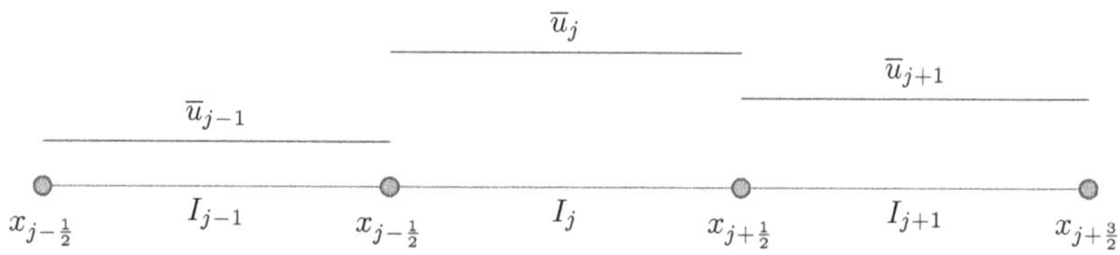


(b) In *Finite Volume* method, piecewise constants are used to approximate a function.

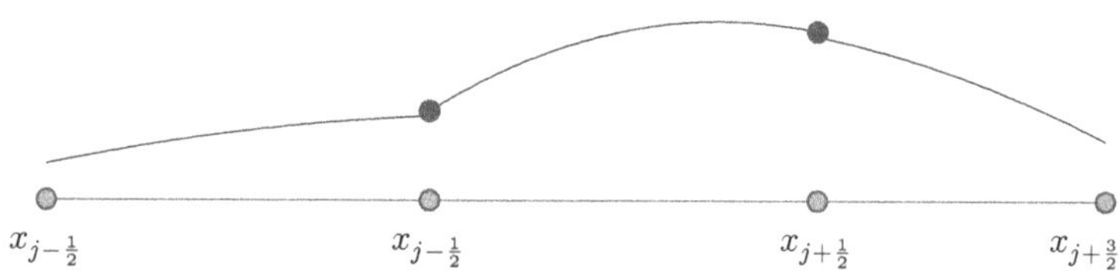


(c) In continuous *Finite Element Method*, a continuous piecewise polynomial is used to approximate a function.

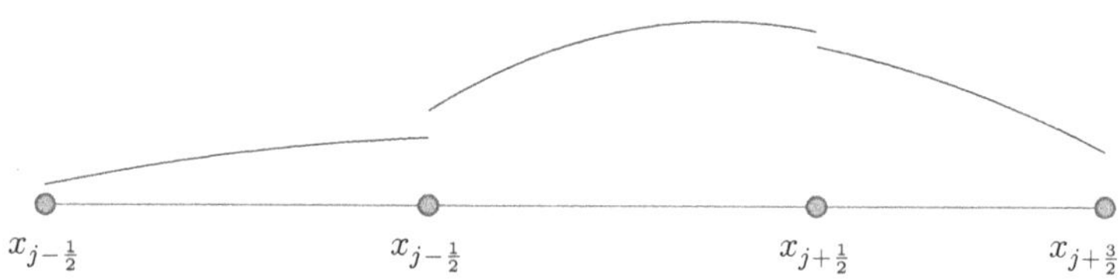


(d) In *discontinuous Galerkin* Finite Element Method, a piecewise polynomial without continuity constraint is used to approximate a function.

Fig. 1.1 An illustration of a few popular methods. The finite difference grid is usually uniform. The meshes for finite volume and finite element methods can be non-uniform.

## 1.4 Examples and applications of the discrete Laplacian

The discrete Laplacian operator, constructed from numerical schemes, is the main focus of this book. It arises from many applications. We list a few quick examples:

(1) The discrete Laplacian operator naturally emerges in solving a parabolic equation like the heat equation.

(2) For solving a more general elliptic equation like $-(a(x)u'(x))' = f(x)$ with a given coefficient function $a(x) > 0$, the discrete Laplacian is often quite useful as a preconditioner, see Appendix B.4.

(3) It naturally emerges in classical fluid dynamics equations. Section 4.1 contains a few fluid dynamics applications. For example, in the incompressible Navier-Stokes equations

$$\mathbf{u}_t + (\mathbf{u} \cdot \nabla)\mathbf{u} + \nabla p = \nu\Delta\mathbf{u} + \mathbf{f}, \quad \nabla \cdot \mathbf{u} = 0,$$

by taking the divergence of the both sides in the momentum conservation equation, we get a Poisson equation $\Delta p = \nabla \cdot (\nu\Delta\mathbf{u} + \mathbf{f} - (\mathbf{u} \cdot \nabla)\mathbf{u})$ for recovering the pressure $p$ from the velocity $\mathbf{u}$. See Section 4.1.2 for a quick heuristic explanation of why the Laplacian operator is hidden in the divergence free constraint.

(4) As will be shown in Section 5.5, the minimization of *Dirichlet Energy* naturally gives the Laplacian operator. This is why the Laplacian operator appears in minimization of many energy models. For example, efficient inversion of the Laplacian operator is quite useful in solving the nonlinear Schrödinger equation for a quantum-mechanical system, with one application shown in Section 4.3.

(5) The discrete Laplacian operator is also related to minimizing the discrete total variation norm (TV norm) used in various models for image processing since 1990s, as will be explained in Section 4.2.

(6) As a core mathematical tool in modern data science, *Optimal Transport* provides a robust way to quantify differences between probability distributions, with applications in machine learning, computer vision, etc. As will be shown in Section 4.4, the Laplacian operator also appears in solving the dynamical formulation of the Optimal Transport problem.

(7) The Laplace–Beltrami operator $\Delta_{\mathcal{M}}$ is the natural extension of the Laplacian operator $\Delta$ from a Euclidean space to a Riemannian manifold $\mathcal{M}$. The TV norm and optimal transport can also be extended from Euclidean spaces to curved domains, for which $\Delta_{\mathcal{M}}$ is needed. These manifold-based extensions and their computational realizations will be covered in Part II, e.g., image processing in Sections 8.1 and optimal transport in Section 8.2.

(8) In the spectral analysis of shapes, the eigenvalue problem for the Laplace–Beltrami operator,

$$\Delta_{\mathcal{M}}\phi_k = -\lambda_k\phi_k,$$

provides a set of intrinsic basis functions $\{\phi_k\}$ that encode the geometry of the manifold $\mathcal{M}$. These eigenfunctions form the foundation of spectral geometry and the functional maps framework used in shape analysis and nonrigid registration, as will be shown in Section 8.3.

(9) In conformal parameterization of genus-0 surfaces, $\Delta_{\mathcal{M}}$ appears in the harmonic map formulation, which seeks a map minimizing the harmonic energy subject to spherical constraints. This approach underlies many conformal mapping and spherical parameterization methods, as discussed in Section 8.3.1.

(10) The Laplace–Beltrami operator also governs committor functions in stochastic processes constrained on manifolds. The committor function $q$ satisfies an elliptic equation of the form

$$-\beta^{-1}\Delta_{\mathcal{M}}q + (\nabla_{\mathcal{M}}U)\cdot\nabla_{\mathcal{M}}q = 0,$$

where $U$ denotes the potential energy function and $\nabla_{\mathcal{M}}$ is the gradient operator on a manifold $\mathcal{M}$. Such formulations arise in modeling transition probabilities and reaction rates in molecular dynamics, which will be demonstrated in Section 8.4.

## 1.5 A reader's guide

Part I of the book focuses on solving $-\Delta u = f$ on a rectangular domain, while Part II of the book covers solving $-\Delta_{\mathcal{M}}u = f$ on a manifold approximated by a triangular mesh or a point cloud. Section titles start with * are materials that may be skipped for beginners. In appendices, we list some tools needed in the book, including the conjugate gradient method for solving a linear system in Appendix B, and ADMM splitting method for convex optimization in Appendix C.

Chapter 2 introduces the traditional finite difference for solving the Poisson equation $-\Delta u = f$ with the second order centered difference derived from formally approximating the second order derivatives, with two and three dimensional schemes included. For inverting the discrete Laplacian matrix, a simple method by eigenvalue decomposition is used, which is not the optimal way but nonetheless efficient enough on modern computers. Readers with college level calculus and linear algebra are fully capable of learning this chapter on their own, with some necessary linear algebra tools and facts listed in Appendix A. Toward a better understanding of the numerical methods, Chapter 2 also provides plenty of problems with some MATLAB codes given for easy implementation, which readers without prior coding experience can follow.

Chapter 3 is an introduction to the continuous finite element method (FEM). Unlike most FEM textbooks, we start with a description of the finite element method only on uniform rectangular meshes, so that beginners may focus more on the key ingredients of FEM and the rationale behind its success, without diving into sophisticated geometry. In this chapter, we introduce and employ more advanced concepts such as weak derivatives and Sobolev spaces, as well as key theorems from functional analysis in a straightforward manner that is accessible to beginners, without presenting their full rigorous versions. We focus on a special implementation of FEM so that it recovers exactly the same scheme as in Chapter 2, with however many advantages summarized in Section 3.15.

In Chapter 4, some classical and modern applications of discrete Laplacian are given, including fluid dynamics such as incompressible Navier-Stokes equations and Cahn-Hilliard phase field equations, TV norm minimization for image denoising, a special nonlinear Schrödinger equation and the optimal transport problem. The projects in Chapter 4 are suitable as research training for undergraduate students and beginning graduate students. For readers interested in trying these projects, it suffices to read only Chapter 2 and appendices, without reading Chapter 3.

Part II continues the progression from Euclidean domains to curved geometric spaces, emphasizing how geometry shapes analysis and computation. The guiding idea is that many physical, biological, and data-driven phenomena evolve intrinsically on manifolds rather than in flat space, and that the Laplace–Beltrami operator provides the unifying bridge between geometry and differential equations. The four chapters in this part follow a natural order, from foundational theory to discrete representations, with applications that connect geometry, analysis, and computation.

Chapter 5 introduces the essential concepts of differential geometry that underpin all subsequent developments. It defines smooth manifolds, tangent spaces, Riemannian metrics, and volume forms, and explains how these structures generalize the familiar notions of differentiation and integration from Euclidean calculus. The central objective is to construct intrinsic differential operators such as the gradient, divergence, and Laplace–Beltrami operator that enable the formulation of partial differential equations on curved surfaces. Readers who have studied vector calculus will find this chapter a conceptual bridge between classical analysis and geometric computation.

Chapter 6 turns to the discrete world of triangular meshes, the standard representation of smooth surfaces in computational differential geometry.

It explains how geometric quantities such as normals, edge lengths, and areas can be computed from mesh elements and how to design discrete analogues of the gradient, divergence, and Laplace–Beltrami operators that respect the intrinsic metric structure. Finite element method type discretizations are presented as core methodologies, together with discussions on convergence, symmetry, and numerical stability. By the end of this chapter, readers will understand how differential geometry, numerical analysis, and scientific computing interact in solving PDEs on mesh-based surfaces.

Chapter 7 extends these ideas to point-cloud representations, where only scattered samples of a manifold are available without explicit connectivity. This setting is particularly relevant for data obtained from 3-D scanners, simulations, or high-dimensional measurements. The chapter develops local approximation techniques estimate tangent spaces and recover intrinsic differential operators directly from the samples. These methods including a moving-least-squares based method and a local mesh method which emphasize geometric consistency and coordinate invariance, ensuring that the numerical schemes depend solely on the manifold's intrinsic geometry. The resulting framework provides a unified approach to solving PDEs on manifolds represented by unstructured data.

Chapter 8 demonstrates how the theoretical and numerical tools developed in the preceding chapters can be applied to a broad range of problems in geometry processing and scientific modeling. It discusses image processing on surfaces, optimal transport constrained by manifold geometry, conformal parameterization and spectral shape analysis, and the computation of committor functions in stochastic dynamics. Each example highlights how intrinsic operators enable computations that respect underlying geometric structure and how the Laplace–Beltrami framework unifies diverse applications across physics, biology, and data science.

# Part I

# Cartesian Grids and Meshes

# Chapter 2

# Finite Difference Method

The Poisson equation (1.1) is one of the most popular linear PDEs for understanding basic concepts of numerical methods. The numerical method introduced in this chapter is the traditional and classical way of constructing a finite difference scheme for approximating (1.1) on a uniform finite difference grid, which was known since no later than 1930s (see [Fox (1947)] and references therein), thrived since 1950s [Collatz (1960); Kantorovich and Krylov (1960)], and still a popular method nowadays [LeVeque (2007)].

Such a method is relatively easy for beginners to follow, since only calculus and linear algebra are needed. But it has many limitations, e.g., difficulties for constructing schemes for problems involving curved domains and variable coefficients, etc. On the other hand, all these difficulties can be dealt with by the finite element method in Chapter 3.

Most schemes in this chapter can be derived from the finite element methods on structured meshes, which will be introduced in Chapter 3. It has been well known that a finite element method can be made equivalent to the same finite difference scheme [Feng (1965)], though such an equivalence is often not emphasized in textbooks for finite difference methods. The two methods will be compared in Chapter 3. To this end, this chapter can be regarded as a warm-up for learning the finite element method in Chapter 3.

## 2.1 Second order finite difference approximations

For a smooth function $u(x)$, define the following finite difference operators approximating $u'(x)$ at the point $\bar{x}$ with a small number $h > 0$ as

- *Forward Difference*: $D_+u(x) = \frac{u(\bar{x}+h)-u(\bar{x})}{h}$;
- *Backward Difference*: $D_-u(\bar{x}) = \frac{u(\bar{x})-u(\bar{x}-h)}{h}$;

- *Centered Difference*: $D_0 u(\bar{x}) = \frac{u(\bar{x}+\frac{h}{2})-u(\bar{x}-\frac{h}{2})}{h}$.

Assume $u(x)$ is a smooth function, then by Taylor expansion,

$$u(\bar{x} \pm \frac{h}{2}) = u(\bar{x}) \pm \frac{h}{2}u'(\bar{x}) + \frac{h^2}{8}u''(\bar{x}) \pm \frac{h^3}{48}u'''(\bar{x}) + \mathcal{O}(h^4),$$

where $\mathcal{O}(h^4)$ denotes the quartic and higher order terms in the remainder, we obtain approximation error

$$D_0 u(\bar{x}) = u'(\bar{x}) + \mathcal{O}(h^2).$$

Similarly, we have

$$D_\pm u(\bar{x}) = u'(\bar{x}) + \mathcal{O}(h).$$

A classical second order finite difference approximation to $u''(x)$ at $\bar{x}$ is,

$$D^2 u(\bar{x}) := D_0[D_0 u(\bar{x})] = \frac{u(\bar{x}+h) - 2u(\bar{x}) + u(\bar{x}-h)}{h^2} = u''(\bar{x}) + \mathcal{O}(h^2).$$

Readers may verify $D^2u(\bar{x}) = D_+[D_- u(\bar{x})]$ as an exercise. Such a finite difference approximation to the second order derivative is often called *centered difference* or *second order finite difference.*

## 2.2 Dirichlet boundary conditions in one dimension

Consider solving the 1D Poisson equation with homogeneous Dirichlet boundary conditions:

$$\begin{cases} -u''(x) = f(x), & x \in (0,1), \\ u(0) = 0, \ u(1) = 0. \end{cases} \tag{2.1}$$

Discretize the domain $[0,1]$ by a uniform grid with spacing $h = \frac{1}{n+1}$ and $n$ interior nodes: $x_j = jh$, $j = 1, 2, \cdots, n$. See Figure 2.1. For convenience, define two ghost points $x_0 = 0$ and $x_{n+1} = 1$. Let $u(x)$ denote the true solution and $f_j = f(x_j)$. Let $u_j$ be the value of the numerical solution at $x_j$, i.e., $u_j$ is an approximation to $u(x_j)$. Since two end values are given as $u(0) = 0, u(1) = 0$, only the interior point values $u_j (j = 1, \cdots, n)$ are unknowns. After approximating $\frac{\mathrm{d}^2}{\mathrm{d}x^2}$ by $D^2$, we get a finite difference scheme

$$-D^2 u_j = \frac{-u_{j-1} + 2u_j - u_{j+1}}{h^2} = f_j, \quad j = 1, 2, \cdots, n \tag{2.2}$$

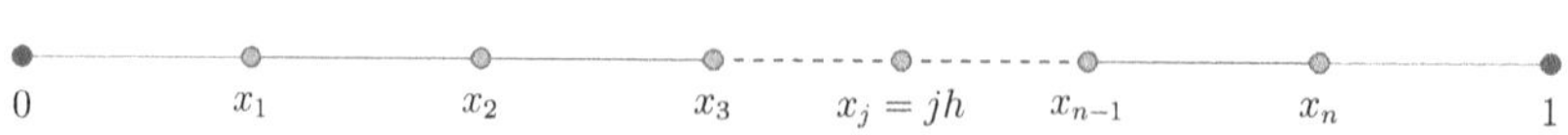


Fig. 2.1 An illustration of the discretized domain.

With the boundary values $u_0 = 0$ and $u_{n+1} = 0$ from the boundary conditions, we can rewrite the finite difference scheme (2.2) in the matrix vector form:

$$KU_h = F,$$

where

$$U_h = \begin{bmatrix} u_1 \\ u_2 \\ \vdots \\ u_n \end{bmatrix}, \quad F = \begin{bmatrix} f_1 \\ f_2 \\ \vdots \\ f_n \end{bmatrix}, \quad K = \frac{1}{h^2}\begin{pmatrix} 2 & -1 & & & \\ -1 & 2 & -1 & & \\ & \ddots & \ddots & \ddots & \\ & & -1 & 2 & -1 \\ & & & -1 & 2 \end{pmatrix}.$$

### 2.2.1 *Consistency, stability and convergence*

Using the scheme (2.2) as an example, we now can understand a few key concepts in numerical methods for PDEs.

The *local truncation error* (LTE) is defined as the residue after replacing the numerical solution by the true solution in the numerical scheme. For instance, the scheme (2.2) can be written as $\frac{-u_{j-1}+2u_j-u_{j+1}}{h^2} - f_j = 0$. With the Taylor expansion of the exact solution $u(x)$ at the location $x_j$, the LTE of this scheme is obtained after replacing $u_j$ by $u(x_j)$ as follows,

$$\begin{aligned}\tau_j &= \frac{-u(x_{j-1}) + 2u(x_j) - u(x_{j+1})}{h^2} - f_j \\ &= -u''(x_j) - \frac{1}{12}h^2 u''''(x_j) + \mathcal{O}(h^4) - f(x_j) \\ &= -\frac{1}{12}h^2 u''''(x_j) + \mathcal{O}(h^4) = \mathcal{O}(h^2),\end{aligned}$$

where $-u''(x_j) = f(x_j)$ is used.

Notice that the LTE is not the true error. Define a vector of the exact solution point values $\widehat{U} = \begin{bmatrix} u(x_1)\ u(x_2) \cdots u(x_n) \end{bmatrix}^\top$ where $\top$ denotes transpose of a vector, then the actual error of the scheme is $E_h = \widehat{U} - U_h$, which is a vector $E_h = \begin{bmatrix} E_1 \cdots E_n \end{bmatrix}^\top$. Define the LTE vector $\tau_h = \begin{bmatrix} \tau_1 \cdots \tau_n \end{bmatrix}^\top$, then it is related to the true error by the following relation if $K$ is invertible:

$$\tau_h = K\hat{U} - F, \quad KU_h = F \Rightarrow E_h = K^{-1}\tau_h.$$

We say the scheme $KU_h = F$ is *stable* if $\|K^{-1}\| \leq C$ for any mesh size $h$ (or equivalently any grid size $n$), where $\|A\|$ denotes the spectral norm of the matrix $A$. In other words, *stability* means that we can find a constant $C$ such that the matrix $K$ of size $n \times n$ satisfies

$$\|K^{-1}\| \leq C, \quad \forall n \in \mathbb{N}.$$

See Appendix A.3 for a review of the matrix norm, which implies the inequality $\|Ax\| \leq \|A\|\cdot\|x\|$ for a vector $x$ and $\|x\|$ denotes the standard vector norm. If $\|K^{-1}\| \leq C$, then $\|E_h\| = \|K^{-1}\tau_h\| \leq \|K^{-1}\|\|\tau_h\| \leq C\|\tau_h\|$.

**Remark 2.1.** *Normal matrices are matrices satisfying $A^*A = AA^*$ with $A^*$ denoting conjugate transpose, e.g., real symmetric matrices, complex Hermitian matrices, see Appendix A.4. For normal matrices, $\|A\| = \max_i |\lambda_i|$ where $\lambda_i$ are eigenvalues of $A$.*

The standard vector norm $\|E_h\|$ is a not a good choice of measuring errors. For instance, assume $E_j = h^2$ for any $j$, then $h = \frac{1}{n+1}$ implies

$$\|E_h\| = \sqrt{\sum_{j=1}^{n} E_j^2} = \sqrt{\sum_{j=1}^{n} h^4} = \mathcal{O}(h^{1.5}).$$

To this end, we define the 2-norm for errors:

$$\|E_h\|_2 = \sqrt{h\sum_{j=1}^{n} E_j^2}. \tag{2.3}$$

**Remark 2.2.** *The 2-norm is a natural discrete version of the function $L^2$-norm. For instance, the $L^2$-norm of a single variable function $f(x)$ on an interval $x \in [0, 1]$ is defined as*

$$\|f\|_{L^2} = \sqrt{\int_0^1 f(x)^2 \mathrm{d}x}.$$

*If we discretize the interval $x \in [0, 1]$ by grid points as in Figure 2.1, and use $\frac{1}{2}[f^2(x_i)+f^2(x_{i+1})]h$ as the simple approximation of integral for each small interval $[x_i, x_{i+1}]$, then for a function $f(x)$ satisfying $f(0) = f(1) = 0$, we get $\|f\|_{L^2} \approx \sqrt{h\sum\limits_{j=1}^{n} f(x_j)^2}$.*

If $\|\tau_h\|_2 \to 0$ when $h \to 0$, we say the scheme is *consistent*. We say the scheme is *convergent* if $\|E_h\|_2 \to 0$ when $h \to 0$. Since

$$\|E_h\| \leq \|K^{-1}\| \cdot \|\tau_h\| \Rightarrow \|E_h\|_2 \leq \|K^{-1}\| \cdot \|\tau_h\|_2,$$

we can see the following classical result for linear problems:

$$\text{Consistency} + \text{Stability} \to \text{Convergence}.$$

### 2.2.2 *Stability by eigenvalues of K*

For the tridiagonal matrix $K$, we derive its eigenvalues which can be used to show stability $\|K^{-1}\| \leq C$ as $h \to 0$.

Any function $u(x)$ satisfying $-u'' = \lambda u, u(0) = u(1) = 0$ is called an *eigenfunction* to the boundary value problem (2.1), and $\lambda$ is the associated *eigenvalue*. It is straightforward to verify that $\sin(m\pi x)$ with integers $m$ are eigenfunctions. We expect that the eigenvectors of $K$ are approximations of $\sin(m\pi x)$ for small $h$ since $K$ approximates the differential operator. With the following trigonometric formulae,

$$\sin(m\pi x_{j+1}) = \sin(m\pi(x_j + h)) = \sin(m\pi x_j)\cos(m\pi h) + \cos(m\pi x_j)\sin(m\pi h),$$
$$\sin(m\pi x_{j-1}) = \sin(m\pi(x_j - h)) = \sin(m\pi x_j)\cos(m\pi h) - \cos(m\pi x_j)\sin(m\pi h),$$

we get

$$-\sin(m\pi x_{j-1}) + 2\sin(m\pi x_j) - \sin(m\pi x_{j+1}) = (2 - 2\cos(m\pi h))\sin(m\pi x_j).$$

Notice the facts that $\sin(m\pi x_0) = 0$ and $\sin(m\pi x_{n+1}) = 0$, we also have

$$2\sin(m\pi x_1) - \sin(m\pi x_2) = (2 - 2\cos(m\pi h))\sin(m\pi x_1),$$

$$-\sin(m\pi x_{n-1}) + 2\sin(m\pi x_n) = (2 - 2\cos(m\pi h))\sin(m\pi x_n).$$

Let $\mathbf{x} = [x_1\, x_2\, \cdots\, x_n]^\top$, then the eigenvectors of $K$ are $\sin(m\pi\mathbf{x})$:

$$K\sin(m\pi\mathbf{x}) = \frac{1}{h^2}(2 - 2\cos(m\pi h))\sin(m\pi\mathbf{x}), \quad m = 1, 2, \cdots, n.$$

Define the eigenvector matrix as $S = [\sin(\pi\mathbf{x}) \quad \sin(2\pi\mathbf{x}) \quad \cdots \quad \sin(n\pi\mathbf{x})]$ and consider the diagonal matrix $\Lambda$ with diagonal entries

$$\frac{1}{h^2}[2 - 2\cos(m\pi h)], \quad m = 1, \cdots, n.$$

Then $K = S\Lambda S^{-1}$. Thus $K^{-1} = S\Lambda^{-1}S^{-1}$ and we get eigenvalues of $K^{-1}$ by L'Hospital's rule:

$$\frac{h^2}{2 - 2\cos(m\pi h)} \to \frac{1}{m^2\pi^2}, \qquad h \to 0.$$

Therefore $\|K^{-1}\| \to \frac{1}{\pi^2}$ as $h \to 0$, which implies the stability (see Prob. 2.1).

Recall that we have derived the truncation error as $\tau_j = -\frac{1}{12}u^{(4)}(x_j)h^2 + \mathcal{O}(h^4)$, if assuming that the problem (2.1) has a smooth solution $u(x)$. Assume $u^{(4)}(x)$ is uniformly bounded, with stability $\|K^{-1}\| \leq C$, we have the convergence

$$\|\tau_h\|_2 = \mathcal{O}(h^2) \Rightarrow \|E_h\|_2 = \mathcal{O}(h^2).$$

**Problem 2.1.** *Show that $\|K^{-1}\| \leq C$ for any $h$, where $C$ is a constant.*

**Problem 2.2.** *Consider* $T(\alpha) = \begin{pmatrix} \alpha & -1 & & & \\ -1 & \alpha & -1 & & \\ & \ddots & \ddots & \ddots & \\ & & -1 & \alpha & -1 \\ & & & -1 & \alpha \end{pmatrix}_{n\times n}$. *Find its eigenvalues and eigenvectors. For what kind of $\alpha$, this matrix will become positive definite?*

### 2.2.3 *An implementation in MATLAB*

For implementing the numerical $KU = F$, all we need to do is to take an inverse of the matrix $K$, of which the cost is always the computational bottleneck. The computational complexity of *Gauss Elimination* for computing inverses of a general matrix of size $n \times n$ is $\mathcal{O}(n^3)$, but for computing $K^{-1}$ it is only $O(n)$ because $K$ is a tridiagonal matrix. In MATLAB, the *mldivide (backslash)* function is one of such efficient solvers for inverting a sparse tridiagonal matrix:

```
U= K \ F;
```

For readers with little coding experience, the following MATLAB code can be copied and used as a quick simple accuracy test of the scheme $KU = F$ solving (2.1) with $f(x) = -\pi^2 \sin(\pi x)$.

```
m=5; % Test the accuracy on m different grids
error_2norm=ones(m,1); error_max=ones(m,1);
exact=@(x) sin(pi*x);  f=@(x) -pi*pi*sin(pi*x);
for k=1:m
    % create the proper uniform grid
    n=10*k; h=1/(n+1); x=[0:h:1]'; x=x(2:end-1);
    % create a sparse tridiagonal matrix K
    e = ones(n,1);
    K=1/h^2*spdiags([-e 2*e -e], [-1 0 1], n, n);
    F=-f(x);
    u=K \ F; % Backslash is efficient for inverting K
    exact_value=exact(x);
    error_2norm(k)=sqrt(h*sum((u-exact_value).^2));
end
hFig=figure; set(hFig, 'Position', [0 800 1500 400])
subplot(1,2,1);
```

```
17 plot(x, exact_value,'+'); hold all; plot(x, u,'o');
18 xlabel('x'); ylabel('u(x)')
19 legend('Exact Solution', 'Numerical Solution')
20 subplot(1,2,2);
21 second_order_slope=(1./(2*[1:m])').^2;
22 loglog(2*[1:m],error_2norm,'o'); hold all
23 loglog(2*[1:m],second_order_slope)
24 legend('Error in 2-norm','Second Order Slope')
25 xlabel('Number of grid points'); ylabel('Error');
```

We can also use the eigenvalue decomposition $K^{-1} = S\Lambda^{-1}S^{-1}$ to invert $K$, which is of course slower than best algorithms for tridiagonal matrices. But the tridiagonal structure of discrete Laplacian matrix $K$ only holds for 1D problems. In 2D and 3D, later we will see that inversion via the eigenvalue decomposition is a convenient and often good enough solver. To implement $U = S\Lambda^{-1}S^{-1}F$, we only need to replace the 11-th line of the code above by the following

```
lambda=1/h^2*(2*ones(n,1)-2*cos([1:n]'*pi*h));
S=sin(x*pi*[1:n]);
u= S*((S \ F)./lambda);
```

**Remark 2.3.** *The matrix* $S = [\sin(\pi\mathbf{x}) \quad \sin(2\pi\mathbf{x}) \quad \cdots \quad \sin(n\pi\mathbf{x})]$ *is the Sine Matrix* $S_{jk} = \sin\left(\frac{jk\pi}{n+1}\right)$ *thus multiplying* $S$ *to a vector* $\mathbf{v}$ *is equivalent to the discrete sine transform (DST) of* $\mathbf{v}$*, which is possible to be computed with* $\mathcal{O}(n\log_2 n)$ *complexity by fast transform implementations. Multiplying* $S^{-1}$ *would be the inverse DST. Run the following code in MATLAB to validate the inversion of* $K$ *by using DST for a random vector:*

```
% difference should be machine zero, i.e., around 10^{-15}
v=randn(n,1); difference = norm(K \ v-dst(idst(v)./lambda))
```

### 2.2.4 *Nonhomogeneous Dirichlet boundary conditions*

Consider solving the same equation with nonhomogeneous Dirichlet boundary conditions:

$$\begin{cases} -u''(x) = f(x), & x \in (0,1), \\ u(0) = a, \ u(1) = b. \end{cases}$$

Then the scheme (2.2) can be changed as

$$\frac{2u_1 - u_2}{h^2} = f_1 + a/h^2,$$

$$\frac{-u_{j-1} + 2u_j - u_{j+1}}{h^2} = f_j, \quad j = 2, \ldots, n-1,$$

$$\frac{-u_{n-1} + 2u_n}{h^2} = f_n + b/h^2.$$

In other words, we can use the same coefficient matrix in the scheme for the nonhomogeneous b.c. with modified right hand side data compensating the nonzero boundary conditions. So from now on we just focus on the homogeneous case.

## 2.3 Dirichlet and Neumann b.c. in one dimension

Consider solving the 1D Poisson equation with homogeneous Dirichlet and Neumann boundary conditions:

$$\begin{cases} -u''(x) = f(x), & x \in (0,1), \\ u'(0) = 0, \ u(1) = 0. \end{cases}$$

Discretize the domain $[0, 1]$ by a uniform grid with spacing $h$. For the interior nodes, we can use the same finite difference scheme (2.2).

### 2.3.1 *A symmetric matrix T*

Discretize the domain $[0, 1]$ by a uniform grid with spacing $h = \frac{1}{n+1}$ and $n$ interior nodes: $x_j = jh$, $j = 1, 2, \cdots, n$. See Figure 2.1. Let $x_0 = 0$ and $x_{n+1} = 1$. For the boundary condition $u'(0) = 0$, we can first consider a simple first order accurate approximation by forward difference:

$$\frac{u_1 - u_0}{h} \approx u'(x_0) \Rightarrow \frac{u_1 - u_0}{h} = 0 \Rightarrow u_0 = u_1,$$

with which the scheme (2.2) at $x_1$ becomes:

$$\frac{u_1 - u_2}{h^2} = f_1.$$

Notice that $U_0$ is not an unknown by this approach, and the unknown numerical solution vector is $U_h = \begin{bmatrix} u_1 & \cdots & u_n \end{bmatrix}^\top$. Then our scheme can be written as $TU_h = F$ with

$$T = \frac{1}{h^2} \begin{pmatrix} 1 & -1 & & & \\ -1 & 2 & -1 & & \\ & \ddots & \ddots & \ddots & \\ & & -1 & 2 & -1 \\ & & & -1 & 2 \end{pmatrix}_{n \times n}.$$

This scheme can be at most first order accurate in the local truncation error since forward difference is used for the boundary condition $u'(0) = 0$.

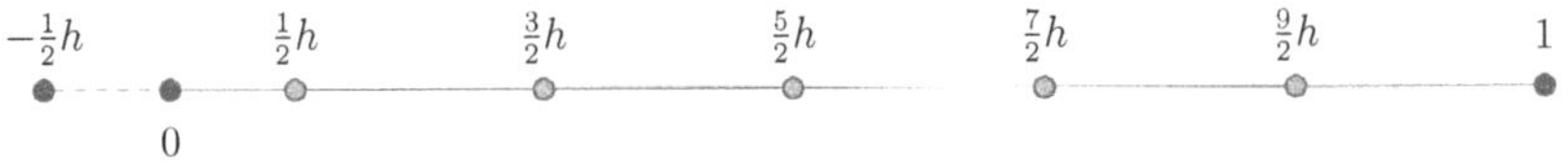


Fig. 2.2 The grid on which $TU_h = F$ is second order accurate.

By considering a new grid $x_j = (j - \frac{1}{2})h$ $(j = 1, 2, \cdots, n)$ with $h = \frac{1}{n+\frac{1}{2}}$, the same scheme $TU_h = F$ becomes second order accurate for approximating the boundary condition. Let $x_0 = -\frac{1}{2}h$ then $\frac{u_1 - u_0}{h}$ is the centered difference approximating $u'(0)$, i.e.,

$$\frac{u(x_1) - u(x_0)}{h} = u'(0) + \mathcal{O}(h^2) \Rightarrow \frac{u_1 - u_0}{h} = 0 \Rightarrow u_0 = u_1.$$

**Remark 2.4.** *For the boundary condition $u'(0) = 0$, instead of considering the ghost point $x_0 = -\frac{h}{2}$ and centered difference $\frac{u_1 - u_0}{h} \approx u'(0)$ for the left boundary, we can obtain the same scheme by approximating the second order derivative $u''(\frac{1}{2}h)$ as follows:*

$$u''(\frac{1}{2}h) \approx \frac{u'(h) - u'(0)}{h} = \frac{u'(h)}{h} \approx \frac{u(\frac{3}{2}h) - u(\frac{1}{2}h)}{h^2} \Rightarrow \frac{u_1 - u_2}{h^2} = f_1,$$

*where we have used the centered difference approximation twice: first for $u''(\frac{1}{2}h)$, then for $u'(h)$.*

However, the approximation error to boundary conditions is NOT the local truncation error (LTE) of the scheme. The LTE of the scheme is actually only first order accurate at the boundary:

$$\frac{u(x_1) - u(x_2)}{h^2} - f(x_1) = \frac{-u(x_0) + 2u(x_1) - u(x_2)}{h^2} - f(x_1) - \frac{u(x_1) - u(x_0)}{h^2}$$

$$= \mathcal{O}(h^2) - \frac{u'(0) + \frac{1}{3}u'''(0)(\frac{h}{2})^3 + \mathcal{O}(h^4)}{h^2} = -\frac{1}{24}u'''(0)h + \mathcal{O}(h^2),$$

where we have used $u'(0) = 0$.

On the other hand, the local truncation error is NOT the true error. The actual order of convergence of this scheme for smooth solutions may or may not be the same as the LTE, which will be further explained in the next two subsections.

**Problem 2.3.** *Let* $\mathbf{x} = [x_1\, x_2\, \cdots\, x_n]^\top$ *with* $x_j = (j - \frac{1}{2})h$ $(j = 1, 2, \cdots, n)$ *and* $h = \frac{1}{n+\frac{1}{2}}$ *in this new grid. Show that the eigenvectors of* $T$ *are* $\cos[(m - \frac{1}{2})\pi\mathbf{x}]$ *with eigenvalues* $\frac{1}{h^2}[2 - 2\cos((m - \frac{1}{2})\pi h)]$, $m = 1, \cdots, n$.

**Problem 2.4.** *Modify the code in Section 2.2.3 to implement the scheme on the grid shown in Figure 2.2 for a properly chosen smooth function* $f(x) = -u''(x)$ *and* $u(x)$ *satisfying the boundary conditions. What order of convergence do you observe?*

### 2.3.2 *Nonsymmetric matrix* $T_2$

We can also construct a scheme on the integer grid $x_j = (j-1)h$ with $h = \frac{1}{n}$, with a second order approximation to the boundary condition $u'(0) = 0$. Let $x_0 = -h$ then $\frac{u_2 - u_0}{2h}$ is the second order centered difference approximation to $u'(0)$. We get a scheme $T_2 U_h = F$ with

$$T_2 = \frac{1}{h^2}\begin{pmatrix} 2 & -2 & & & \\ -1 & 2 & -1 & & \\ & \ddots & \ddots & \ddots & \\ & & -1 & 2 & -1 \\ & & & -1 & 2 \end{pmatrix}.$$

Unfortunately, $T_2$ is no longer symmetric, not even a normal matrix. But we can convert it to $T$ by dividing the first equation by 2, then the scheme becomes

$$TU_h = \begin{pmatrix} f_1/2 \\ f_2 \\ \vdots \\ f_n \end{pmatrix}. \tag{2.4}$$

If we perform local truncation error analysis at the first grid point, then we will find out that it is only first order in LTE even though it is obtained by a second order approximation to the Neumann boundary condition. On the other hand, the scheme is actually second order accurate not only in the discrete 2-norm but also in the maximum norm. See next subsection for explanation.

**Problem 2.5.** *Find the eigenvalues and eigenvectors for* $T_2$.

**Problem 2.6.** *Find the local truncation error of the scheme* (2.4) *at the left boundary.*

**Problem 2.7.** *Prove the stability and convergence of the scheme* $T_2U_h = F$*. Notice that* $T_2$ *is not normal thus eigenvalues of* $T_2$ *do not give its spectral norm. Instead, find a diagonal matrix* $D$ *s.t.* $T_2 = DT$ *where* $T$ *matrix is symmetric (thus normal and its spectral norm can be estimated by eigenvalues, see Appendix A.4). Use eigenvalues of* $T$ *and the inequality* $\|DT\| \leq \|D\|\|T\|$ *to prove the stability.*

**Problem 2.8.** *Consider the following problem:*

$$\begin{cases} -u''(x) = 25\pi^2 \sin(5\pi x) - 6x, \; x \in (0,1), \\ u'(0) = 5\pi, \; u(1) = 1. \end{cases}$$

*The exact solution is* $u(x) = \sin(5\pi x) + x^3$*. Implement the finite difference scheme on the grid* $x_j = (j-1)h$ *with* $h = \frac{1}{n}$ *using the nonsymmetric matrix* $T_2$*. The matrix-vector forms of the scheme is* $T_2U = \widehat{F}$ *where* $\widehat{F}$ *is the modified right hand side data due to the nonzero boundary conditions. Implement this scheme and test its order of convergence. You should observe second order convergence for errors in both 2-norm and max norm.*

## 2.4 *Convergence in maximum norm

For beginners, this subsection should be skipped. First, review the definitions of vector norms and matrix norms in Appendix A. In particular, the matrix $\infty$-norm is $\|A\|_\infty = \max_i \sum_{j=1}^{n} |a_{ij}|$, and it satisfies the inequality:

$$E_h = K^{-1}\tau_h \Rightarrow \|E_h\|_\infty \leq \|K^{-1}\|_\infty \|\tau_h\|_\infty.$$

Since truncation error is usually straightforward to estimate, to get true errors in $\infty$-norm, we need $\infty$-norm estimates for the matrices. In general, this task is much harder than estimating the spectral norm. However, the matrices $K$, $T$ and $T_2$ are very special, and they belong to a class of matrices called *M-matrices*. The sharp characterization of an M-matrix [Plemmons (1977)] is complicated and we only need a convenient sufficient condition here:

**Theorem 2.1.** *For a real square matrix* $A$ *with positive diagonal entries and non-positive off-diagonal entries,* $A$ *is a nonsingular M-matrix if all the row sums of* $A$ *are non-negative and at least one row sum is positive.*

We need the following well-known but nontrivial result to proceed:

**Theorem 2.2.** *The inverse matrix of a nonsingular M-matrix has non-negative entries.*

By this theorem, we have $K^{-1} \geq 0, T^{-1} \geq 0, T_2^{-1} \geq 0$, where the inequalities are entry-wise inequalities, i.e., these matrices have non-negative entries. If a matrix has an inverse with non-negative entries, then we call it *monotone*. So $K$, $T$ and $T_2$ are monotone matrices.

Usually monotonicity of the matrix $A$ can help to establish the estimate on $\infty$-norm of $A^{-1}$. If we can find a vector $\mathbf{v}$ such that $A\mathbf{v} = \mathbf{1}$ where $\mathbf{1}$ is a vector with each entry being 1, then

$$A^{-1} \geq 0 \Rightarrow \|A^{-1}\|_\infty = \|A^{-1}\mathbf{1}\|_\infty = \|\mathbf{v}\|_\infty.$$

### 2.4.1 *Dirichlet boundary conditions*

For the $K$ matrix, in order to find $\mathbf{v}$ such that $K\mathbf{v} = \mathbf{1}$, first think about the exact solution to the problem $-u'' = 1, u(0) = u(1) = 0$, which is $v(x) = \frac{1}{2}x(1-x)$. Let $\mathbf{v} = v(\mathbf{x})$ where $\mathbf{x}$ is the grid points for the corresponding scheme, i.e., $\mathbf{x} = \begin{bmatrix} h \ 2h \cdots nh \end{bmatrix}^\top$ with $h = \frac{1}{n+1}$. It is straightforward to verify that $K\mathbf{v} = \mathbf{1}$. On the other hand, since $0 \leq v(x) \leq \frac{1}{8}$ for $x \in (0,1)$, we have $\|\mathbf{v}\|_\infty \leq \frac{1}{8}$, thus $\|K^{-1}\|_\infty = \|\mathbf{v}\|_\infty \leq \frac{1}{8}$, with which we can easily establish the second order convergence in maximum norm for the scheme $KU_h = F$ solving a smooth problem.

### 2.4.2 *Dirichlet and Neumann b.c.*

The exact solution to the problem $-u'' = 1, u'(0) = u(1) = 0$, which is $v(x) = \frac{1}{2} - \frac{1}{2}x^2$. Let $\mathbf{v} = v(\mathbf{x})$ where $\mathbf{x}$ is the grid points for the corresponding scheme to $T_2$, i.e., $\mathbf{x} = \begin{bmatrix} 0 \ h \ 2h \cdots (n-1)h \end{bmatrix}^\top$ with $h = \frac{1}{n}$. It is straightforward to verify that $A\mathbf{v} = \mathbf{1}$. Thus similarly as in previous subsection, $\|T_2^{-1}\|_\infty = \|\mathbf{v}\|_\infty \leq \frac{1}{8}$.

### 2.4.3 *Why the scheme $T_2U_h = F$ is second order accurate*

Let $D = diag\{\frac{1}{2}, 1, \cdots, 1\}$ be a diagonal matrix. Recall that the local truncation error is only first order at the left boundary. In order to establish second order convergence, we need the estimate for the first column of $T_2^{-1}$. For convenience, let $A = T_2$. If we can find a vector $\mathbf{w}$ such that $A\mathbf{w} \geq \begin{bmatrix} 1 \ 0 \cdots 0 \end{bmatrix}^\top$. Then the first column of $A^{-1}$ is given by $A^{-1}\begin{bmatrix} 1 \ 0 \cdots 0 \end{bmatrix}^\top \leq \mathbf{w}$, where inequality holds because $A^{-1}$ has non-negative entries.

Let $w(x) = h(\frac{1}{2} - \frac{1}{2}x)$ and $\mathbf{w} = w(\mathbf{x})$, then it is straightforward to verify $T_2\mathbf{w} = \begin{bmatrix} 1 \ 0 \cdots 0 \end{bmatrix}^\top$. Let $A = [a_{ij}]$ and $A^{-1} = [a^{ij}]$. Then $\mathbf{w} = A^{-1}\begin{bmatrix} 1 \ 0 \cdots 0 \end{bmatrix}^\top$ implies that $\max_i |a^{i1}| = \max_i |w_i| \leq \frac{1}{2}h$.

So the true error is $(U_h - \hat{U})_i = (A^{-1}\tau_h)_i = a^{i1}\tau_1 + \sum_{j=2}^{n} a^{ij}\tau_j$ where $\tau_1 = \mathcal{O}(h)$ and $\tau_j = \mathcal{O}(h^2), j = 2, 3, \cdots, n$. Thus we get

$$
\begin{aligned}
|(U_h - \hat{U})_i| &\le \frac{1}{2}h\tau_1 + \max_{j\ge 2}\tau_j \sum_{j=2}^{n} a^{ij} \\
\le \frac{1}{2}h\tau_1 + \max_{j\ge 2}\tau_j \sum_{j=1}^{n} a^{ij} &\le \frac{1}{2}h\tau_1 + \max_{j\ge 2}\tau_j \|A^{-1}\|_\infty = \mathcal{O}(h^2).
\end{aligned}
$$

**Remark 2.5.** *Here $w(x)$ is not the exact solution to $-u'' = \delta(0), u'(0) = u(1) = 0$. See Section 3.14.1 for the meaning of the function $\delta(x)$. But to find such a vector, we can simply ask MATLAB to compute $T_2^{-1}\begin{bmatrix}1 & 0 & \cdots & 0\end{bmatrix}^\top$, which gives this vector* $\mathbf{w}$.

We have explained why the true error of the scheme $T_2 U_h = F$ is second order accurate despite its first order LTE at the boundary. But the explanation here has a severe limitation: extension of this argument to multiple dimensions is intractable. Instead, in Chapter 3, techniques from the finite element method can be used to give a much better explanation of the second order accuracy of the same scheme.

## 2.5 Neumann boundary conditions in one dimension

### 2.5.1 *The matrix $B$ on the one-half grid*

Consider solving the 1D Poisson equation with homogeneous Neumann boundary conditions:

$$
\begin{cases}
-u''(x) = f(x), & x \in (0,1), \\
u'(0) = 0, \ u'(1) = 0.
\end{cases}
$$

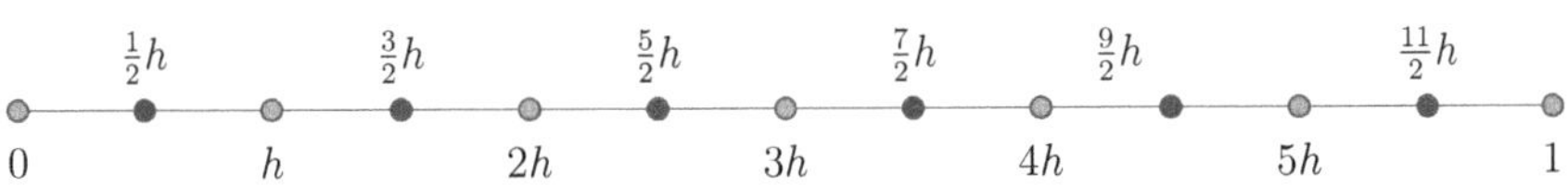


Fig. 2.3 The integer grid and the one-half grid.

Following Section 2.3.1, we use the grid $x_j = (j - \frac{1}{2})h$, $j = 1, \cdots, n$ with $h = \frac{1}{n}$ and obtain a second order scheme $BU_h = F$ with

$$B = \frac{1}{h^2} \begin{pmatrix} 1 & -1 & & & \\ -1 & 2 & -1 & & \\ & \ddots & \ddots & \ddots & \\ & & -1 & 2 & -1 \\ & & & -1 & 1 \end{pmatrix}.$$

The eigenvectors of $B$ are $\cos(m\pi\mathbf{x})$ with eigenvalues $\frac{1}{h^2}[2 - 2\cos(m\pi h)]$, $m = 0, \cdots, n-1$, and $\mathbf{x}$ must consist of $x_j = (j - \frac{1}{2})h$, $j = 1, \cdots, n$ with $h = \frac{1}{n}$.

**Problem 2.9.** *We cannot choose* $\cos(m\pi\mathbf{x})$ *with integers* $m = 1, \cdots, n$ *as eigenvectors for* $B$*. Why?*

### 2.5.2 *The matrix B on the integer grid*

Following Section 2.3.2, on the grid $x_j = (j-1)h$, $j = 1, \cdots, n$ with $h = \frac{1}{n-1}$, we have a second order scheme

$$BU_h = \begin{pmatrix} f_1/2 \\ f_2 \\ \vdots \\ f_n/2 \end{pmatrix}, \quad B = \frac{1}{h^2} \begin{pmatrix} 1 & -1 & & & \\ -1 & 2 & -1 & & \\ & \ddots & \ddots & \ddots & \\ & & -1 & 2 & -1 \\ & & & -1 & 1 \end{pmatrix}.$$

Notice that the eigenvectors and eigenvalues of $B$ are the same as in the previous subsection because it is the same matrix. However, be aware that the $\mathbf{x}$ and $h$ in the eigenvector and eigenvalue formulae are different from the integer grid used in the scheme in this subsection. The advantage of $B$ is its symmetry.

### 2.5.3 *The matrix $B_2$ on the integer grid*

On the grid $x_j = (j-1)h$, $j = 1, \cdots, n$ with $h = \frac{1}{n-1}$, the scheme in Section 2.5.2 is equivalent to $B_2 U_h = F$ with

$$B_2 = \frac{1}{h^2} \begin{pmatrix} 2 & -2 & & & \\ -1 & 2 & -1 & & \\ & \ddots & \ddots & \ddots & \\ & & -1 & 2 & -1 \\ & & & -2 & 2 \end{pmatrix}.$$

The eigenvectors of this matrix are $\cos(\mathbf{x}m\pi)$ with eigenvalues $\frac{1}{h^2}[2 - 2\cos(m\pi h)], m = 0, 1, 2, ..., n-1$, where $\mathbf{x}$ must be $x_j = (j-1)h$, $j = 1, \cdots, n$ with $h = \frac{1}{n-1}$. Notice that the eigenvectors are not the same as those for the $B$ matrix.

### 2.5.4 *Compatibility condition of Neumann b.c.*

Consider a purely Neumann boundary value problem

$$\begin{cases} -u''(x) = f(x), & x \in (0,1), \\ u'(0) = \sigma_0, \; u'(1) = \sigma_1. \end{cases}$$

Integrating the equation, we get

$$\int_0^1 f(x)\mathrm{d}x = \int_0^1 -u''(x)\mathrm{d}x = -u'(1) + u'(0) = \sigma_0 - \sigma_1. \tag{2.5}$$

For a given $f(x)$, the boundary data $\sigma_0, \sigma_1$ must satisfy the compatibility condition (2.5) so that there is a solution. When (2.5) is satisfied, the solution is unique up to any constant. For instance, if $u(x)$ is a solution, then so is $u(x) + c$ for any constant $c$.

Next we consider the scheme in Section 2.5.3:

$$\frac{1}{h^2}\begin{pmatrix} 2 & -2 & & & \\ -1 & 2 & -1 & & \\ & \ddots & \ddots & \ddots & \\ & & -1 & 2 & -1 \\ & & & -2 & 2 \end{pmatrix}\begin{pmatrix} u_1 \\ u_2 \\ \vdots \\ u_{n-1} \\ u_n \end{pmatrix} = \begin{pmatrix} f_1 - 2\sigma_0/h \\ f_2 \\ \vdots \\ f_{n-1} \\ f_n + 2\sigma_1/h \end{pmatrix}.$$

For convenience, we denote it as $B_2 U_h = F$. Since each row of $B_2$ sums to zero, the null space of $B_2$ is nontrivial thus $B_2$ is not invertible. The linear system $B_2 U_h = F$ has a solution only if $F$ is in the column space of $B_2$, which is not necessarily true even if (2.5) is satisfied. In other words, we still need a discrete compatibility condition so that the linear system $B_2 U_h = F$ has a solution.

By the relations of column space and left null space of a matrix, $F \in \mathrm{Col}(B_2)$ is equivalent to $F \perp \mathrm{Null}(B_2^\top)$. Since all columns of $B_2$ are orthogonal to $\left[\frac{1}{2}\; 1 \cdots 1\; \frac{1}{2}\right]^\top$, $\mathrm{Null}(B_2^\top)$ is spanned by this vector. So we obtain the following discrete compatibility condition

$$F \in \mathrm{Col}(B_2) \Leftrightarrow \frac{h}{2}f_1 + hf_2 + \cdots + hf_{n-1} + \frac{h}{2}f_n = \sigma_0 - \sigma_1. \tag{2.6}$$

Notice that $\frac{h}{2}f_1 + hf_2 + \cdots + hf_{n-1} + \frac{h}{2}f_n = \int_0^1 f(x)\mathrm{d}x + \mathcal{O}(h^2)$ by the trapezoidal quadrature (numerical integration) rule for each interval $[x_i, x_{i+1}]$, thus (2.6) is indeed a discrete version of (2.5).

On the other hand, given a problem satisfying (2.5), (2.6) is not necessarily satisfied for $B_2U_h = F$. We would like to have a second order scheme $B_2U_h = \bar{F}$ by generating a new right hand side vector $\bar{F}$ belonging to the column space of $B_2$, which can be obtained by projecting $F$ to $\mathrm{Col}(B_2)$. In general, such a projection might be nontrivial to obtain. However, the rank of $B_2$ is $n-1$ and we know the orthogonal complement of $\mathrm{Col}(B_2)$ is spanned by $\left[\frac{1}{2}\ 1 \cdots 1\ \frac{1}{2}\right]^\top$, thus it is straightforward to obtain this projection. First we have the compatibility error as

$$c = \frac{h}{2}f_1 + hf_2 + \cdots + hf_{n-1} + \frac{h}{2}f_n - \sigma_0 + \sigma_1 = \mathcal{O}(h^2).$$

Second, the projection $\bar{F}$ can be written as

$$\bar{F} = \begin{pmatrix} f_1 - 2\sigma_0/h \\ f_2 \\ \vdots \\ f_{n-1} \\ f_n + 2\sigma_1/h \end{pmatrix} + a \begin{pmatrix} \frac{1}{2} \\ 1 \\ \vdots \\ 1 \\ \frac{1}{2} \end{pmatrix},$$

where $a$ is a parameter determined by requiring $\bar{F}$ is orthogonal to $\left[\frac{1}{2}\ 1 \cdots 1\ \frac{1}{2}\right]^\top$. So we obtain $a = \frac{-c}{h(n-\frac{3}{2})}$, thus

$$\bar{F} = \begin{pmatrix} f_1 - 2\sigma_0/h - c/h/(n-3/2)/2 \\ f_2 - c/h/(n-3/2) \\ \vdots \\ f_{n-1} - c/h/(n-3/2) \\ f_n + 2\sigma_1/h - c/h/(n-3/2)/2 \end{pmatrix}. \tag{2.7}$$

Finally, if we prefer to solve a symmetric system, we can symmetrize the system by dividing the first and last row by two:

$$\frac{1}{h^2}\begin{pmatrix} 1 & -1 & & & \\ -1 & 2 & -1 & & \\ & \ddots & \ddots & \ddots & \\ & & -1 & 2 & -1 \\ & & & -1 & 1 \end{pmatrix} \begin{pmatrix} u_1 \\ u_2 \\ \vdots \\ u_{n-1} \\ u_n \end{pmatrix} = \begin{pmatrix} \frac{1}{2}f_1 - \sigma_0/h - c/h/(n-3/2)/4 \\ f_2 - c/h/(n-3/2) \\ \vdots \\ f_{n-1} - c/h/(n-3/2) \\ \frac{1}{2}f_n + \sigma_1/h - c/h/(n-3/2)/4 \end{pmatrix}.$$

**Problem 2.10.** *Suppose $B_2U_h = F$ does not satisfy the discrete compatibility condition. The least square solution $\hat{U}_h$ for the inconsistent linear system $B_2U_h = F$ is defined as the minimizer for the function $\|\frac{1}{h^2}B_2U_h - F\|^2$ under the standard vector norm. Show that $\hat{U}_h$ is a solution to $B_2U_h = \bar{F}$.*

### 2.5.5 *Inverting B*

Notice that $B$ is a singular matrix. In other words, there are infinitely many numerical solutions, which corresponds to the fact that the true solutions to the purely Neumann b.c. are not unique. To *invert* $B$, we can do the following by choosing a particular solution for a consistent linear system $BU_h = F$:

```
% eigenvectors and eigenvalues of B
h=1/n; x=[\frac{h}{2}:h:1-\frac{h}{2}]';
lambda=(2*ones(n,1)-2*cos([0:n-1]'*pi*h))/h/h;
S=cos(x*pi*[0:n-1]);
% Multiply the "inverse" of B  to a vector F
U=(inv(S)*F)./lambda;
U(1)=0;  % special treatment for the zero eigenvalue
U=S*U;
```

Let $\lambda_i = 2-2\cos((i-1)\pi h)$ then $\lambda_1 = 0$. Define $\Lambda^{-1}$ as a diagonal matrix with diagonal entries $0, \frac{1}{\lambda_2}, \frac{1}{\lambda_3}, \cdots, \frac{1}{\lambda_n}$. Let $S$ denote the eigenvector matrix then we set $U_h = S\Lambda^{-1}S^{-1}F$. Such a procedure is equivalent to redefine $\frac{1}{\lambda_1} = 0$, which gives a zero sum solution, explained as follows.

Since $B$ is a real symmetric matrix, the eigenvector matrix $S$ has orthogonal columns. If we further choose orthonormal eigenvectors, e.g., normalize each column of $S$ to be a unit vector, then $S^\top = S^{-1}$. Then its eigendecomposition $B = S\Lambda S^{-1}$ is also its singular value decomposition $B = S\Lambda S^\top$ (see Appendix (A)). By setting $\frac{1}{\lambda_1} = 0$, the numerical solution has zero component along the first eigenvector which is $[1 \quad 1 \quad \cdots \quad 1]^\top$. This means that the numerical solution $U_h$ is perpendicular to $[1 \quad 1 \quad \cdots \quad 1]^\top$ thus $U_h = \begin{bmatrix} u_1 \cdots u_n \end{bmatrix}^\top$ satisfies $\sum_j u_j = 0$.

**Remark 2.6.** *Since the exact solution is not unique, we should compare the zero sum numerical solution above with a zero sum exact solution: suppose* $\widehat{U} = \begin{bmatrix} u(x_1) \cdots u(x_n) \end{bmatrix}^\top$ *where* $u(x)$ *is any exact solution, then we should compute the error as* $E_h = U - \widetilde{U}$ *with* $\widetilde{U} = \begin{bmatrix} u(x_1) - \bar{U} \cdots u(x_n) - \bar{U} \end{bmatrix}^\top$ *where* $\bar{U} = \frac{1}{n}\sum_j u(x_j)$.

**Problem 2.11.** *Use SVD to show that setting* $\frac{1}{\lambda_1} = 0$ *gives the least square solution for* $BU_h = F$.

**Problem 2.12.** *For the* $B_2$ *matrix, we can use the same procedure as above by setting* $\frac{1}{\lambda_1} = 0$. *However, it no longer gives a solution satisfying* $\sum_j u_j = 0$. *Why?*

### 2.5.6 *Inverting $B_2$*

For the $B_2$ matrix, let $B_2 = S\Lambda S^{-1}$ be its eigendecomposition with $\lambda_1 = 0$. Recall that the eigenvectors and eigenvalues of $B_2$ are: $\cos(\mathbf{x}m\pi)$, $2 - 2\cos\left(m\pi\frac{1}{n-1}\right), m = 0, 1, 2, ..., n-1$, where $\mathbf{x} = \begin{bmatrix} 0 & \frac{1}{n-1} & \frac{2}{n-1} & \cdots & 1 \end{bmatrix}$. Let $\mathbf{w} = \begin{bmatrix} \frac{1}{2} & 1 & \cdots & 1 & \frac{1}{2} \end{bmatrix}^\top$. Let $\mathbf{v}_i$ be the eigenvector for $\lambda_i$. Then $0 = \mathbf{w}^\top B_2 \mathbf{v}_i = \lambda_i \mathbf{w}^\top \mathbf{v}_i$ implies $\mathbf{w}^\top \mathbf{v}_i = 0$ if $\lambda_i \neq 0$. Thus the eigenvectors associated with nonzero eigenvalues are orthogonal to $\mathbf{w}$, i.e., the orthogonal complement of the column space of $B_2$.

For the $B_2$ matrix, and a vector $F$ which does not satisfy the discrete compatibility condition (2.6), i.e., $F$ is not in the column space of $B_2$, the vector $\hat{U}_h = S\Lambda^{-1}S^{-1}F$ (with $\frac{1}{\lambda_1}$ being defined as zero) is the least square solution to $B_2 U_h = F$. In other words, $B_2\hat{U}_h = B_2 S\Lambda^{-1}S^{-1}F$ is the projection of $F$ onto the column space of $B_2$. To see why it is true, assume $\tilde{F}$ is the projection of $F$ onto the column space of $B_2$, then Problem 2.13 implies

$$B_2\hat{U}_h = B_2 S\Lambda^{-1}S^{-1}(F-\tilde{F}) + S\begin{pmatrix} 0 & & & \\ & 1 & & \\ & & \ddots & \\ & & & 1 \end{pmatrix} S^{-1}\tilde{F} = B_2 S\Lambda^{-1}S^{-1}(F-\tilde{F}) + \tilde{F}.$$

Problem 2.14 gives $B_2 S\Lambda^{-1}S^{-1}(F-\tilde{F}) = 0$ thus $B_2\hat{U}_h = \tilde{F}$. By the definition of $\tilde{F}$, $\hat{U}_h$ is the least square solution to $B_2 U_h = F$.

Notice that $F - \tilde{F}$ should be exactly the shift we added to generate $\bar{F}$ in (2.7).

**Problem 2.13.** *For any vector $\bar{F}$ satisfying the discrete compatibility condition (2.6), $\bar{F}$ is orthogonal to $\mathbf{w}$. Show that $U_h = S\Lambda^{-1}S^{-1}\bar{F}$ (with $\frac{1}{\lambda_1}$ being defined as zero) is a solution to $B_2 U_h = \bar{F}$. In other words, show that*

$$S\begin{pmatrix} 0 & & & \\ & 1 & & \\ & & \ddots & \\ & & & 1 \end{pmatrix} S^{-1}\bar{F} = \bar{F}.$$

*Hint: show that $\bar{F}$ lives in the eigenspace for $\lambda_2, \cdots, \lambda_n$ thus* $\bar{F} = S\begin{pmatrix} 0 \\ d_2 \\ \vdots \\ d_n \end{pmatrix}$.

**Problem 2.14.** *Show that $B_2 S\Lambda^{-1}S^{-1}(F-\tilde{F}) = 0$.*

## 2.6 Periodic boundary conditions in one dimension

Consider solving the periodic boundary conditions:

$$\begin{cases} -u''(x) = f(x), x \in (0,1), \\ u(0) = u(1). \end{cases}$$

We use the grid $x_j = (j-1)h$, $j = 1, \cdots, n$ with $h = \frac{1}{n}$ and obtain a second order scheme $CU_h = F$ with

$$C = \frac{1}{h^2}\begin{pmatrix} 2 & -1 & & & -1 \\ -1 & 2 & -1 & & \\ & \ddots & \ddots & \ddots & \\ & & -1 & 2 & -1 \\ -1 & & & -1 & 2 \end{pmatrix}.$$

The eigenvectors of $C$ are $\exp(-\mathbb{i}m2\pi\mathbf{x})$ with imaginary unit $\mathbb{i}$, and eigenvalues are $\frac{1}{h^2}[2 - 2\cos(m2\pi h)]$, $m = 0, \cdots, n-1$. By choosing these eigenvectors, the eigenvector matrix is precisely the discrete Fourier transform matrix, i.e., dftmtx(n) in MATLAB. Multiplying the DFT matrix is equivalent to the fast Fourier transform.

**Remark 2.7.** *The matrix $C$ is circulant. The columns of DFT matrix are eigenvectors to any circulant matrix thus the DFT matrix can diagonalize any circulant matrix.*

The matrix $C$ is singular. To *invert* the matrix $C$, we can do the same thing as previously for the matrix $B$: set the component along the zero eigenvector to be zero, which gives a numerical solution summing to zero.

## 2.7 Dirichlet boundary conditions in two dimensions

Consider solving the 2D Poisson equation with homogeneous Dirichlet boundary conditions on a rectangular domain $\Omega = (0,1) \times (0,2)$ with its boundary $\Gamma$:

$$\begin{cases} -u_{xx}(x,y) - u_{yy}(x,y) = f(x,y), \quad (x,y) \in \Omega, \\ u(x,y)|_\Gamma = 0. \end{cases}$$

Consider a Cartesian grid $x_i = i\Delta x$, $i = 1, \cdots, N_x$ with $\Delta x = \frac{1}{N_x+1}$ and $y_j = j\Delta y$, $j = 1, \cdots, N_y$ with $\Delta y = \frac{2}{N_y+1}$. Let $U_{ij}$ be the numerical solution at $(x_i, y_j)$, and $f_{ij} = f(x_i, y_j)$. The second order finite difference scheme for the interior nodes is given by

$$\frac{-U_{i-1,j} + 2U_{ij} - U_{i+1,j}}{\Delta x^2} + \frac{-U_{i,j-1} + 2U_{ij} - U_{i,j+1}}{\Delta y^2} = f_{ij}. \tag{2.8}$$

In order to write out the matrix vector form of the scheme (2.8), we will need two operators:

- Kronecker product of two matrices: if $A$ is $m \times n$ and $B$ is $p \times q$, then $A \otimes B$ is $mp \times nq$ given by

$$A \otimes B = \begin{pmatrix} a_{11}B & \cdots & a_{1n}B \\ \vdots & \vdots & \vdots \\ a_{m1}B & \cdots & a_{mn}B \end{pmatrix}.$$

- For a 2D array or matrix $X$ of size $m \times n$, $vec(X)$ denotes the vectorization of $X$, i.e., rearranging $X$ into a big vector in a column by column fashion. In MATLAB the *reshape* function can act as the inverse of the *vec* operator: if $v = vec(X)$, then $reshape(v, m, n)$ recovers $v$ back into $X$.

The following properties for matrices $A, B, C, D$ and $X$ with proper sizes will be used:

(1) $(A \otimes B)(C \otimes D) = AC \otimes BD$.
(2) $(A \otimes B)^{-1} = A^{-1} \otimes B^{-1}$.
(3) $(B^\top \otimes A) vec(X) = vec(AXB)$.
(4) $(A \otimes B)^\top = A^\top \otimes B^\top$.

Next, we need to use some less intuitive symbols. We define $U$ as a 2D array of size $N_y \times N_x$, with entries $U(j, i) = U_{ij}$, which is the numerical solution at location $(x_i, y_j)$. Let $F$ be a $N_y \times N_x$ array with entries $F(j, i) = f(x_i, y_j)$. The matrix vector form of the scheme (2.8) can be written as

$$(K_x \otimes I_y + I_x \otimes K_y)\, vec(U) = vec(F),$$

where $K_x$ is the $K$ matrix defined in Section 2.2 of size $N_x \times N_x$ with $h = \Delta x$, $I_x$ is the identity matrix of size $N_x \times N_x$, $K_y$ is the $K$ matrix of size $N_y \times N_y$ with $h = \Delta y$, and $I_y$ is the identity matrix of size $N_y \times N_y$.

**Remark 2.8.** *If we choose a more natural symbol by defining $U$ as a 2D array of size $N_x \times N_y$, with entries $U(i, j) = U_{ij}$, which is the numerical solution at location $(x_i, y_j)$. The matrix form of scheme becomes*

$$(I_y \otimes K_x + K_y \otimes I_x)\, vec(U) = vec(F).$$

*Either choice of symbols is fine. We choose the symbols such that the discrete Laplacian matrix is $-(K_x \otimes I_y + I_x \otimes K_y)$ for 2D problems in this chapter.*

Let $K2D$ denote the big matrix $K_x \otimes I_y + I_x \otimes K_y$. Since we know the eigenvectors of $K_x$ and $K_y$, we can find the inverse of the matrix $K2D$ by the following *eigenvector method.* Suppose the eigendecompositions of $K_x$ and $K_y$ are $K_x = S_x\Lambda_x S_x^{-1}$ and $K_y = S_y\Lambda_y S_y^{-1}$, then

$$\begin{aligned}K_x \otimes I_y &= S_x\Lambda_x S_x^{-1} \otimes I_y = S_x\Lambda_x S_x^{-1} \otimes S_y I_y S_y^{-1}\\ &= (S_x \otimes S_y)(\Lambda_x S_x^{-1} \otimes I_y S_y^{-1}) = (S_x \otimes S_y)(\Lambda_x \otimes I_y)(S_x^{-1} \otimes S_y^{-1}),\end{aligned}$$

where the first property of the Kronecker product is used twice. Similarly,

$$I_x \otimes K_y = (S_x \otimes S_y)(I_x \otimes \Lambda_y)(S_x^{-1} \otimes S_y^{-1}).$$

Thus we get the eigenvectors and eigenvalues of $K2D$:

$$K2D = (S_x \otimes S_y)(\Lambda_x \otimes I_y + I_x \otimes \Lambda_y)(S_x^{-1} \otimes S_y^{-1}),$$

which gives the inverse matrix as

$$K2D^{-1} = (S_x \otimes S_y)(\Lambda_x \otimes I_y + I_x \otimes \Lambda_y)^{-1}(S_x^{-1} \otimes S_y^{-1}).$$

Notice that $\Lambda_x \otimes I_y + I_x \otimes \Lambda_y$ is a diagonal matrix with diagonal entries $\Lambda_x(i,i) + \Lambda_y(j,j)$. Thus, to compute $K2D^{-1}vec(F)$, we only need to consider three steps:

Step I. Compute $(S_x^{-1} \otimes S_y^{-1})vec(F) = vec(S_y^{-1}FS_x^{-\top})$ where $S_x^{-\top}$ is the transpose of $S_x^{-1}$.

Step II. Define a 2D array $\Lambda$ with entries $\Lambda(j,i) = \Lambda_x(i,i)+\Lambda_y(j,j)$. Then we have

$$(\Lambda_x \otimes I_y + I_x \otimes \Lambda_y)^{-1}vec(S_y^{-1}FS_x^{-\top}) = vec(S_y^{-1}FS_x^{-\top}./\Lambda),$$

where ./ denotes the component-wise division of two arrays.

Step III. $(S_x \otimes S_y)vec(S_y^{-1}FS_x^{-\top}./\Lambda) = vec(S_y(S_y^{-1}FS_x^{-\top}./\Lambda)S_x^{\top})$.

If using array notation, the three steps above simply writes as

$$U = S_y(S_y^{-1}FS_x^{-\top}./\Lambda)S_x^{\top}. \tag{2.9}$$

We briefly discuss the computational complexity, which quantifies the number of operations needed. Since multiplication/division is more expensive than addition/subtraction, usually the number of operations refer to number of multiplication/division when counting complexity. In (2.9), it has $\mathcal{O}(N^3)$ complexity if $N = N_x = N_y$. If fast discrete sine transform is used for multiplying $S_x$ and $S_x^{-1}$, it reduces to $\mathcal{O}(N^2 \log_2 N)$. As a comparison, the Gaussian elimination of $K2D$ costs $\mathcal{O}(N^4)$ (costs $\mathcal{O}(N^7)$ in 3D) because its bandwidth is $N$ ($N^2$ in 3D).

The implementation of (2.9) in MATLAB is given as follows (notice that the big matrix $K2D$ should never be constructed, and only small matrices like $S_x$ and arrays like $U$ are used):

```
m=5; err=ones(m,1);
for n=1:m
dx=1/(10*n);x=[0:dx:1]';x=x(2:end-1);Nx=prod(size(x));
dy=2/(10*n);y=[0:dy:2]';y=y(2:end-1);Ny=prod(size(y));
% -u_xx-u_yy=f on (x,y) in [0,1]*[0,2]
f=5/4*pi^2*sin(pi/2*y)*sin(pi*x');
exact_sol=sin(pi/2*y)*sin(pi*x');
% define eigenvectors and eigenvalues of K2D
lambda_x=[2*ones(Nx,1)-2*cos([1:Nx]'*pi*dx)]/dx/dx;
Sx=sin(x*pi*[1:Nx]);iSx=inv(Sx);
lambda_y=[2*ones(Ny,1)-2*cos([1:Ny]'*pi*dy/2)]/dy/dy;
Sy=sin(y/2*pi*[1:Ny]);iSy=inv(Sy);
lambda=lambda_y*ones(1,Nx)+ones(Ny,1)*lambda_x';
% The eigenvector method to "invert" the matrix K2D
sol=Sy*((iSy*f*iSx')./lambda)*Sx';
error=sol-exact_sol;
err(n)=sqrt(sum(error(:).^2)*dx*dy);
end
hFig=figure;
set(hFig, 'Position', [400 800 1500 400])
set(0,'DefaultTextFontSize',12,'DefaultAxesFontSize', 12)
subplot(1,3,1)
loglog(10*[1:m],err,'o'); hold all
loglog(10*[1:m],(1./(2*[1:m])').^2,'-')
xlabel('1/dx');ylabel('Error')
legend('Error in 2 norm','Second Order Slope')
subplot(1,3,2)
contour(x,y,exact_sol);colorbar;axis equal
xlabel('X');ylabel('Y')
title('The contour of the exact solution')
subplot(1,3,3)
contour(x,y,sol);colorbar;axis equal
xlabel('X');ylabel('Y')
title('The contour of the numerical solution')
```

**Remark 2.9.** *Since $K$ is real symmetric, it has orthonormal eigenvectors. If we normalize each eigenvector to obtain orthonormal eigenvectors for constructing $S$, then $S^{-1} = S^{\top}$. The computation of $S^{-1}$ can be avoided.*

**Problem 2.15.** *Find $S_y$ and $\Lambda_y$. Notice that $y \in [0, 2]$, which might make a difference in the formula.*

**Problem 2.16.** *Show that $K2D$ is symmetric.*

**Problem 2.17.** *Consider a matrix given in the form of $A \otimes B + B \otimes A$ where $A$ and $B$ are matrices of size $n \times n$. Can it be inverted using a similar procedure as described in this subsection through eigen-decomposition of* *small matrices of size $n \times n$?*

*Hint:* $A \otimes B + B \otimes A = (B \otimes B)(B^{-1}A \otimes I + I \otimes B^{-1}A)$.

## 2.8 Neumann boundary conditions in two dimensions

Consider solving the 2D Poisson equation with homogeneous Neumann boundary conditions:

$$\begin{cases} -u_{xx}(x,y) - u_{yy}(x,y) = f(x,y),\ (x,y) \in \Omega = (0,1) \times (0,2), \\ \frac{\partial}{\partial \mathbf{n}} u(x,y)|_\Gamma = 0, \end{cases}$$

where $\frac{\partial}{\partial \mathbf{n}} u(x,y)|_\Gamma$ denotes the directional derivative of $u$ along the direction normal to the domain boundary $\Gamma$. The normal derivatives at four corner points of a rectangle are not well defined, so we should understand the boundary condition as unspecified at four corners.

We have two choices of 1D methods to construct the scheme in 2D:

(1) Use the one-half grid and the $B$ matrix in Section 2.5.1.
(2) Use matrix $B$ on an integer grid in Section 2.5.2, which can also be equivalently written using $B_2$ in Section 2.5.3.

If we just want to implement a finite difference scheme, then either choice is fine. But the second choice allows a rigorous explanation of the second order accuracy. As we have seen in Section 2.3.1, the local truncation error of these schemes is only first order accurate at the boundary point, thus it is quite difficult to explain why these schemes are second order accurate schemes for smooth solutions. The special discussion in Section 2.4.3 does not generalize to even 2D. On the other hand, as will be explained in the next chapter (Section 3.9.5), the scheme in Section 2.5.2 and the one in Section 2.5.3 are equivalent to a finite element method, thus their accuracy can be proven via more advanced techniques for the finite element method.

To this end, we will only use the second approach. In other words, we consider a finite difference scheme that is also equivalent to the finite element method in Section 3.9.5. Consider the grid $x_i = (i-1)\Delta x$, $i = 1, \cdots, N_x$ with $\Delta x = \frac{1}{N_x - 1}$ and $y_j = (j-1)\Delta y$, $j = 1, \cdots, N_y$ with $\Delta y = \frac{2}{N_y - 1}$, let $B_x$ and $B_y$ be the matrix $B_2$ in Section 2.5.3 of suitable sizes with $h = \Delta x$ and $h = \Delta y$ respectively on this integer grid. Following similar derivations as in Section 2.7, the scheme has the following matrix vector form

$$(B_x \otimes I_y + I_x \otimes B_y)\, vec(U) = vec(F).$$

Such a scheme can be implemented following the same three steps in Section 2.7, with one more complication: the smallest eigenvalue of the big matrix $B_x \otimes I_y + I_x \otimes B_y$ is zero. Let $B_x = S_x \Lambda_x S_x^{-1}$ and $B_y = S_y \Lambda_y S_y^{-1}$ with eigenvalues ordered from the smallest to the largest. Define a 2D array $\Lambda$ with entries $\Lambda(j,i) = \Lambda_x(i,i) + \Lambda_y(j,j)$. Then $\Lambda(1,1) = 0$. By redefining $\frac{1}{\Lambda(1,1)} = 0$, the scheme can be implemented by

$$U = S_y (S_y^{-1} F S_x^{-\top} ./\Lambda) S_x^{\top}$$

By setting the $(1,1)$ entry in $S_y^{-1} F S_x^{-\top} ./\Lambda$ to be zero, it returns a numerical solution with zero average.

**Problem 2.18.** *We have used the fact that $B_2$ matrix in Section 2.5.3 is diagonalizable, i.e., $B_2 = S\Lambda S^{-1}$ with diagonal matrix $\Lambda$. The $B$ matrix in Section 2.5.2 is diagonalizable because it is real symmetric. How do we show that the nonsymmetric $B_2$ is still diagonalizable? Hint: $B_2 = DB$ with a diagonal matrix $D$, thus $B_2 = D^{\frac{1}{2}}(D^{\frac{1}{2}} S D^{\frac{1}{2}}) D^{-\frac{1}{2}}$. Notice that eigenvalue decomposition of $D^{\frac{1}{2}} S D^{\frac{1}{2}}$ gives eigenvalue decomposition of $B_2$.*

**Problem 2.19.** *Design a second order finite difference scheme for solving*

$$-u_{xx}(x,y) - u_{yy}(x,y) = \frac{5}{4}\pi^2 \sin\left(\frac{1}{2}\pi x\right)\cos(\pi y)$$

*on the rectangle $(x,y) \in (0,3) \times (0,1.5)$ with the following boundary conditions and the exact solution is $u(x,y) = \sin\left(\frac{1}{2}\pi x\right)\cos(\pi y)$:*

- *Dirichlet b.c. on the left and right boundaries: $u(0,y) = 0$, $u(3,y) = \sin\left(\frac{3}{2}\pi\right)\cos(\pi y)$.*
- *Neumann b.c. on the top and bottom boundaries: $\frac{\partial}{\partial y}u(x,0) = 0$, $\frac{\partial}{\partial y}u(x,\frac{3}{2}) = -\pi \sin\left(\frac{1}{2}\pi x\right)\sin\left(\frac{3}{2}\pi\right)$.*

*The scheme should have the form*

$$(K_x \otimes I_y + I_x \otimes B_y)\, vec(U) = vec(\hat{F}),$$

*where $\hat{F}$ contains the nonzero boundary data. Implement it by $U = S_y(S_y^{-1} F S_x^{-\top} ./\Lambda) S_x^{\top}$. Unlike the purely Neumann b.c., there is no zero eigenvalue in 2D array $\Lambda$ due to presence of the Dirichlet boundary conditions. Set $\Delta x = \Delta y$ and test the max norm and 2-norm error on a few different grids and compare it with the second order slope.*

## 2.9 Efficient implementations in three dimensions

For simplicity, we only consider solving $-u_{xx} - u_{yy} - u_{zz} = f$ on a cube $\Omega = [0,1]^3$ with homogeneous Dirichlet boundary conditions.

Consider a uniform Cartesian mesh in 3D with grid points $(x_i, y_j, z_k)$ $(i = 1, \cdots, N_x; j = 1, \cdots, N_y; k = 1, \cdots, N_z)$.

Following our choice of notation for 2D problems, for representing numerical solutions, let $U$ be a 3D array of size $N_z \times N_y \times N_x$ with $(k, j, i)$-th entry $U(k, j, i)$ denoting the point value at $(x_i, y_j, z_k)$.

For a 3D array $U$, we define a page as the matrix obtained by fixing the last index of $U$. Namely, $U(:,:,i)$ for any fixed $i$ is a page of $U$. For a matrix $U(:,:,i)$ of size $N_z \times N_y$, recall that $vec(U(:,:,i))$ is a column vector of size $N_z N_y$. We define $\hat{U}$ as the following matrix of size $N_z N_y \times N_x$ obtained by reshaping $U$:

$$\hat{U} = \left[vec(U(:,:,1))\ vec(U(:,:,2)) \cdots vec(U(:,:,N_x))\right].$$

Then we define $vec(U)$ as the vector of size $N_z N_y N_x \times 1$ by reshaping $\hat{U}$ in a column by column order.

With the notation above, it is straightforward to verify that

$$(A_1^\top \otimes A_2^\top \otimes A_3) vec(U) = vec((A_2^\top \otimes A_3)\hat{U} A_1). \tag{2.10}$$

Let $Y$ be a 3D array of size $N_z \times N_y \times N_x$ defined by

$$vec(Y) = (A_1^\top \otimes A_2^\top \otimes A_3) vec(U). \tag{2.11}$$

With the simple property (2.10), among other choices, one simple and efficient implementation of multiplying the big matrix $A_1^\top \otimes A_2^\top \otimes A_3$ in (2.11) in MATLAB is to use two functions *tensorprod* and *pagemtimes*:

```
% Computing a 3D array Y of the same size as U defined above
Y = tensorprod(U,A1,3,1);
Y = pagemtimes(Y,A2);
Y = squeeze(tensorprod(A3,Y,2,1));
```

Let $F$ be a 3D array with $F(k, j, i) = f(x_i, y_j, z_k)$, then the second order finite difference scheme can be written as

$$(K_x \otimes I_y \otimes I_z + I_x \otimes K_y \otimes I_z + I_x \otimes I_y \otimes K_z)\, vec(U) = vec(F).$$

With the eigenvalue decomposition $K = S\Lambda S^{-1}$, similar to the derivation 2D problems, the scheme is equivalent to

$$(S_x \otimes S_y \otimes S_z)\,(\Lambda_x \otimes I_y \otimes I_z + I_x \otimes \Lambda_y \otimes I_z + I_x \otimes I_y \otimes \Lambda_z)\,(S_x^{-1} \otimes S_y^{-1} \otimes S_z^{-1}) vec(U) = vec(F).$$

Define a 3D array $\Lambda$ with its $(k, j, i)$-th entry being equal to $\Lambda_x(i,i) + \Lambda_y(j,j) + \Lambda_z(k,k)$, then it can be implemented efficiently as follows in MATLAB:

```
% InvS denotes the inverse matrix of S
U = tensorprod(F,InvSx',3,1);
U = pagemtimes(U,InvSy');
U = squeeze(tensorprod(InvSz,U,2,1));
U = U./Lambda;
U = tensorprod(U,Sx',3,1);
U = pagemtimes(U,Sy');
U = squeeze(tensorprod(Sz,U,2,1));
```

Recall that $S^{-1} = S$ if we use orthonormal eigenvectors, as pointed out in Remark 2.9. In *Python*, with package *Jax*, it can be similarly implemented using the function *jax.numpy.einsum*:

```
U = jnp.einsum('ijk,kl->ijl',F,invSx.transpose())
U = jnp.einsum('ijk,jl->ilk',U,invSy.transpose())
U = jnp.einsum('li,ijk->ljk',invSz,U)
U = U/Lambda
U = jnp.einsum('ijk,kl->ijl',U,Sx.transpose())
U = jnp.einsum('ijk,jl->ilk',U,Sy.transpose())
U = jnp.einsum('li,ijk->ljk',Sz,U)
```

**Remark 2.10.** *See [Liu et al. (2024b)] for how to use the MATLAB and Python scripts above on a GPU device. In particular, on one Nvidia A100 80G GPU card, it costs around* 0.8 *seconds for inverting one 3D discrete Laplacian of size* $1000^3$.

**Problem 2.20.** *Derive the exact number of operations (only counting the number of multiplication/division) needed for the method in this section inverting* $K \otimes I \otimes I + I \otimes K \otimes I + I \otimes I \otimes K$ *where* $K$ *and* $I$ *are matrices of size* $N \times N$.

**Problem 2.21.** *Write a MATLAB code for solving* $-u_{xx} - u_{yy} - u_{zz} = 14\pi^2 \sin(\pi x)\sin(2\pi y)\sin(3\pi z)$ *with homogeneous Dirichlet b.c. on the domain* $[0,1]^3$. *Test your code for the exact solution* $u(x,y,z) = \sin(\pi x)\sin(2\pi y)\sin(3\pi z)$. *Test the convergence rate on several different grids. On a modern desktop computer, the implementation in this section should be quite efficient for a grid of size* $200^3$ *or even* $300^3$. *If your computer has large enough memory, try to push the grid size to as large as it allows.*

**Problem 2.22.** *If you have access to computers on which you can run MATLAB with GPU, follow the script Poisson3D.m in [Liu et al. (2024b),*

*Appendix] to test your MATLAB code on GPU. Notice that Poisson3D.m in [Liu et al. (2024b), Appendix] first creates arrays on CPU memory thus it needs to move data to GPU memory, e.g.,*

```
% create an array then move it to GPU memory
a=rand(3,3);a=gpuArray(a);
```

*Moving data around can be slow for large problems. Instead, you can create arrays on GPU directly, to achieve a better implementation on GPU:*

```
 % create an array directly on GPU memory
a=rand(3,3,"gpuArray");
```

## 2.10 Higher order accuracy: the 9-point discrete Laplacian

The scheme (2.8) for the solving the 2D Poisson equation is second order accurate and it is often called *5-point discrete Laplacian* due to its 5-point stencil. If we use $\Delta_5$ to denote the 5-point discrete Laplacian, then

$$-\Delta_5 U_{i,j} = \frac{-U_{i-1,j} + 2U_{ij} - U_{i+1,j}}{\Delta x^2} + \frac{-U_{i,j-1} + 2U_{ij} - U_{i,j+1}}{\Delta y^2},$$

and the matrix representation of the operator $-\Delta_5$ for the homogeneous Dirichlet boundary condition is

$$K2D = K_x \otimes I_y + I_x \otimes K_y.$$

Now consider the following 9-point Laplacian for $h = \Delta x = \Delta y$:

$$\begin{aligned}\Delta_9 U_{i,j} =& \frac{1}{6h^2}(4U_{i-1,j} + 4U_{i+1,j} + 4U_{i,j-1} + 4U_{i,j+1} \\ &+ U_{i-1,j-1} + U_{i+1,j-1} + U_{i+1,j+1} + U_{i-1,j+1} - 20U_{i,j}).\end{aligned}$$

For convenience, let $u_{i,j}$ denote the value of a smooth function $u(x,y)$ at the point $(x_i, y_j)$, then by the Taylor expansion for the smooth function around the point $(x_i, y_j)$, for the case $h = \Delta x = \Delta y$, we get

$$\begin{aligned}\Delta_9 u(x_i, y_j) =& u_{xx} + u_{yy} + \frac{1}{12}h^2(u_{xxxx} + 2u_{xxyy} + u_{yyyy}) \\ &+ \frac{1}{360}h^4(\partial_x^2 + \partial_y^2)(u_{xxxx} + 4u_{xxyy} + u_{yyyy}) + \mathcal{O}(h^6).\end{aligned}$$

which implies the scheme $-\Delta_9 U_{i,j} = f(x_i, y_j)$ is only second order accurate. However, the second order error term is precisely $\Delta^2 u = \Delta(\Delta u) = u_{xxxx} +$

$2u_{xxyyy} + u_{yyyy}$, where $\Delta^2$ is the biharmonic operator. If $u$ is the solution to $-\Delta u = -u_{xx} - u_{yy} = f$, then $\Delta^2 u = -\Delta f$. Therefore, we can modify it to achieve fourth-order accuracy by considering the scheme:

$$-\Delta_9 U_{i,j} = f(x_i, y_j) + \frac{h^2}{12}\Delta_5 f(x_i, y_j).$$

Let $K2D_9$ be the matrix representing the 9-point Laplacian operator $-\Delta_9$ for the homogeneous Dirichlet boundary condition, then it can be written as

$$K2D_9 = -\frac{1}{6h^2}[H_x \otimes H_y - 36 I_x \otimes I_y],$$

where $H$ is a symmetric tridiagonal matrix

$$H = \begin{pmatrix} 4 & 1 & & & \\ 1 & 4 & 1 & & \\ & \ddots & \ddots & \ddots & \\ & & 1 & 4 & 1 \\ & & & 1 & 4 \end{pmatrix}.$$

For the 9-point Laplacian, the size of the matrices $H_x$ and $H_y$ do not need to be the same, and we do not need $\Delta x = \Delta y$. As a matter of fact the 9-point discrete Laplacian for the Poisson equation can be also written as

$$\frac{1}{12}\left[\frac{1}{\Delta x^2}\begin{pmatrix} 1 & -2 & 1 \\ 10 & -20 & 10 \\ 1 & -2 & 1 \end{pmatrix} + \frac{1}{\Delta y^2}\begin{pmatrix} 1 & 10 & 1 \\ -2 & -20 & -2 \\ 1 & 10 & 1 \end{pmatrix}\right] \circ \bar{U} = \frac{1}{12}\begin{pmatrix} 0 & 1 & 0 \\ 1 & 8 & 1 \\ 0 & 1 & 0 \end{pmatrix} \circ \bar{F},$$

which reduces to the following under the assumption $\Delta x = \Delta y = h$,

$$\frac{1}{6h^2}\begin{pmatrix} 1 & 4 & 1 \\ 4 & -20 & 4 \\ 1 & 4 & 1 \end{pmatrix} \circ \bar{U} = \frac{1}{12}\begin{pmatrix} 0 & 1 & 0 \\ 1 & 8 & 1 \\ 0 & 1 & 0 \end{pmatrix} \circ \bar{F}.$$

Here $\circ$ denotes the Hadamard product of matrices, and $\bar{U}$ and $\bar{F}$ are matrices:

$$\bar{U} = \begin{pmatrix} U_{i-1,j+1} & U_{i,j+1} & U_{i+1,j+1} \\ U_{i-1,j} & U_{i,j} & U_{i+1,j} \\ U_{i-1,j-1} & U_{i,j-1} & U_{i+1,j-1} \end{pmatrix}, \quad \bar{F} = \begin{pmatrix} F_{i-1,j+1} & F_{i,j+1} & F_{i+1,j+1} \\ F_{i-1,j} & F_{i,j} & F_{i+1,j} \\ F_{i-1,j-1} & F_{i,j-1} & F_{i+1,j-1} \end{pmatrix}.$$

**Remark 2.11.** *The 9-point discrete Laplacian can be regarded as a compact finite difference scheme. Thus fourth-order compact finite difference schemes can be used to extend it to Neumann boundary conditions, see [Li and Zhang (2024), Appendix].*

**Remark 2.12.** *We have seen in Section 2.4 that the $K$ matrix has an M-matrix structure (see Theorem 2.1), which can be used for $\infty$-norm estimate. It is straightforward to see that both matrices for the two operators $-\Delta_5$ and $-\Delta_9$ also satisfy Theorem 2.1 thus are M-matrices. Such a structure can also be used for constructing positivity-preserving schemes since M-matrix structure implies that $K^{-1}$ has positive entries (see Theorem 2.2). See [Li* et al. *(2018); Li and Zhang (2024)] for how to take advantage of similar structure in high order compact finite difference schemes for constructing bound-preserving high order schemes.*

**Problem 2.23.** *Explain why the matrix representation of the 9-point Laplacian operator is given as the matrix $K2D_9$.*

**Problem 2.24.** *Find the eigenvectors and eigenvalues of $K2D_9$ and use them to establish the stability thus convergence of the scheme.*

**Problem 2.25.** *Derive the local truncation error in the 9-point scheme for the case $h = \Delta x = \Delta y$.*

**Problem 2.26.** *Prove that the accuracy of the 9-point scheme becomes sixth order for solving the Laplace equation $\Delta u = 0$ with Dirichlet boundary conditions. Solutions to the 2D Laplace equations $\Delta u = 0$ are harmonic functions, thus this is a simple very high order accurate scheme for finding harmonic functions on rectangular domains.*

## 2.11 Variable coefficient problems

### 2.11.1 *Dirichlet b.c. in one dimension (variable coefficients)*

We first consider a 1D variable coefficient problem:

$$-(a(x)u'(x))' = f(x), \quad x \in [0, 1],$$

with homogeneous Dirichlet boundary conditions. A conservative discretization should be used:

$$\frac{1}{\Delta x^2}[-a_{j-\frac{1}{2}}u_{j-1} + (a_{j-\frac{1}{2}} + a_{j+\frac{1}{2}})u_j - a_{j+\frac{1}{2}}u_{j+1}] = f_j,$$

where $a_{j-\frac{1}{2}} = a(x_j - \frac{1}{2}\Delta x)$. The matrix vector form of this scheme is $CU_h = F$ where $C$ is a real symmetric tridiagonal matrix:

$$C = \frac{1}{\Delta x^2}\begin{pmatrix} a_{\frac{1}{2}} + a_{\frac{3}{2}} & -a_{\frac{3}{2}} & & \\ -a_{\frac{3}{2}} & a_{\frac{3}{2}} + a_{\frac{5}{2}} & -a_{\frac{5}{2}} & \\ & \ddots & \ddots & \ddots \end{pmatrix}. \tag{2.12}$$

Notice that $x_j = \frac{j}{n+1}$ for $j = 1, \cdots, n$ are the $n$ grid points and the coefficient function $a(x)$ is sampled at $n+1$ points $x_{j-\frac{1}{2}} = (j - \frac{1}{2})\frac{1}{n+1}$ for $j = 1, \cdots, n+1$. Let $D$ denote the $(n+1) \times n$ matrix:

$$D = \frac{1}{\Delta x}\begin{bmatrix} 1 & & & \\ -1 & 1 & & \\ 0 & \ddots & \ddots & \\ & & -1 & 1 \\ & & & -1 \end{bmatrix},$$

and $A$ be a diagonal matrix with diagonal entries $a_{\frac{1}{2}}, \cdots, a_{n+\frac{1}{2}}$. Then we have $D^\top D = K$ and $C = D^\top A D$.

It is difficult to compute eigenvalues of the matrix $C$. However, it is still possible to have a useful estimate of the smallest eigenvalues of the matrix $C$, as will be shown in Chapter 3 (Section 3.6.3), so that we can establish the stability thus the convergence of the scheme.

### 2.11.2 *Dirichlet b.c. in two dimensions (variable coefficients)*

Now consider the following 2D problem

$$\nabla(a(x,y)\nabla u) = f(x,y),$$

with homogeneous Dirichlet b.c. on a rectangular domain, where $a(x,y) > 0$ is some known coefficient function. If we use the same conservative centered difference discretization as above, then we obtain

$$\left[(D_x^\top \otimes I_y)A_1(D_x \otimes I_y) + (I_x \otimes D_y^\top)A_2(I_x \otimes D_y)\right] vec(U) = vec(F),$$

where $A_1$ and $A_2$ are two diagonal matrices defined as follows.

Let $a_1$ be a 2D array of size $N_y \times (N_x+1)$ satisfying $a_2(j,i) = a(x_{i-\frac{1}{2}}, y_j)$ and $a_2$ be a 2D array of size $(N_y+1) \times N_x$ satisfying $a_1(j,i) = a(x_i, y_{j-\frac{1}{2}})$. The diagonal entries of $A_1$ are

$$reshape(a_1, (N_x+1)N_y, 1),$$

and the diagonal entries of $A_2$ are

$$reshape(a_2, (N_y+1)N_x, 1).$$

**Problem 2.27.** *Show that the matrix $(D_x^\top \otimes I_y)A_1(D_x \otimes I_y) + (I_x \otimes D_y^\top)A_2(I_x \otimes D_y)$ is symmetric.*

### 2.11.3 *Neumann b.c. in one dimension (variable coefficients)*

Next we consider the Neumann boundary conditions:

$$-(a(x)u'(x))' = f(x), \quad x \in [0,1], \quad u'(0) = \sigma_0, u'(1) = \sigma_1.$$

The compatibility condition for this problem is

$$\int_0^1 f(x)\mathrm{d}x = -a(1)\sigma_1 + a(0)\sigma_0.$$

In this section we use the grid points $x_j = j\frac{1}{n+1}$ for $j = 0, \cdots, n+1$. The half grid points are $x_{j+\frac{1}{2}} = (j+\frac{1}{2})\frac{1}{n+1}$. For approximating the boundary conditions, the Taylor expansions

$$\frac{1}{\Delta x}\left(-\frac{3}{2}u(0) + 2u(\Delta x) - \frac{1}{2}u(2\Delta x)\right) = u'(0) + \mathcal{O}(\Delta x^2),$$

$$\frac{1}{\Delta x}\left(\frac{1}{2}u(1-2\Delta x) - 2u(1-\Delta x) + \frac{3}{2}u(1)\right) = u'(1) + \mathcal{O}(\Delta x^2),$$

imply a second order one-sided difference:

$$-\frac{3}{2}u_0 + 2u_1 - \frac{1}{2}u_2 = h\sigma_0,$$

$$\frac{1}{2}u_{n-1} - 2u_n + \frac{3}{2}u_{n+1} = h\sigma_1.$$

Therefore, we get

$$u_0 = \frac{4}{3}u_1 - \frac{1}{3}u_2 - \frac{2}{3}h\sigma_0,$$

$$u_{n+1} = -\frac{1}{3}u_{n-1} + \frac{4}{3}u_n + \frac{2}{3}h\sigma_1.$$

So we get a second order approximation to the first order derivative:

$$\begin{pmatrix} u'(x_{\frac{1}{2}}) \\ u'(x_{\frac{3}{2}}) \\ \vdots \\ u'(x_{\frac{n+1}{2}}) \end{pmatrix} \approx \frac{1}{h}\begin{bmatrix} -1 & 1 & & \\ 0 & \ddots & \ddots & \\ & & -1 & 1 \end{bmatrix} \begin{pmatrix} u_0 \\ u_1 \\ \vdots \\ u_{n-1} \\ u_{n+1} \end{pmatrix} = \frac{1}{h}\begin{bmatrix} -\frac{1}{3} & \frac{1}{3} & & \\ -1 & 1 & & \\ 0 & \ddots & \ddots & \\ & & -1 & 1 \\ & & -\frac{1}{3} & \frac{1}{3} \end{bmatrix} \begin{pmatrix} u_1 \\ u_2 \\ \vdots \\ u_{n-1} \\ u_n \end{pmatrix} + \begin{pmatrix} \frac{2}{3}\sigma_0 \\ 0 \\ \vdots \\ 0 \\ \frac{2}{3}\sigma_1 \end{pmatrix}.$$

Thus we get a second order approximation for $(a(x)u')'$:

$$\frac{1}{h}\begin{bmatrix} -1 & 1 & & \\ 0 & \ddots & \ddots & \\ & & -1 & 1 \end{bmatrix} \begin{bmatrix} a_{\frac{1}{2}} & & & \\ & a_{\frac{3}{2}} & & \\ & & \ddots & \\ & & & a_{n+\frac{1}{2}} \end{bmatrix} \left( \frac{1}{h}\begin{bmatrix} -\frac{1}{3} & \frac{1}{3} & & \\ -1 & 1 & & \\ 0 & \ddots & \ddots & \\ & & -1 & 1 \\ & & -\frac{1}{3} & \frac{1}{3} \end{bmatrix} \begin{pmatrix} u_1 \\ u_2 \\ \vdots \\ u_{n-1} \\ u_n \end{pmatrix} + \begin{pmatrix} \frac{2}{3}\sigma_0 \\ 0 \\ \vdots \\ 0 \\ \frac{2}{3}\sigma_1 \end{pmatrix} \right).$$

A second order scheme for $-(a(x)u')' = f$ can be written as

$$\frac{1}{h^2}\begin{pmatrix} a_{\frac{3}{2}} - \frac{1}{3}a_{\frac{1}{2}} & -a_{\frac{3}{2}} + \frac{1}{3}a_{\frac{1}{2}} & & & \\ -a_{\frac{3}{2}} & a_{\frac{3}{2}} + a_{\frac{5}{2}} & -a_{\frac{5}{2}} & & \\ & \ddots & \ddots & \ddots & \\ & & -a_{n-\frac{3}{2}} & a_{n-\frac{3}{2}} + a_{n-\frac{1}{2}} & -a_{n-\frac{1}{2}} \\ & & & \frac{1}{3}a_{n+\frac{1}{2}} - a_{n-\frac{1}{2}} & a_{n-\frac{1}{2}} - \frac{1}{3}a_{n+\frac{1}{2}} \end{pmatrix}\begin{pmatrix} u_1 \\ u_2 \\ \vdots \\ u_{n-1} \\ u_n \end{pmatrix}$$

$$= \begin{pmatrix} f_1 - \frac{1}{h}a_{\frac{1}{2}}\frac{2}{3}\sigma_0 \\ f_2 \\ \vdots \\ f_{n-1} \\ f_n + \frac{1}{h}a_{n+\frac{1}{2}}\frac{2}{3}\sigma_1 \end{pmatrix},$$

which can be denoted as

$$AU_h = F.$$

The discrete compatibility condition is

$$\frac{a_{\frac{3}{2}}}{a_{\frac{3}{2}} - \frac{1}{3}a_{\frac{1}{2}}}(hf_1 - a_{\frac{1}{2}}\frac{2}{3}\sigma_0) + hf_2 + \cdots + hf_{n-1} + \frac{a_{n-\frac{1}{2}}}{a_{n-\frac{1}{2}} - \frac{1}{3}a_{n+\frac{1}{2}}} \times(hf_n + a_{n+\frac{1}{2}}\frac{2}{3}\sigma_1) = 0. \tag{2.13}$$

**Problem 2.28.** *Show that the discrete compatibility condition is indeed* (2.13) *and show it is a second order accurate approximation to the compatibility condition for $a(x)$ being a constant.*

To obtain a modified right hand side $\bar{F}$ so that $AU_h = F$ has a solution if $F$ does not satisfy (2.13), we can project $F$ to the column space of $A$. Similar to the discussions in Section 2.5.4, we have

$$\bar{F} = \begin{pmatrix} f_1 - 2a_0/h \\ f_2 \\ \vdots \\ f_{n-1} \\ f_n + 2a_1/h \end{pmatrix} + b\mathbf{v}, \quad \mathbf{v} := \begin{pmatrix} a_{\frac{3}{2}}/(a_{\frac{3}{2}} - \frac{1}{3}a_{\frac{1}{2}}) \\ 1 \\ \vdots \\ 1 \\ a_{n-\frac{1}{2}}/(a_{n-\frac{1}{2}} - \frac{1}{3}a_{n+\frac{1}{2}}) \end{pmatrix},$$

where $b$ is defined by requiring $\bar{F}$ is orthogonal to the vector $\mathbf{v}$.

Sometimes we may prefer to solve a symmetric system, especially for iterative solvers, which will be briefly explained in Appendix B. For this matter, we can symmetrize the system by dividing proper numbers in the first and the last row of the system to obtain an equivalent symmetric one:

$$\frac{1}{h^2}\begin{pmatrix} a_{\frac{3}{2}} & -a_{\frac{3}{2}} & & & \\ -a_{\frac{3}{2}} & a_{\frac{3}{2}}+a_{\frac{5}{2}} & -a_{\frac{5}{2}} & & \\ & \ddots & \ddots & \ddots & \\ & & -a_{n-\frac{3}{2}} & a_{n-\frac{3}{2}}+a_{n-\frac{1}{2}} & -a_{n-\frac{1}{2}} \\ & & & -a_{n-\frac{1}{2}} & a_{n-\frac{1}{2}} \end{pmatrix}\begin{pmatrix} u_1 \\ u_2 \\ \vdots \\ u_{n-1} \\ u_n \end{pmatrix}$$

$$= \begin{pmatrix} \frac{a_{\frac{3}{2}}}{a_{\frac{3}{2}}-\frac{1}{3}a_{\frac{1}{2}}}\left(f_1 - \frac{1}{h}a_{\frac{1}{2}}\frac{2}{3}\sigma_0\right) + \left(\frac{a_{\frac{3}{2}}}{a_{\frac{3}{2}}-\frac{1}{3}a_{\frac{1}{2}}}\right)^2 b \\ f_2 + b \\ \vdots \\ f_{n-1} + b \\ \frac{a_{n-\frac{1}{2}}}{a_{n-\frac{1}{2}}-\frac{1}{3}a_{n+\frac{1}{2}}}\left(f_n + \frac{1}{h}a_{\frac{n+1}{2}}\frac{2}{3}\sigma_1\right) + \left(\frac{a_{n-\frac{1}{2}}}{a_{n-\frac{1}{2}}-\frac{1}{3}a_{n+\frac{1}{2}}}\right)^2 b \end{pmatrix}.$$

Chapter 3

# Finite Element Method on Rectangular Meshes

## 3.1 Motivation and overview

Finite difference schemes can only be used on a rectangular domain or a domain which can be transformed to a rectangle, e.g., a disk via polar coordinates. For complicated problems in real applications, the finite element method is the most successful approach due to its rich theories and flexibility with geometries. In Part I of this book, we focus on rectangular domains, on which the finite difference method is the most convenient choice. On the other hand, the finite element method on a rectangular domain can be implemented as a finite difference method when integrals are replaced with quadrature. For instance, there is no essential difference between grid point values and piece-wise linear polynomials represented by its grid point values. Even on unstructured triangular meshes, a linear polynomial on a triangle can be represented by its point values on three vertices of the triangle, which is often called *nodal representation.*

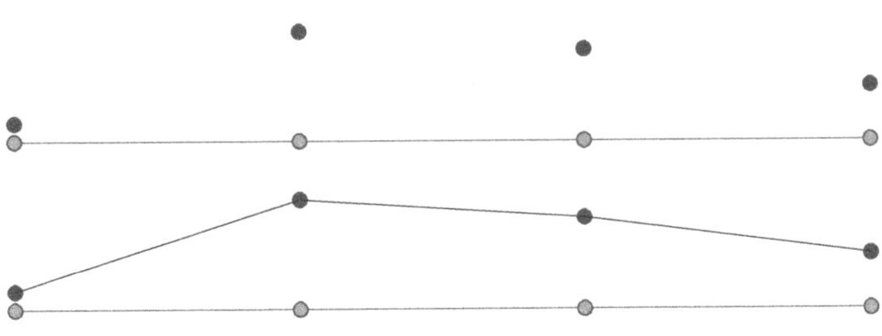

The conventional approach of constructing a finite difference method that we have seen in Chapter 2 includes two crucial steps: first, develop a consistent discretization or approximation to the differential operator and the boundary conditions then try to establish the stability, i.e., try to show $\|A^{-1}\| \leq C$ if the matrix-vector form of the scheme is $A\mathbf{u} = \mathbf{f}$. There are severe drawbacks and practical concerns:

- It is not elegant or convenient to design a high order accurate scheme to use Taylor expansion. For instance, do you actually enjoy solving Problem 2.25? It also becomes harder to deal with boundary conditions in high order schemes.
- Estimating the inverse of a matrix is always hard. If using only linear algebra, singular values and eigenvalues are impossible to estimate for more general schemes, e.g., stability of a high order accurate scheme for $-(a(x)u')' = f$.
- Loss of accuracy on non-uniform meshes: if the local truncation error is obtained by Taylor expansion on uniform grids, the proof of order of accuracy breaks down.
- Loss of symmetry in the matrix $A$: the matrix in general is not symmetric and hard to symmetrize, e.g., $-\nabla(a(x,y)\nabla u) = f$ with Neumann boundary conditions. See remarks in Section 3.9.7 for why symmetry can be useful and desired.

All these concerns and difficulties can be solved by using the finite element method! What is even better is that a finite element method on rectangular meshes (or regular triangular meshes) implemented by a suitable numerical integration is exactly a finite difference method. In this chapter, we will first see how a finite element method is defined then implement it as a finite difference method.

**Caution to readers**: this is NOT a conventional introduction to the finite element method because we focus mainly on rectangular domains and rectangular meshes. Moreover, we will give up certain math rigor such as complete definitions of distribution and Sobolev spaces, proof of existence and uniqueness of variational formulation and important estimates. Instead, they will be given and stated as facts. Despite these simplifications, the key ingredients of the finite element method will be presented in this chapter.

## 3.2 Preliminaries

### 3.2.1 *Weak derivatives and Sobolev spaces*

Let $C_0^\infty(\mathbb{R})$ be the set of all infinitely differentiable functions which can be nonzero only on a finite interval.

If a function $f(x)$ is differentiable, then after integration by parts, for any smooth function $v(x) \in C_0^\infty(\mathbb{R})$ , we have

$$\int_{-\infty}^{+\infty} f(x)v'(x)\,\mathrm{d}x = -\int_{-\infty}^{+\infty} f'(x)v(x)\,\mathrm{d}x. \tag{3.1}$$

The fact (3.1) will be used to define *weak derivatives* or *generalized derivative* for non-differentiable but continuous functions like $f(x) = |x|$. For example, intuitively, we may understand the derivative of $f(x) = |x|$ as the step function $g(x) = \begin{cases} 1 & x \geq 0 \\ -1 & x < 0 \end{cases}$, and $g(x)$ is indeed the *weak derivative* of $f(x)$ by the following definition:

**Definition 3.1.** *A function $g(x)$ is defined to be the weak or generalized derivative of $f(x)$ if it satisfies*

$$\int_{-\infty}^{+\infty} f(x)v'(x)\,\mathrm{d}x = -\int_{-\infty}^{+\infty} g(x)v(x)\,\mathrm{d}x, \quad \forall v(x) \in C_0^\infty(\mathbb{R}).$$

**Example 3.1.** *It is a straightforward exercise to verify that the step function is the weak derivative of the absolute value function, i.e.,*

$$\int_{-\infty}^{+\infty} |x|v'(x)\,\mathrm{d}x = -\int_{0}^{+\infty} v(x)\,\mathrm{d}x + \int_{-\infty}^{0} v(x)\,\mathrm{d}x, \quad \forall v(x) \in C_0^\infty(\mathbb{R}).$$

If a function is differentiable, then its weak derivative is simply the classical derivative. **From now on, in this book, all derivatives are understood as weak or generalized derivatives.**

Next we need to define a few spaces:

- The $L^2$ function space is also called $H^0$ function space:

  $$L^2(0,1) = H^0(0,1) = \left\{ f(x) : \int_0^1 f(x)^2\,\mathrm{d}x < \infty \right\}.$$

  For a general domain $\Omega$, $L^2(\Omega)$ is similarly defined. The integral here is the Lebesgue integral since $f(x)$ may not be continuous. The $L^2(\Omega)$-norm or $H^0(\Omega)$-norm will be denoted as

  $$\|f\|_{0,\Omega} = \|f\|_{L^2(\Omega)} = \left( \int_\Omega f(x)^2\,\mathrm{d}x \right)^{\frac{1}{2}}.$$

When there is no confusion, we will drop $\Omega$ in the subscript, e.g., $\|f\|_0$ simply denotes the $L^2$-norm.

- The $H^1$ function space is

$$H^1(0,1) := \left\{ f(x), f'(x) \in L^2 : \int_0^1 [f(x)^2 + f'(x)^2]\, \mathrm{d}x < \infty \right\}.$$

For a domain $\Omega$ in multiple dimensions, the derivative is replaced by the gradient operator, and the $H^1(\Omega)$-norm will be denoted as

$$\|f\|_{1,\Omega} = \|f\|_{H^1(\Omega)} = \left( \int_\Omega [f(x)^2 + |\nabla f(x)|^2]\, \mathrm{d}x \right)^{\frac{1}{2}}.$$

We also define a seminorm:

$$|f|_{1,\Omega} = |f|_{H^1(\Omega)} = \left( \int_\Omega |\nabla f(x)|^2\, \mathrm{d}x \right)^{\frac{1}{2}}.$$

When there is no confusion, we will drop $\Omega$ in the subscript, e.g., $\|f\|_1$ simply denotes the $H^1$-norm.

- $H_0^1(0,1) := \{f \in H^1(0,1) : f(0) = f(1) = 0\}$, i.e., it is the subset of $H^1(0,1)$ with the property of vanishing at the boundary.
- $H^2$ space is similarly defined:

$$H^2([0,1]) = \left\{ f(x), f'(x), f''(x) \in L^2 : \int_0^1 [f(x)^2 + f'(x)^2 + f''(x)^2]\, \mathrm{d}x < \infty \right\}.$$

Norm and semi-norm for $\Omega = (0,1)$ are

$$\|f\|_{2,\Omega} = \|f\|_{H^2(\Omega)} = \left( \int_\Omega [f(x)^2 + f'(x)^2 + f''(x)^2]\, \mathrm{d}x \right)^{\frac{1}{2}},$$

$$|f|_{2,\Omega} = |f|_{H^2(\Omega)} = \left( \int_\Omega f''(x)^2\, \mathrm{d}x \right)^{\frac{1}{2}}.$$

- $H^3$ space and its norm are also similarly defined: just add $f'''(x)$.

We list a few facts about continuity of these functions:

- The most general statement is from general Sobolev inequalities [Evans (2012)], which imply that

$$f(\mathbf{x}) \in H^k(\Omega),\ k > \frac{n}{2} \Longrightarrow f(\mathbf{x}) \in C(\bar{\Omega}),$$

where $\Omega \subset \mathbb{R}^n$ is a bounded open set with a $C^1$ boundary.

- The special case for one dimension: $H^1(-1,1) \subset C[-1,1]$. In other words, $H^1$ functions in one dimension are continuous.

- Two dimensions: $f(x,y) \in H^2(\Omega) \Longrightarrow f(x,y) \in C(\bar{\Omega})$.
- Three dimensions: $f(x,y,z) \in H^2(\Omega) \Longrightarrow f(x,y,z) \in C(\bar{\Omega})$.
- In two dimensions, $H^1$ is not enough for continuity.

**Example 3.2.** *Here is a function that is $H^1$ but not continuous in two dimensions: let $\Omega \subset \mathbb{R}^2$ be a disk centered at the origin with radius $R = \frac{1}{2}$, then the following function is not continuous at the origin even if redefining $f(0,0) = b$ for any number $b$:*

$$f(x,y) = \left(-\log(x^2+y^2)\right)^\alpha \in H^1(\Omega),$$

*where $\alpha \in (0, \frac{1}{2})$ is a constant. Next, we can show $f \in H^1(\Omega)$. Let $r = \sqrt{x^2+y^2}$, we first have*

$$\iint_\Omega |f|^2 \, \mathrm{d}x\mathrm{d}y = \int_{r=0}^{\frac{1}{2}} \int_{\theta=0}^{2\pi} [-\log r^2]^{2\alpha} r \, \mathrm{d}r\mathrm{d}\theta \le C$$

*because $[-\log r^2]^{2\alpha} r$ is bounded and continuous on $r \in [0, \frac{1}{2}]$. Then*

$$|\nabla f| = 2\alpha \left(-\log r^2\right)^{\alpha-1} \frac{1}{r} = C(-\log r)^{\alpha-1} r^{-1}.$$

$$\iint_\Omega |\nabla f|^2 \, \mathrm{d}x\mathrm{d}y = \int_{r=0}^{\frac{1}{2}} \int_{\theta=0}^{2\pi} C(-\log r)^{2\alpha-2} r^{-2} r \, \mathrm{d}r\mathrm{d}\theta$$

$$= C \int_{r=0}^{\frac{1}{2}} (-\log r)^{2\alpha-2} r^{-1} \, \mathrm{d}r = -C \int_{t=-\log\frac{1}{2}}^{+\infty} t^{2\alpha-2} \, \mathrm{d}t < +\infty,$$

*where $t = -\log r$.*

Finally, we state two versions of the Poincaré inequalities that we need in this chapter, and more general versions can be found in classical references such as [Evans (2012)].

**Theorem 3.1** (Poincaré inequalities). *Assume $\Omega \subset \mathbb{R}^n$ is a bounded connected open set and its boundary $\partial\Omega$ is $C^1$. Then there exists a constant $C$, depending only on $\Omega$, such that for any $u \in H^1(\Omega)$,*

$$\|u - \bar{u}\|_{L^2(\Omega)} \le C\|\nabla u\|_{L^2(\Omega)} = C|u|_1,$$

*where $\bar{u} = \frac{1}{|\Omega|}\int_\Omega u \, \mathrm{d}x$ is the average of $u$. For the $H_0^1(\Omega)$ space, there exists a constant $C$ depending only on $\Omega$ such that*

$$\|u\|_{L^2(\Omega)} \le C|u|_1, \quad \forall u \in H_0^1(\Omega).$$

### 3.2.2 *Interpolation and quadrature*

Finite element methods are built upon basic tools including interpolation and quadrature (numerical integration).

- *Lagrange interpolation* is a convenient polynomial approximation to a function through its point values: given $k+1$ point values of $f(x)$ at $k+1$ grid points $x_i$ $(i=1,2,\cdots,k+1)$, there is a unique polynomial $p(x)$ of degree $k$ satisfying $p(x_i)=f(x_i)$ $(i=1,2,\cdots,k+1)$.
  The linear interpolation at $x_i, x_{i+1}$ for $f(x)$ is given by
  $$\frac{x-x_{i+1}}{x_i-x_{i+1}}f_i+\frac{x-x_i}{x_{i+1}-x_i}f_{i+1}.$$
  The quadratic interpolation at $x_{i-1}, x_i, x_{i+1}$ for a function $f(x)$ is given by
  $$\tfrac{(x-x_i)(x-x_{i+1})}{(x_{i-1}-x_i)(x_{i-1}-x_{i+1})}f_{i-1}+\tfrac{(x-x_{i-1})(x-x_{i+1})}{(x_i-x_{i-1})(x_i-x_{i+1})}f_i+\tfrac{(x-x_i)(x-x_{i-1})}{(x_{i+1}-x_i)(x_{i+1}-x_{i-1})}f_{i+1}.$$
- *Quadrature* rules are numerical integration formulae, for approximating integrals.
  $$\text{Trapezoidal rule}: \int_{-1}^{1} f(x)\,\mathrm{d}x \approx f(-1)+f(1).$$
  $$\text{Simpson's rule}: \int_{-1}^{1} f(x)\,\mathrm{d}x \approx \frac{1}{3}f(-1)+\frac{4}{3}f(0)+\frac{1}{3}f(1).$$
  Trapezoidal rule is exact if $f(x)$ is a linear polynomial. Simpson's rule is also 3-point Gauss-Lobatto rule or 3-point Newton-Cotes rule, which is exact if $f(x)$ is a cubic polynomial.

Consider a uniform mesh with grids $0=x_0<x_1<\cdots<x_N<x_{N+1}=1$ with spacing $h=\frac{1}{N+1}$ for the interval $[0,1]$, which consists of $N+1$ intervals $I_k=[x_{k-1},x_k]$ $(k=1,\cdots,N+1)$. Then for each interval we can use a linear polynomial to approximate $f(x)$ if given $f_i=f(x_i)$. Let $\Pi_1 f(x)$ denote such a piecewise linear polynomial function.

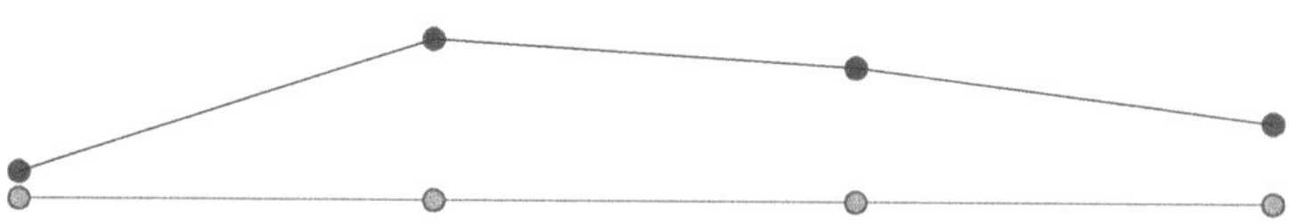

Fig. 3.1 Four grid points and three intervals. For each interval, a linear polynomial is interpolated.

Next consider a uniform mesh with grids $0=x_0<x_1<\cdots<x_N<x_{N+1}=1$ with spacing $h=\frac{1}{N+1}$ for the interval $[0,1]$. And this time we

assume $N = 2n-1$ is odd. Then there are $n$ small intervals $I_k = [x_{2k-2}, x_{2k}]$ $(k = 1, \cdots, n)$, on which we can define a piecewise quadratic interpolation polynomial, denoted by $\Pi_2 f(x)$.

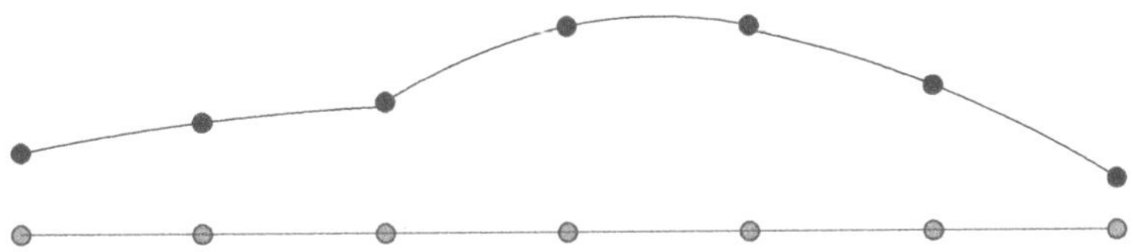

Fig. 3.2 Seven grid points and three intervals. For each interval, a quadratic polynomial is interpolated.

Here are the facts that we will use without any proof first: for a smooth enough function $f(x)$, the interpolation error and quadrature error are given as follows.

- $L^2$ and $H^1$ errors of piecewise linear interpolation:

$$\|f - \Pi_1 f\|_0 \leq Ch^2|f|_2, \quad \|f - \Pi_1 f\|_1 \leq Ch|f|_2. \tag{3.2}$$

- $L^2$ and $H^1$ errors of piecewise quadratic interpolation:

$$\|f - \Pi_2 f\|_0 \leq Ch^3|f|_3, \quad \|f - \Pi_2 f\|_1 \leq Ch^2|f|_3. \tag{3.3}$$

- Quadrature error of trapezoidal rule for intervals in Figure 3.1 :

$$\left| \int_0^1 f(x)\,\mathrm{d}x - \sum_{k=1}^{N+1} \frac{1}{2} h[f(x_{k-1}) + f(x_k)] \right| \leq Ch^2|f|_2.$$

- Quadrature error of Simpson's rule for intervals in Figure 3.2:

$$\left| \int_0^1 f(x)\,\mathrm{d}x - \sum_{k=1}^{n} h[\frac{1}{3} f(x_{2k-1}) + \frac{4}{3} f(x_{2k}) + \frac{1}{3} f(x_{2k+1})] \right| \leq Ch^4|f|_4.$$

Notice that the estimate above only needs the minimal assumption on the function, e.g., for (3.2) we only need to assume $f(x) \in H^2(\Omega)$, which ensures that $f(x)$ is at least a continuous function in 1D/2D/3D so that $f(x_i)$ is well defined. The same order can be obtained by Taylor expansion, but obviously we need the derivatives to exist in the classical sense. All these estimates above can be derived from the Bramble-Hilbert Lemma in Section 3.7.2.

**Problem 3.1.** *Show that the piecewise interpolation polynomial function $\Pi_1 f(x)$ is in the Sobolev space $H^1(0, 1)$. Find its (weak) derivative.*

**Problem 3.2.** *Consider a continuous function $f(x)$ defined on the interval $[0,1]$ and a uniform mesh with integer grids $x_j = jh$ so that the piecewise quadratic interpolation polynomial function $\Pi_2 f(x)$ is well defined. Pick a smooth function $f(x)$ and write a code in MATLAB to*

*(1) evaluate the interpolation errors $|f(x) - \Pi_2 f(x)|$ at the half grids $x_{j+\frac{1}{2}} = x_j + \frac{1}{2}h$ as shown in Figure 2.3.*
*(2) evaluate the quadrature error:*

$$\left| \int_0^1 f(x)\,\mathrm{d}x - \sum_{k=1}^{n} h[\frac{1}{3}f(x_{2k-1}) + \frac{4}{3}f(x_{2k}) + \frac{1}{3}f(x_{2k+1})] \right|.$$

*Follow the MATLAB code in Section 2.2.3 to compute errors on a few different sets of grids with smaller and smaller $h$, and show that the interpolation error is third order in 2-norm and also maximum norm, and show that the quadrature error is fourth order.*

## 3.3 Homogeneous Dirichlet b.c. in one dimension

### 3.3.1 *Variational formulation*

Given a function $f(x) \in L^2(0,1)$, consider solving $-u''(x) = f(x)$ on the interval $(0,1)$ with boundary conditions $u(0) = 0, \quad u(1) = 0$. Multiplying a test function $v \in H_0^1(0,1)$, after integration by parts, we get

$$\int_0^1 u'(x)v'(x)\,\mathrm{d}x = \int_0^1 f(x)v(x)\,\mathrm{d}x. \tag{3.4}$$

Define

$$(f,g) := \int_0^1 f(x)g(x)\,\mathrm{d}x,$$

then (3.4) can be denoted as

$$(u',v') = (f,v).$$

It can be shown that the solution to the PDE $-u''(x) = f(x)$ is equivalent to the solution to the following variational formulation

$$\text{seek} \quad u \in H_0^1(0,1), \quad \text{satisfying} \quad (u',v') = (f,v), \quad \forall v \in H_0^1(0,1). \tag{3.5}$$

The following theorem only explains why a smooth solution to (3.5) is also a solution to PDE.

**Theorem 3.2.** *Assume $f(x) \in C[0,1]$ and $u(x) \in C^2[0,1]$ satisfies (3.5), then $-u''(x) = f(x)$.*

*Proof.* After integration by parts in (3.5), we get

$$0 = (f, v) - (u', v') = (f, v) + (u'', v) = (f + u'', v) = \int_0^1 [u''(x) + f(x)]v(x)\,dx.$$

If $u''(x) + f(x) \neq 0$, then due to continuity, $u''(x) + f(x)$ is either positive or negative on an interval $[x_0, x_1] \subset [0, 1]$. Without loss of generality, assume $u''(x) + f(x) > 0$ on $[x_0, x_1] \subset [0, 1]$. Consider a test function

$$v(x) = \begin{cases} 0, & x < x_0 \\ (x - x_0)^2(x - x_1)^2, & x \in [x_0, x_1] \\ 0, & x > x_1 \end{cases},$$

and we have

$$\int_0^1 [u''(x) + f(x)]v(x)\,dx = \int_{x_0}^{x_1} [u''(x) + f(x)]v(x)\,dx > 0,$$

which is a contradiction. □

Of course, the solution of (3.5) is not necessarily a $C^2$ function. The *Elliptic Regularity Theorem* states that the solution of (3.5) has second order derivative, i.e., (3.5) on a suitable domain $\Omega$ has the property

$$f(x) \in L^2(\Omega) \Rightarrow u(x) \in H^2(\Omega).$$

For a function $u(x)$ that is only in $H^2$ but not in $C^2$, its second order derivative in the weak derivative, i.e., its second order derivative in the boundary value problem for the PDE $-u''(x) = f(x)$ with $u(0) = u(1) = 0$ should be understood as (3.4) for any $v \in C_0^\infty(\mathbb{R})$.

Thus the variational form is a more general form to write out the PDE $-u''(x) = f(x)$. From now on, we simply assume that *Elliptic Regularity Theorem* holds for the domains that we study, thus solutions to the variational form is in $H^2$ thus continuous in 1D/2D/3D.

### 3.3.2 *Bilinear form*

Now consider a more general variable coefficient problem

$$-(a(x)u')' = f, \quad x \in (0, 1), \quad u(0) = u(1) = 0, \tag{3.6}$$

where $a(x) > 0$ is a given continuous coefficient function.

We can introduce a new notation called bilinear form

$$\mathcal{A}(u, v) := \int_0^1 a(x)u'(x)v'(x)\,dx,$$

and $\mathcal{A}(\cdot, \cdot)$ is linear with respect to each of its arguments. Then the equivalent variational formulation of (3.6) is

$$\text{seek} \quad u \in H_0^1(0, 1), \quad \text{satisfying} \quad \mathcal{A}(u, v) = (f, v), \quad \forall v \in H_0^1(0, 1). \tag{3.7}$$

### 3.3.3 The abstract finite element method

Given a mesh with $n$ intervals $I_j$ $(j = 1, \cdots, n)$, let $P^k(I_j)$ denote the set of polynomials of degree $k$ defined on $I_j$. We will mainly consider $k = 1$ or $k = 2$, i.e., linear or quadratic polynomial approximation. In general, these small intervals $I_j$ do not have to be uniform. But for convenience and also for the sake of constructing a finite difference scheme on a uniform mesh, let us assume they have a uniform interval size. We first define a continuous piecewise polynomial space:

$$V^h := \{v_h(x) \in C(0,1) : v_h \text{ is a polynomial of degree } k \text{ on each interval } I_j\},$$

which is an approximation to the function space $H^1(0,1)$. In other words, $v_h(x) \in V^h$ restricted to each interval $I_j$ is a polynomial, and $v_h(x)$ is continuous on the whole domain $(0,1)$. See Figures 3.1 and 3.2 for illustrations of $v_h(x)$.

The space $V_0^h$ is similarly defined as an approximation to $H_0^1(0,1)$:

$$V_0^h := \{v_h(x) \in V^h : v_h(0) = v_h(1) = 0\}.$$

A continuous piecewise polynomial can have a weak derivative as defined by Definition 3.1, which is the piecewise derivative inside each interval, just like that the weak derivative of $f(x) = |x|$ is the step function. Thus we have the following fact:

$$V^h \subset H^1(0,1), \quad V_0^h \subset H_0^1(0,1).$$

Given $V_0^h$, the abstract finite element method for approximating (3.7) is defined as

$$\text{seek} \quad u_h \in V_0^h, \quad \text{satisfying} \quad \mathcal{A}(u_h, v_h) = (f, v_h), \quad \forall v_h \in V_0^h. \tag{3.8}$$

We call it the abstract finite element method because it can never be exactly implemented. For example, the right hand side integral $(f, v_h)$ can never be computed exactly, unless $f(x)$ is a special function like polynomials.

### 3.3.4 The abstract implementation

Assume we know how to compute all integrals in (3.8), e.g., if the coefficient $a(x) \equiv 1$ and $f(x)$ is a polynomial in (3.6), then all integrands are polynomials. Then let us think about how the scheme (3.8) should be implemented, e.g., in the scheme (3.8) what does arbitrariness of the test function $v_h$ mean?

First of all, once the mesh is fixed and polynomial degree is fixed, the piecewise polynomial space $V_0^h$ is a finite dimensional vector space. Assume it is $N$-dimensional with basis functions $\{\phi_i(x) : i = 1, \cdots, N\}$.

Second, in the scheme (3.8), both the left hand side and the right hand side are linear operators with respect to the test function $v_h$. Therefore, $A(u_h, v_h) = (f, v_h)$ for an arbitrary test function $v_h$ in an $N$-dimensional vector space $V_0^h$ is equivalent to $A(u_h, v_h) = (f, v_h)$ for $v_h$ being all basis functions $\phi_i(x)$. Namely, (3.8) is equivalent to

$$\mathcal{A}(u_h, \phi_i) = (f, \phi_i), \quad i = 1, \cdots, N.$$

Third, $u_h \in V_0^h$ implies that $u_h$ is a linear combination of the basis functions:

$$u_h(x) = \sum_{j=1}^{N} u_j \phi_j(x).$$

Plugging it in and using the linearity of the bilinear form $\mathcal{A}$, we get that

$$\sum_{j=1}^{N} u_j \mathcal{A}(\phi_j, \phi_i) = (f, \phi_i), \quad i = 1, \cdots, N,$$

which is a system of $N$ linear equations.

The last step is to solve a linear system $S\mathbf{u} = \mathbf{f}$ where the *stiffness* matrix $S$ has entries $S_{ij} = \mathcal{A}(\phi_j, \phi_i)$, and

$$\mathbf{u} = \begin{pmatrix} u_1 \\ u_2 \\ \vdots \\ u_N \end{pmatrix}, \quad \mathbf{f} = \begin{pmatrix} (f, \phi_1) \\ (f, \phi_2) \\ \vdots \\ (f, \phi_N) \end{pmatrix}.$$

### 3.3.5 *The simple implementation on uniform meshes*

To implement the scheme (3.8), we need to compute integrals. One convenient choice is to use quadrature. Let us use trapezoidal rule for $P^1$ method (and Simpson's rule for $P^2$ method). Let $\mathcal{A}_h(\cdot, \cdot)$ and $\langle f, v_h \rangle_h$ denote the quadrature approximation to $\mathcal{A}(\cdot, \cdot)$ and $(f, v_h)$ respectively. Then we get a new scheme

$$\text{seek} \quad u_h \in V_0^h, \quad \text{satisfying} \quad \mathcal{A}_h(u_h, v_h) = \langle f, v_h \rangle_h, \quad \forall v_h \in V_0^h. \tag{3.9}$$

Assume that there are $N$ interior grid points in the $P^1$ mesh in Figure 3.1 (or $P^2$ mesh in Figure 3.2). Let $\phi_i(x)$ $(i = 1, \cdots, N)$ denote the basis functions in $V_0^h$ satisfying

$$\phi_i(x_j) = \delta_{ij} = \begin{cases} 1 & i = j, \\ 0 & i \neq j, \end{cases}, \quad \forall j = 1, \cdots, N.$$

This kind of basis is often called *Lagrange basis* or *nodal basis.* For instance, $\phi_i(x)$ for $P^1$ method is given as the *hat function*:

$$\phi_i(x) = \begin{cases} \frac{1}{h}(x - x_{i-1}), & x \in [x_{i-1}, x_i], \\ \frac{1}{h}(x_{i+1} - x), & x \in [x_i, x_{i+1}], \\ 0, & \text{otherwise,} \end{cases}$$

and its weak derivative is

$$\phi_i'(x) = \begin{cases} \frac{1}{h}, & x \in [x_{i-1}, x_i], \\ -\frac{1}{h}, & x \in [x_i, x_{i+1}], \\ 0, & \text{otherwise.} \end{cases}$$

The $P^2$ Lagrange basis can be similarly written out using Lagrange interpolation by quadratic polynomials. For Lagrange basis $\phi_i(x)$, we have $u_i = u_h(x_i)$ because

$$\phi_j(x_i) = \delta_{ij} \Rightarrow u_h(x_i) = \sum_{j=1}^{N} u_j \phi_j(x_i) = \sum_{j=1}^{N} u_j \delta_{ij} = u_i,$$

thus $u_h$ can also be denoted as a vector of point values:

$$\mathbf{u} = \begin{bmatrix} u_1 \\ u_2 \\ \vdots \\ u_N \end{bmatrix}.$$

Since it suffices to ask $\mathcal{A}_h(u_h, v_h)$ to hold for $v_h = \phi_i$ for all $i$, plugging $u_h = \sum\limits_{j=1}^{N} u_j \phi_j(x)$ into the bilinear form, the scheme (3.9) is equivalent to a linear system for the unknown vector of point values $\mathbf{u} \in \mathbb{R}^N$ satisfying

$$\sum_{j=1}^{N} \mathcal{A}_h(\phi_j(x), \phi_i(x)) u_j = \langle f, \phi_i(x) \rangle_h, \quad \forall i = 1, \cdots, N. \tag{3.10}$$

Let $f_i = f(x_i)$, then the right hand side for $P^1$ element can be explicitly written as

$$\langle f, \phi_i(x) \rangle_h = \sum_{k=0}^{N} \frac{1}{2} h[f(x_k)\phi_i(x_k) + f(x_{k+1})\phi_i(x_{k+1})] = f_i h,$$

where we have used the fact that $\phi_i \in V_0^h \Rightarrow \phi_i(0) = 0, \phi_i(1) = 0$. So the matrix-vector form of (3.10) is $S\mathbf{u} = h\mathbf{f}$ where $\mathbf{f}$ is now the vector consisting of point values $f(x_i)$, and the stiffness matrix $S$ has its $(i, j)$-th entry as

$$S_{ij} = \mathcal{A}_h(\phi_j(x), \phi_i(x)).$$

Consider the simplest constant coefficient case $a(x) \equiv 1$ (i.e., approximating Laplacian), then

$$S_{ij} = \mathcal{A}_h(\phi_j(x), \phi_i(x)) = \langle \phi_j'(x), \phi_i'(x)\rangle_h = \begin{cases} \frac{2}{h} & i = j \\ -\frac{1}{h} & i = j \pm 1 \\ 0 & \text{otherwise.} \end{cases}$$

In other words, for solving $-u'' = f, u(0) = u(1) = 0$, the matrix-vector form of the $P^1$ finite element method with trapezoidal quadrature is precisely the second order centered difference in the previous chapter:

$$\frac{1}{h}\begin{pmatrix} 2 & -1 & & & & \\ -1 & 2 & -1 & & & \\ & -1 & 2 & -1 & & \\ & & \ddots & \ddots & \ddots & \\ & & & -1 & 2 & -1 \\ & & & & -1 & 2 \end{pmatrix}\begin{bmatrix} u_1 \\ u_2 \\ u_3 \\ \vdots \\ u_{N-1} \\ u_N \end{bmatrix} = h\begin{bmatrix} f_1 \\ f_2 \\ f_3 \\ \vdots \\ f_{N-1} \\ f_N \end{bmatrix}.$$

For the variable coefficient problem $-(au')' = f, u(0) = u(1) = 0$, similarly we can derive the matrix-vector form for the scheme (3.10) with piecewise linear basis:

$$\frac{1}{h}\frac{1}{2}\begin{pmatrix} a_0 + 2a_1 + a_2 & -a_1 - a_2 & \\ -a_1 - a_2 & a_1 + 2a_2 + a_3 & -a_2 - a_3 \\ \ddots & \ddots & \ddots \end{pmatrix}\begin{bmatrix} u_1 \\ u_2 \\ \vdots \end{bmatrix} = h\begin{bmatrix} f_1 \\ f_2 \\ \vdots \end{bmatrix} \tag{3.11}$$

**Remark 3.1.** *If we compare the scheme* (3.11) *to the abstract finite element method* (3.8)*, one key difference is that we need continuous functions* $a(x)$ *and* $f(x)$ *in* (3.11) *such that* $a_i$ *and* $f_i$ *are well defined. In contrast,* (3.8) *involves only integration of* $a(x)$ *and* $f(x)$ *thus it allows more general functions such as* $f(x) \in L^2[0,1]$*, e.g.,* $f(x)$ *can be a step function. For implementing* (3.8) *with discontinuous* $a(x)$ *and* $f(x)$*, one needs to choose or design a more sophisticated quadrature rule so that their integrals can be accurately approximated. Here we only consider continuous* $a(x)$ *and* $f(x)$*.*

Recall that the traditional finite difference scheme (2.12) in Chapter 2 is given as

$$\frac{1}{\Delta x^2}\begin{pmatrix} a_{\frac{1}{2}} + a_{\frac{3}{2}} & -a_{\frac{3}{2}} & \\ -a_{\frac{3}{2}} & a_{\frac{3}{2}} + a_{\frac{5}{2}} & -a_{\frac{5}{2}} \\ \ddots & \ddots & \ddots \end{pmatrix}\begin{bmatrix} u_1 \\ u_2 \\ \vdots \end{bmatrix} = \begin{bmatrix} f_1 \\ f_2 \\ \vdots \end{bmatrix},$$

and the matrix can be easily written as $B = D^\top AD$, where $A$ be a diagonal matrix with diagonal entries $a_{\frac{1}{2}}, \cdots, a_{n+\frac{1}{2}}$ and

$$D = \frac{1}{\Delta x}\begin{bmatrix} 1 & & & \\ -1 & 1 & & \\ 0 & \ddots & \ddots & \\ & & -1 & 1 \\ & & & -1 \end{bmatrix}_{(n+1)\times n}. \tag{3.12}$$

Notice that the two schemes (2.12) and (3.11) would be exactly the same if we use an approximation $a_{j+\frac{1}{2}} \approx \frac{a_j+a_{j+1}}{2}$ in (2.12). For smooth $a(x)$, the approximation $a_{j+\frac{1}{2}} \approx \frac{a_j+a_{j+1}}{2}$ is second order accurate by Taylor expansion. Because of this, the stiffness matrix $S$ in the finite element method (3.11) can be easily written as

$$S = \frac{\Delta x}{2} D^\top AD$$

where $A$ is a diagonal matrix with diagonal entries $a_0 + a_1, a_1 + a_2, a_2 + a_3, \cdots$.

**Remark 3.2.** *Sometimes, there are alternatives to compute or approximate integrals in* (3.8) *without using quadrature. For example, we can approximate* $a(x)$ *by its piecewise Lagrange interpolation* $a_h(x)$*, then we can evaluate the integral* $\int_0^1 a_h(x)u_h'(x)v_h'(x)\,\mathrm{d}x$ *exactly. See [Li and Zhang (2020b)] for such an implementation.*

**Problem 3.3.** *For the variable coefficient problem* $-(au')' = f, u(0) = u(1) = 0$*, derive the equivalent matrix-vector form* (3.11) *for the scheme* (3.10) *with piecewise linear basis.*

**Problem 3.4.** *On the mesh with points* $x_0, x_1, \cdots, x_6$ *shown in Figure 3.2, derive the Lagrange basis functions* $\phi_i(x), i = 0, 1, \cdots, 6$ *for the* $P^2$ *method and find the explicit matrix-vector form of the scheme* (3.10) *for the* $-u'' = f, u(0) = u(1) = 0$*. Hint:* $\phi_i(x_j) = \delta_{ij}$ *and it is a quadratic polynomial on each interval* $[x_{2k}, x_{2k+2}]$*. For instance, by the quadratic interpolation formula for the function values* $0, 1, 0$ *at* $x_0, x_1, x_2$*, we have*

$$\phi_1(x) = \begin{cases} 0, & x \notin [x_0, x_2] \\ \frac{(x-x_0)(x-x_2)}{(x_1-x_0)(x_1-x_2)}, & x \in [x_0, x_2] \end{cases}.$$

**Problem 3.5.** *Implement both schemes* (2.12) *and* (3.11) *for a problem with a smooth solution* $u(x)$ *for a smooth coefficient* $a(x)$, *by modifying the MATLAB code in Section 2.2.3. Does one scheme have smaller errors than the other, or do the two schemes have similar errors?*

**Problem 3.6.** *Implement the Preconditioned Conjugate Gradient method in Appendix B for the scheme* (3.11) *using the* $K$ *matrix in Chapter 2 as the preconditioner. Plot three figures for relative error V.S. iteration number, relative residue V.S. iteration number, and cost function value* ($f(x) = \frac{1}{2}x^\top Ax - b^\top x$ *for solving* $Ax = b$) *V.S. iteration number. The exact solution to the linear system can be computed by direct inversion of the matrix.*

## 3.4 Basic properties of the bilinear form

We consider the bilinear form $\mathcal{A}(u,v) = \int_0^1 au'v' \,\mathrm{d}x$ with a continuous coefficient $a(x)$ satisfying $0 \leq C_1 \leq \min_{x\in[0,1]} a(x) \leq a(x) \leq \max_{x\in[0,1]} a(x) \leq C_2$.

### 3.4.1 *Continuity*

The *continuity* of $\mathcal{A}$ means that there is a constant $C$ such that

$$\forall u, v \in H^1(\Omega), \quad |\mathcal{A}(u,v)| \leq C\|u\|_1\|v\|_1,$$

which is implied by the Cauchy-Schwarz inequality:

$$\begin{aligned}\mathcal{A}(u,v) &= \int_0^1 au'v' \,\mathrm{d}x \leq \max_x a(x) \int_0^1 u'v' \,\mathrm{d}x \\ &\leq \max_x a(x) \sqrt{\int_0^1 [u']^2 \,\mathrm{d}x} \sqrt{\int_0^1 [v']^2 \,\mathrm{d}x} \leq C\|u\|_1\|v\|_1.\end{aligned}$$

A continuous bilinear form is also called *bounded*.

### 3.4.2 *Coercivity*

One key concept is called *coercivity* of the bilinear form:

$$\forall v \in H_0^1(\Omega), \quad \mathcal{A}(v,v) = \int_0^1 a(x)v'v' \,\mathrm{d}x \geq \left(\min_x a(x)\right)|v|_1^2 \geq C\|v\|_1^2,$$

where $C > 0$ is another constant, and we have used the fact that the $H^1$-seminorm $|\cdot|_{1,\Omega}$ and the $H^1$-norm $\|\cdot\|_{1,\Omega}$ are *equivalent* in the special function space $H_0^1(\Omega)$. Two norms of the same space are equivalent if they can be controlled by each other with constants. In particular, for $H_0^1(\Omega)$, there is a constant $C > 0$ depending only on $\Omega$ such that

$$\forall v \in H_0^1(\Omega), \quad C\|v\|_{1,\Omega}^2 \leq |v|_{1,\Omega}^2 \leq \|v\|_{1,\Omega}^2. \tag{3.13}$$

The second inequality in (3.13) is trivial. The first inequality in (3.13) simply says that the function value can be controlled by the gradient, which is in general not true. For example, if $v(x) \equiv 1$ on $\Omega = [0, 1]$, then $\|v\|_0 = 1$ and $|v|_1 = \|v'(x)\|_0 = 0$ thus the first inequality cannot hold, unless $v \in H_0^1(\Omega)$.

For $\Omega = (0, 1)$, here are some quick arguments to see why it is even possible to control function values by derivatives for $v(x) \in H_0^1(0, 1)$. If $v(0) = 0$ and $v'(x)$ exists everywhere in the classical sense, then by the Mean Value Theorem we have $\frac{v(x)-v(0)}{x-0} = v'(y)$ for some $y \in (0, x)$, thus $v(x) = xv'(y)$ and $|v(x)| \leq |v'(y)|$ for any $x \in [0, 1]$. The general case of (3.13) can be proven by using *Poincaré inequality* in Theorem 3.1.

**Remark 3.3.** *The estimates in* (3.13) *hold even for a function* $v(x)$ *which vanishes only along a part or a very small part of the boundary of* $\Omega$ *in two dimensions.*

### 3.4.3 *Coercivity is stability*

Recall that the abstract finite element method gives a linear system:

$$\sum_{j=1}^{N} u_j \mathcal{A}(\phi_j, \phi_i) = (f, \phi_i), \quad i = 1, \cdots, N,$$

where $u_j = u_h(x_j)$ if using Lagrange basis $\phi_i(x)$. Whenever a scheme is given as a linear system for an elliptic equation, the first question should be whether the linear system has a solution. In other words, we need to show the solvability, i.e., the invertibility of the *stiffness* matrix $S$ with entries $S_{ij} = \mathcal{A}(\phi_j(x), \phi_i(x))$. In Chapter 2, we computed eigenvalues of $K$ matrix for constant coefficient problems so that we can show the nonsingularity of the matrix $K$.

Obviously, for a variable coefficient problem, e.g., the scheme (3.11), the eigenvalues of the stiffness matrix should be estimated rather than computed because it is nearly impossible to compute. Such an eigenvalue estimate can be done by the coercivity. In this section we only focus on the bilinear form $\mathcal{A}$ and we will discuss $\mathcal{A}_h$ later. For any $v_h \in V_0^h$, we have

$$\mathcal{A}(v_h, v_h) = \mathcal{A}(\sum_{i=1}^{n} v_i\phi_i, \sum_{j=1}^{n} v_j\phi_j) = \sum_{i=1}^{n}\sum_{j=1}^{n} \mathcal{A}(\phi_j, \phi_i)v_jv_i = \mathbf{v}^\top S\mathbf{v}.$$

Let $\|v\|_0$ denote the $H^0$-norm, i.e., the $L^2$-norm of $v$. The coercivity implies

$$\mathcal{A}(v,v) \geq C_1\|v\|_1^2 \geq C_1\|v\|_0^2, \quad \forall v(x) \in H_0^1(\Omega),$$

where the constant $C_1$ only depends on the domain $\Omega$. Thus

$$v_h(x) \in V_0^h \subset H_0^1(\Omega) \Rightarrow \mathcal{A}(v_h, v_h) \geq C_1\|v_h\|_0^2.$$

*Equivalence of Norms Theorem* states that any two norms of a finite dimensional vector space are *equivalent*, i.e., they can be controlled by each other via a constant. Notice that $\|v_h\|_0$ and $\|\mathbf{v}\|$ are both norms of the same finite dimensional vector space $V_0^h$, thus they are equivalent:

$$C_2\|\mathbf{v}\|^2 \leq \|v_h\|_0^2 \leq C_3\|\mathbf{v}\|^2,$$

where the constants $C_2, C_3$ depends on the vector space $V_0^h$. Thus

$$\mathbf{v}^\top S\mathbf{v} = \mathcal{A}(v_h, v_h) \geq C_1\|v_h\|_0^2 \geq C_1C_2\|\mathbf{v}\|^2 \Rightarrow \frac{\mathbf{v}^\top S\mathbf{v}}{\mathbf{v}^\top\mathbf{v}} \geq C_1C_2.$$

Recall that $\mathcal{A}(u,v) = \int_0^1 a(x)u'(x)v'(x)\,\mathrm{d}x$, thus $\mathcal{A}(u,v) = \mathcal{A}(v,u)$ implies that $S$ is real symmetric. By the Courant-Fisher-Weyl min-max principle (see Appendix A.1), $\frac{\mathbf{v}^\top S\mathbf{v}}{\mathbf{v}^\top\mathbf{v}} \geq C_1C_2$ implies that the smallest eigenvalue of $S$ is greater than or equal to $C_1C_2 > 0$. Therefore $S$ is invertible. In Section 3.6.3, we will further show that we can get a useful estimate for $\|S^{-1}\|$ from the coercivity.

## 3.5 Error estimates of the abstract finite element method

We have seen that the finite element method with linear basis and quadrature recovers the $K$ matrix for approximating second order derivative. Even though it gives the same finite difference scheme, it is much more useful to understand the finite difference scheme from the finite element perspective.

We will first show the error estimates for (3.8), i.e., the finite element method without quadrature. We will focus on the analysis for $P^1$ finite element method for the one-dimensional problem with homogeneous Dirichlet boundary conditions to understand the key components in the analysis. Readers should keep in mind that all discussions in this section apply to much more general cases such as solving variable coefficient elliptic equations by $P^k$ polynomial finite element method with Neumann boundary conditions on unstructured meshes for a curved domain in multiple dimensions.

### 3.5.1 $H^1$-norm estimate and convergence

The continuous piecewise polynomial has weak derivatives, thus we have $V_0^h \subset H_0^1(\Omega)$, and such a finite element method is called *conforming*. Here is one simple example of nonconforming finite element space for the Poisson equation: the discontinuous piecewise polynomial space is not a subspace of the $H_0^1(\Omega)$ function space.

**Theorem 3.3** (Galerkin orthogonality). *Let $u$ be the solution to* (3.7). *The solution $u_h$ to the conforming finite element method* (3.8) *satisfies:*

$$\mathcal{A}(u - u_h, w_h) = 0, \quad \forall w_h \in V_0^h.$$

*Proof.* The exact solution $u$ satisfies

$$\mathcal{A}(u, w) = (f, w), \quad \forall w \in H_0^1(\Omega)$$

thus

$$\mathcal{A}(u, w_h) = (f, w_h), \quad \forall w_h \in V_0^h \subset H_0^1(\Omega).$$

The numerical solution $u_h$ satisfies

$$\mathcal{A}(u_h, w_h) = (f, w_h), \quad \forall w_h \in V_0^h.$$

Subtracting these two equations, we get Galerkin orthogonality, which is implied by the choice of approximation space $V_0^h \subset H_0^1(\Omega)$. □

Galerkin orthogonality simply says that the true error $u - u_h$ is somehow "orthogonal" to any test function $w_h$ in $V_0^h$ through the bilinear form $\mathcal{A}$. More importantly, Galerkin orthogonality is the analogue of the consistency or truncation error in Chapter 2. Since $\mathcal{A}(u_h, w_h) = (f, w_h)$, we have

$$\mathcal{A}(u - u_h, w_h) = \mathcal{A}(u, w_h) - \mathcal{A}(u_h, w_h) = \mathcal{A}(u, w_h) - (f, w_h).$$

So Galerkin orthogonality is the same as

$$\mathcal{A}(u, w_h) - (f, w_h) = 0, \quad \forall w_h \in V_0^h,$$

which is nothing but replacing $u_h$ by $u$ in the numerical scheme. On the other hand, it seems that the "truncation error" is zero here, which is due to the direct approximation of the variational form. Notice that the "truncation error" $\mathcal{A}(u, w_h) - (f, w_h)$ is zero only when the test function $w_h$ is in the approximated space $V_0^h$.

**Theorem 3.4** (Céa's Lemma). *Let $u$ be the solution to* (3.7). *The solution $u_h$ to the conforming finite element method* (3.8) *satisfies:*

$$\|u - u_h\|_1 \le C \inf_{w_h \in V_0^h} \|u - w_h\|_1.$$

*Proof.* First of all, we have $u_h \in V_0^h \subset H_0^1(\Omega)$ thus $u - u_h \in H_0^1(\Omega)$. The coercivity implies

$$C_1\|u-u_h\|_1^2 \le \mathcal{A}(u-u_h, u-u_h).$$

Next, we have

$$\mathcal{A}(u-u_h, u-u_h) = \mathcal{A}(u-u_h, w_h-u_h) + \mathcal{A}(u-u_h, u-w_h).$$

Galerkin orthogonality implies $\mathcal{A}(u-u_h, w_h-u_h) = 0$. So we get

$$\mathcal{A}(u-u_h, u-u_h) = \mathcal{A}(u-u_h, u-w_h) \le C_2\|u-u_h\|_1\|u-w_h\|_1,$$

where continuity is used. Finally, coercivity and continuity imply

$$C_1\|u-u_h\|_1^2 \le \mathcal{A}(u-u_h, u-u_h) = \mathcal{A}(u-u_h, u-w_h) \le C_2\|u-u_h\|_1\|u-w_h\|_1,$$

thus

$$\|u-u_h\|_1^2 \le C_2\|u-u_h\|_1\|u-w_h\|_1.$$

So we have $\|u-u_h\|_1 \le C_2\|u-w_h\|_1$ for any $w_h \in V_0^h$. □

Céa's Lemma says that the finite element solution error is controlled by the best piecewise polynomial approximation error, which we do not know. On the other hand, we do know polynomial interpolation error estimates (3.2) and (3.3). Assuming $u \in H^2(\Omega)$ or $u \in H^3(\Omega)$ (i.e., assuming $u$ is smooth enough), we easily obtain error estimate in $H^1$-norm for the finite element method using $P^k$ basis:

$$\|u-u_h\|_1 \le C \inf_{w_h\in V_0^h} \|u-w_h\|_1 \le C\|u-\Pi_k u\|_1 = \begin{cases} Ch|u|_2, & k=1 \\ Ch^2|u|_3, & k=2 \end{cases}. \tag{3.14}$$

Céa's Lemma gives (3.14), which is the *convergence* of the finite element method. On the other hand, Céa's Lemma is implied by both Galerkin orthogonality (consistency) and Coercivity (stability). So we get the same conclusion as in Chapter 2 for linear schemes solving linear PDEs:

$$\text{consistency} + \text{stability} \longrightarrow \text{convergence}.$$

Recall that $u_h \in V_0^h \subset H_0^1(\Omega)$ and $u \in H_0^1(\Omega)$, thus the error function $u - u_h$ is an element in the $H_0^1(\Omega)$ space, in which $H^1$-norm is equivalent to the $H^1$-seminorm. So the $H^1$-norm estimate above simply implies that $P^1$ finite element method generates a numerical solution satisfying that

$$\sqrt{\int_0^1 |u'(x) - u_h'(x)|^2 \, \mathrm{d}x} = \mathcal{O}(h).$$

For function values, we will get one order higher, explained in the next subsection.

### 3.5.2 * $L^2$-norm estimate

Beginners may skip this subsection and only read it after getting more familiar with the finite element method.

Notice that $H^1$ estimate cannot explain why $P^1$ method gives a second order accurate scheme. Recall that $H^1$-norm, i.e., $\|u - u_h\|_1$, measures the error both in the function value and the first order derivative. The $L^2$-norm, i.e., $\|u - u_h\|_0$ measures the error in the function value. For example, we already know that the $P^1$ finite element method on a uniform mesh gives exactly the standard centered finite difference, which is second order accurate for the function value.

The $L^2$ estimate will be one order higher than $H^1$ estimate:

**Theorem 3.5** (Aubin-Nitsche Lemma). *Let $u$ be the solution to* (3.7). *The solution $u_h$ to the conforming finite element method* (3.8) *satisfies:*

$$\|u - u_h\|_0 \leq Ch\|u - u_h\|_1,$$

*where $h$ is the mesh size.*

For proving the Aubin-Nitsche Lemma, we need a basic fact about the Poisson equation, which is called *elliptic regularity*: the solution to (3.7) satisfies $\|u\|_2 \leq C\|f\|_0$, which simply says that the second order derivative of $u$ and lower order ones are controlled by function value of $f(x)$.

Even though we only seek $u(x) \in H_0^1(\Omega)$ in (3.7), the elliptic regularity theorem guarantees that $f(x) \in L^2(\Omega) \Rightarrow u(x) \in H^2(\Omega)$. In particular, if $f(x)$ is infinitely differentiable, then so is $u(x)$. The elliptic regularity can be proven under certain assumptions for the domain $\Omega$.

We also need a *dual* problem to help us here. A dual problem of (3.7) is to find $w \in H_0^1(\Omega)$ satisfying

$$\mathcal{A}(w, v) = (u - u_h, v), \quad \forall v \in H_0^1(\Omega). \tag{3.15}$$

The equivalent PDE form of the dual problem above is

$$-w''(x) = u(x) - u_h(x),$$

if the original PDE we want to solve is $-u''(x) = f(x)$.

The elliptic regularity theorem on the dual problem gives the following

$$\|w\|_2 \leq C\|u - u_h\|_0.$$

For the dual problem, its finite element solution for finding $w_h \in V_0^h$ s.t.

$$\mathcal{A}(v_h, w_h) = (u - u_h, v_h), \quad \forall v_h \in V_0^h,$$

where we have used the symmetry of the bilinear form $\mathcal{A}(w,v) = \mathcal{A}(v,w)$. By Céa's Lemma and $H^1$-estimate applied to the finite element solution $w_h$ for approximating the dual problem, we have

$$\|w - w_h\|_1 \leq Ch|w|_2 \leq Ch\|w\|_2,$$

where we have used the interpolation error estimate. With elliptic regularity for the dual problem $\|w\|_2 \leq C\|u - u_h\|_0$, we get

$$\|w - w_h\|_1 \leq Ch\|w\|_2 \leq Ch\|u - u_h\|_0.$$

Now we can understand how $L^2$-norm estimate in Aubin-Nitsche Lemma is proven:

*Proof.* Let $w_h$ be the finite element solution for the dual problem. Then Galerkin orthogonality implies $\mathcal{A}(u - u_h, w_h) = 0$, thus

$$\mathcal{A}(u - u_h, w) = \mathcal{A}(u - u_h, w - w_h) + \mathcal{A}(u - u_h, w_h) = \mathcal{A}(u - u_h, w - w_h).$$

Continuity implies

$$\mathcal{A}(u - u_h, w - w_h) \leq C\|u - u_h\|_1\|w - w_h\|_1.$$

Recall that $w$ is the solution to the dual problem thus plugging in $v = u - u_h \in H_0^1(\Omega)$ in (3.15), we get

$$\mathcal{A}(u - u_h, w) = \mathcal{A}(w, u - u_h) = (u - u_h, u - u_h) = \|u - u_h\|_0^2.$$

Finally, putting everything together

$$\|u-u_h\|_0^2 = \mathcal{A}(u-u_h, w) \leq C\|u-u_h\|_1\|w-w_h\|_1 \leq C\|u-u_h\|_1 h\|u-u_h\|_0,$$

which gives

$$\|u - u_h\|_0 \leq Ch\|u - u_h\|_1.$$

□

With the $H^1$-norm estimate (3.14), the Aubin-Nitsche Lemma gives us the $L^2$-norm error estimates:

$$\|u - u_h\|_0 \leq Ch\|u - u_h\|_1 = \begin{cases} Ch^2|u|_2, & k = 1 \\ Ch^3|u|_3, & k = 2 \end{cases}. \tag{3.16}$$

This is consistent with what we already know: $P^1$ finite element method gives a second order accurate scheme for function values. The $P^2$ finite element method gives a third order accurate scheme for function values, which is consistent with the interpolation error order (3.3). However, if we implement $P^2$ finite element method as a finite difference scheme, we can actually get a fourth order accurate finite difference scheme, which is called *superconvergence*. It will be explained in Section 3.14.

**Remark 3.4.** *In estimates like* (3.14) *and* (3.16), *it is already assumed that* $u$ *should be smooth enough such that either* $u \in H^2(\Omega)$ *or* $u \in H^3(\Omega)$. *The elliptic regularity theorem implies that* $f(x) \in L^2(\Omega) \Rightarrow u \in H^2(\Omega)$ *and* $f(x) \in H^1(\Omega) \Rightarrow u \in H^3(\Omega)$.

### 3.5.3 *Summarization and comparison*

Now let us just focus on the $P^1$ finite element method and think about how the second order accuracy is proven differently from the one we did in Chapter 2. In Chapter 2, we computed the eigenvalues of the $K$ matrix for proving stability $\|A^{-1}\| \leq C$ in a matrix-vector form of the scheme $A\mathbf{u} = \mathbf{f}$. On the one hand, it only requires simpler knowledge of linear algebra. On the other hand, it is highly restrictive because we cannot even compute eigenvalues for a one-dimensional variable coefficient problem.

The discussion in this section obviously applies to the variable coefficient problem, but we need so many much more advanced tools such as Sobolev spaces and elliptic regularity. Recall how exactly we can prove the second order error in $P^1$ finite element method for function values:

(1) The homogeneous Dirichlet boundary condition is built into the function space $H_0^1(\Omega)$, which in return gives the *Poincaré inequality*:

$$\int_0^1 |v'(x)|^2 \, \mathrm{d}x \geq C \left[ \int_0^1 |v'(x)|^2 \, \mathrm{d}x + \int_0^1 |v(x)|^2 \, \mathrm{d}x \right], \quad \forall v(x) \in H_0^1(\Omega).$$

(2) The *Poincaré inequality* gives the *coercivity*

$$\mathcal{A}(v_h, v_h) \geq C\|v_h\|_0^2,$$

which is the *stability.*

(3) From the fact that it is conforming $V_0^h \subset H_0^1(\Omega)$, *Galerkin orthogonality* is easily obtained:

$$\mathcal{A}(u - u_h, v_h) = 0, \quad \forall v_h \in V_0^h.$$

*Galerkin orthogonality* is the *consistency.*

(4) With *Galerkin orthogonality* and *coercivity*, we get Céa's Lemma, which says the finite element solution error is controlled by the best piecewise polynomial approximation error:

$$\|u - u_h\|_1 \leq C \inf_{w_h \in V_0^h} \|u - w_h\|_1.$$

In other words, consistency and stability imply convergence.

(5) We know the interpolation error using $P^k$ polynomials, so the $H^1$-estimate is simply by Céa's Lemma:

$$\|u-u_h\|_1 \leq C \inf_{v_h\in V_0^h} \|u-v_h\|_1 \leq C\|u-\Pi_k u\|_1 \leq \begin{cases} Ch|u|_2, & k=1 \\ Ch^2|u|_3, & k=2 \end{cases}.$$

(6) Finally, with the elliptic regularity on a dual problem, we get the Aubin-Nitsche Lemma for error in $L^2$-norm (the $H^0$-norm):

$$\|u-u_h\|_0 \leq Ch\|u-u_h\|_1 \leq \begin{cases} Ch^2|u|_2, & k=1 \\ Ch^3|u|_3, & k=2 \end{cases}.$$

## 3.6 Properties of the bilinear form with quadrature

Since in practice quadrature is used to implement the finite element method, we also need to know whether coercivity and continuity hold for $\mathcal{A}_h$. Usually the discrete continuity can be easily derived from Cauchy-Schwarz inequality. The discrete coercivity is called *$V^h$-ellipticity.*

We only consider the continuous piecewise linear space $V_0^h$ as an example in this section. Let $x_i$ $(i=0,1,\cdots,N+1)$ be a uniform grid for the whole interval $[0,1]$, where $x_0=0$ and $x_{N+1}=1$ are boundary points. The grid spacing is $h=\frac{1}{N+1}$.

### 3.6.1 *Equivalent norms of the piecewise linear polynomial*

Everything in this subsection can be derived by abstract arguments. But instead we use some explicit elementary tools to derive what we need for coercivity.

For any $v_h\in V_0^h$, let $v_i=v_h(x_i)$ and $\mathbf{v}=\begin{bmatrix}v_1 & \cdots & v_N\end{bmatrix}^\top$. So $\|v_h\|_0$ and $\|\mathbf{v}\|$ are both norms of $V_0^h$, thus by *Equivalence of Norms Theorem*, they satisfy

$$C_2\|\mathbf{v}\|^2 \leq \|v_h\|_0^2 \leq C_3\|\mathbf{v}\|^2,$$

where $C_2, C_3$ depend on the dimension $N$ of the vector space $V_0^h$.

It is useful to figure out the exact dependence of these constants on the dimension $N$ or the mesh size $h$. For the one-dimensional problem continuous piecewise linear polynomial space $V_0^h$ on a uniform mesh with mesh size $h$, we have

$$\|v_h\|_0^2 = \sum_{j=0}^{N}\int_{x_j}^{x_{j+1}} |v_h(x)|^2\,\mathrm{d}x = \sum_{j=0}^{N}\int_0^h \left[\frac{v_{j+1}-v_j}{h}x+v_j\right]^2 \mathrm{d}x$$

$$= \frac{1}{3}h\sum_{j=0}^{N}\left(v_{j+1}^2 + v_j^2 + v_{j+1}v_j\right) = \frac{2}{3}h\sum_{j=1}^{N} v_j^2 + \frac{1}{3}h\sum_{j=0}^{N} v_{j+1}v_j,$$

where the last step is implied by $v_h(x) \in V_0^h \Rightarrow v_0 = v_{N+1} = 0$. With two simple inequalities

$$-\frac{1}{2}v_{j+1}^2 - \frac{1}{2}v_j^2 \leq v_{j+1}v_j \leq \frac{1}{2}v_{j+1}^2 + \frac{1}{2}v_j^2,$$

we can derive

$$\frac{1}{3}h\|\mathbf{v}\|^2 \leq \|v_h\|_0^2 \leq h\|\mathbf{v}\|^2. \tag{3.17}$$

### 3.6.2 *Coercivity and continuity of the bilinear form $\mathcal{A}_h$*

Recall that $v_h(x) \in V_0^h$ is a piecewise polynomial thus weakly differentiable, and its weak derivative $v_h'(x)$ is a piecewise constant function:

$$v_h'(x) = \frac{v_{j+1} - v_j}{h}, \quad x \in [x_j, x_{j+1}], \quad j = 1, \cdots, N.$$

Thus $v_h'(x_j)$ is double valued at interior nodes $x_j$. Let $v_h'(x_j)^-$ and $v_h'(x_j)^+$ denote two values obtained by taking derivatives in the intervals $[x_{j-1}, x_j]$ and $[x_j, x_{j+1}]$ respectively. In other words,

$$v_h'(x_{j+1})^- = \frac{v_{j+1} - v_j}{h} = v_h'(x_j)^+. \tag{3.18}$$

For integrating a piecewise constant function $v_h'(x)$, we have

$$\int_0^1 [v_h'(x)]^2 \, \mathrm{d}x = h\sum_{j=0}^{N} \frac{(v_j - v_{j+1})^2}{h^2}.$$

If using trapezoidal rule for $P^1$ finite element method in each cell $I_j = [x_j, x_{j+1}]$ in Figure 3.1, then for any $v_h(x) \in V_0^h$ we have

$$\begin{aligned}
\mathcal{A}_h(v_h, v_h) &= \sum_{j=0}^{N} \frac{h}{2}\left[a(x_j)[v_h'(x_j)^+]^2 + a(x_{j+1})[v_h'(x_{j+1})^-]^2\right] \\
&= \sum_{j=0}^{N} \frac{h}{2}\left[a(x_j)\frac{(v_j - v_{j+1})^2}{h^2} + a(x_{j+1})\frac{(v_j - v_{j+1})^2}{h^2}\right] \\
&\geq \min_j a(x_j) h \sum_{j=0}^{N} \frac{(v_j - v_{j+1})^2}{h^2} \\
&= \min_j a(x_j) \int_0^1 [v_h'(x)]^2 \, \mathrm{d}x = \min_j a(x_j) |v_h'(x)|_1^2.
\end{aligned}$$

With (3.13), we obtain coercivity of the bilinear form $\mathcal{A}_h$:

$$\mathcal{A}_h(v_h, v_h) = \min_j a(x_j)|v_h(x)|_1^2 \geq C \min_x a(x)\|v_h(x)\|_1^2,$$

where the constant $C$ is independent of $h$ or $N$.

For continuity of $\mathcal{A}_h$, consider any $w_h, v_h \in V_0^h$, then with (3.18) we have

$$\begin{aligned}
&\mathcal{A}_h(w_h, v_h)\\
&= \sum_{j=0}^{N} \frac{h}{2}\left(a(x_j)[w_h'(x_j)^+][v_h'(x_j)^+] + a(x_{j+1})[w_h'(x_{j+1})^-][v_h'(x_{j+1})^-]\right)\\
&\leq \max_j a(x_j) h \sum_{j=0}^{N} \frac{w_{j+1}-w_j}{h}\frac{v_{j+1}-v_j}{h}\\
&\leq \max_j a(x_j) h \sqrt{\sum_{j=0}^{N} \frac{(w_{j+1}-w_j)^2}{h^2}}\sqrt{\sum_{j=0}^{N} \frac{(v_{j+1}-v_j)^2}{h^2}}\\
&= \max_j a(x_j)|w_h|_1|v_h|_1 \leq \max_x a(x)||w_h||_1||v_h||_1,
\end{aligned}$$

where we have used the Cauchy-Schwarz inequality for vectors

$$\sum_i a_i b_i \leq \sqrt{\sum_i a_i^2}\sqrt{\sum_i b_i^2}.$$

### 3.6.3 *Coercivity implies stability of the finite difference scheme*

Recall that in Section 3.4.3 we have shown the nonsingularity of the *stiffness* matrix for the abstract finite element method without any quadrature.

Now we are ready to discuss how the $V^h$-ellipticity or the discrete coercivity can imply nonsingularity of the *stiffness* matrix $S$ with entries $S_{ij} = \mathcal{A}_h(\phi_j, \phi_i)$ for the finite element method with quadrature. In particular, for $P^1$ finite element method with trapezoidal rule solving a variable coefficient problem, from (3.11) we know the stiffness matrix can be written as

$$S = \frac{1}{h}\frac{1}{2}\begin{pmatrix} a_0 + 2a_1 + a_2 & -a_1 - a_2 & & \\ -a_1 - a_2 & a_1 + 2a_2 + a_3 & -a_2 - a_3 & \\ & & & \\ & \ddots & \ddots & \ddots \end{pmatrix}$$

Since $v_h(x) = \sum_{j=1}^{N} v_j \phi_j(x)$, we have

$$\mathcal{A}_h(v_h, v_h) = \sum_{i=1}^{N}\sum_{j=1}^{N} \mathcal{A}_h(\phi_j(x), \phi_i(x)) v_i v_j = \mathbf{v}^\top S \mathbf{v}.$$

With the coercivity in Section 3.6.2 and (3.17), we have

$$\mathcal{A}_h(v_h, v_h) \geq C\|v_h\|_1^2 \geq C\|v_h\|_0^2 \geq Ch\|\mathbf{v}\|^2$$

So we have $\mathbf{v}^\top S\mathbf{v} \geq Ch\|\mathbf{v}\|^2$ for any $\mathbf{v} \in \mathbb{R}^N$, which implies $S$ is positive definite. The symmetry of $S$ is implied by $\mathcal{A}_h(w_h, v_h) = \mathcal{A}_h(v_h, w_h)$. By the Courant-Fisher-Weyl min-max principle (see Appendix A.1), $\frac{\mathbf{v}^\top S\mathbf{v}}{\mathbf{v}^\top\mathbf{v}} \geq Ch$ implies that the smallest eigenvalue of $S$ is greater than or equal to $Ch > 0$. Therefore $S$ is invertible.

The matrix-vector form of (3.11) is $S\mathbf{u} = h\mathbf{f}$, thus $\mathbf{u} = hS^{-1}\mathbf{f}$. Since we have shown $S$ is real symmetric positive definite, its singular values are also eigenvalues for both $S$ and $S^{-1}$. So $\|S^{-1}\|$ is simply the reciprocal of the smallest eigenvalue of $S$. Therefore we get $\|hS^{-1}\| = h\|S^{-1}\| \leq C$, which is precisely the stability in the sense of traditional finite difference method in Chapter 2.

**Problem 3.7.** *Recall that the traditional finite difference scheme* (2.12) *in Chapter 2 is given as*

$$\frac{1}{\Delta x^2}\begin{pmatrix} a_{\frac{1}{2}} + a_{\frac{0}{2}} & a_{\frac{3}{2}} & & \\ -a_{\frac{3}{2}} & a_{\frac{3}{2}} + a_{\frac{5}{2}} & -a_{\frac{5}{2}} & \\ & \ddots & \ddots & \ddots \end{pmatrix}\begin{bmatrix} u_1 \\ u_2 \\ \vdots \end{bmatrix} = \begin{bmatrix} f_1 \\ f_2 \\ \vdots \end{bmatrix}.$$

*Apply the discussion in this section to prove the stability of this scheme. The consistency or the truncation error of the scheme* (2.12) *is straightforward to derive. Once we have the stability, we have its convergence following Chapter 2. Hint: we can get its stability if we can define some $b_i$ so that it has an equivalent form*

$$\frac{1}{h}\frac{1}{2}\begin{pmatrix} b_0 + 2b_1 + b_2 & -b_1 - b_2 & & \\ -b_1 - b_2 & b_1 + 2b_2 + b_3 & -b_2 - b_3 & \\ & \ddots & \ddots & \ddots \end{pmatrix}\begin{bmatrix} u_1 \\ u_2 \\ \vdots \end{bmatrix} = h\begin{bmatrix} f_1 \\ f_2 \\ \vdots \end{bmatrix}.$$

*So how do we define $b_i$ so that they are equivalent?*

## 3.7 *Error estimates of the finite element method with quadrature

In order to derive the error estimates of the finite element method with quadrature (3.9), we need to show all the lemmas and theorems in Section 3.5 also hold when $\mathcal{A}(\cdot,\cdot)$ is replaced by $\mathcal{A}_h(\cdot,\cdot)$. If this is your first time reading this chapter, you can assume that this is true and skip this section.

### 3.7.1 *First Strang Lemma*

The First Strang Lemma is the Céa's Lemma for the scheme (3.9).

**Theorem 3.6** (First Strang Lemma).

$$\|u-u_h\|_1 \le C \inf_{v_h\in V_0^h}\left\{\|u-v_h\|_1 + \sup_{w_h\in V_0^h}\frac{|\mathcal{A}(v_h,w_h)-\mathcal{A}_h(v_h,w_h)|}{\|w_h\|_1}\right\}$$

$$+C\sup_{w_h\in V_0^h}\frac{|\langle f,w_h\rangle_h-(f,w_h)|}{\|w_h\|_1}.$$

**Remark 3.5.** *Compared to Céa's Lemma, the extra terms in the First Strang Lemma is nothing but quadrature error terms.*

*Proof.* First, we can rewrite the bilinear form

$$\begin{aligned}\mathcal{A}_h(u_h-v_h,u_h-v_h) &= \mathcal{A}_h(u_h,u_h-v_h)-\mathcal{A}_h(v_h,u_h-v_h)\\ &\quad+\mathcal{A}(u-v_h,u_h-v_h)-\mathcal{A}(u-v_h,u_h-v_h)\\ &= \mathcal{A}_h(u_h,u_h-v_h)-\mathcal{A}_h(v_h,u_h-v_h)\\ &\quad+\mathcal{A}(u-v_h,u_h-v_h)+\mathcal{A}(v_h,u_h-v_h)-\mathcal{A}(u,u_h-v_h).\end{aligned}$$

By coercivity of $\mathcal{A}_h$, and the facts $\mathcal{A}_h(u_h,u_h-v_h)=\langle f,u_h-v_h\rangle_h$ and $\mathcal{A}(u,u_h-v_h)=(f,u_h-v_h)$, we get

$$\begin{aligned}C\|u_h-v_h\|_1^2 \le \mathcal{A}_h(u_h-v_h,u_h-v_h)&=\mathcal{A}(u-v_h,u_h-v_h)+\mathcal{A}(v_h,u_h-v_h)\\ &-\mathcal{A}_h(v_h,u_h-v_h)+\langle f,u_h-v_h\rangle_h-(f,u_h-v_h),\quad \forall v_h\in V_0^h.\end{aligned}$$

With $\mathcal{A}(u-v_h,u_h-v_h)\le C_2\|u-v_h\|_1\|u_h-v_h\|_1$, we have

$$\begin{aligned}C\|u_h-v_h\|_1 \le \quad & C_2\|u-v_h\|_1+\frac{\mathcal{A}(v_h,u_h-v_h)-\mathcal{A}_h(v_h,u_h-v_h)}{\|u_h-v_h\|_1}\\ &+\frac{\langle f,u_h-v_h\rangle_h-(f,u_h-v_h)}{\|u_h-v_h\|_1}\end{aligned}$$

thus

$$\begin{aligned}\|u_h - v_h\|_1 \leq \quad & C\|u - v_h\|_1 + C \sup_{w_h \in V_0^h} \frac{|\mathcal{A}(v_h, w_h) - \mathcal{A}_h(v_h, w_h)|}{\|w_h\|_1} \\ & + C \sup_{w_h \in V_0^h} \frac{|\langle f, w_h \rangle_h - (f, w_h)|}{\|w_h\|_1}.\end{aligned}$$

The proof is done after using the triangle inequality: $\|u - u_h\|_1 \leq \|u - v_h\|_1 + \|u_h - v_h\|_1$. □

### 3.7.2 *Quadrature estimate: Bramble-Hilbert Lemma*

The first Strang Lemma means that the Céa's Lemma holds up to the quadrature error, which can be estimated by the Bramble-Hilbert Lemma for a continuous linear form (i.e., a continuous/bounded linear mapping from function space to $\mathbb{R}$):

**Theorem 3.7** (Bramble-Hilbert Lemma). *For some integer $k \geq 0$ and a suitable domain $\Omega$, let $\mathcal{L}$ be a continuous linear form on the space $H^{k+1}(\Omega)$ with the property that $\forall p(x) \in P^k(\Omega)$ (all polynomials of degree $k$), $\mathcal{L}[p(x)] = 0$. Then*

$$|\mathcal{L}(f)| \leq C\|\mathcal{L}\|_{k+1}^* |f|_{k+1},$$

*where $\|\cdot\|_{k+1}^*$ is the operator norm, and $|f|_{k+1} = \sqrt{\int_\Omega |f^{(k+1)}(x)|^2 \, dx}$ for $\Omega = [0, 1]$.*

The notation in the Bramble-Hilbert Lemma are abstract but one typical example of such a continuous linear form (i.e., a bounded linear functional) is the interpolation error operator. For instance, given point values $f_i$ of a continuous function $f(x)$ on a uniform mesh, we can do a piecewise linear polynomial interpolation as in Section 3.2.2. The interpolation error $\mathcal{L}(f) = f(x_0) - \Pi_1 f(x_0)$ at a given location $x_0$ is a linear operator with respect to $f(x)$, and the interpolation error is always zero if $f(x)$ is a linear polynomial. Then the Bramble-Hilbert Lemma implies that this piecewise linear interpolation error is controlled by $|f|_2$, which contains the second order (weak) derivative. On the other hand, we can also get a similar conclusion that the piecewise linear interpolation error is dominated by or related to $f''(x)$ through Taylor expansion. So if you prefer, you can think of the Bramble-Hilbert Lemma as the better alternative as opposed to performing Taylor expansion.

The power of the abstraction in the Bramble-Hilbert Lemma lies in the fact that we can use it to easily extend the interpolation and quadrature error estimates in Section 3.2.2 to unstructured meshes on any shape of domain. Recall that the $H^1$-norm error estimate of the abstract finite element method is built upon the interpolation error estimate. This is why the arguments for deriving error estimates in this chapter also apply to any general setup such as problems in multiple dimensions.

**Example 3.3.** *Consider the quadrature error operator, which is linear and also zero for polynomials of certain degree. For instance, if considering the trapezoidal rule for each interval in Figure 3.1, then*

$$\int_0^1 f(x)\,\mathrm{d}x - \sum_{i=0}^{N} \frac{1}{2}h[f(x_i)+f(x_{i+1})] = \sum_{i=0}^{N}\left(\int_{x_i}^{x_{i+1}} f(x)\,\mathrm{d}x - \frac{1}{2}h[f(x_i)+f(x_{i+1})]\right).$$

*Consider a mapping from the small cell $[x_i, x_{i+1}]$ to the reference cell $[0,1]$ by*

$$x = h\hat{x} + x_i, \quad \hat{f}(\hat{x}) = f(h\hat{x}+x_i).$$

*Let*

$$E_i(f) = \int_{x_i}^{x_{i+1}} f(x)\,\mathrm{d}x - \frac{1}{2}h[f(x_i)+f(x_{i+1})]$$

*be the quadrature error on a small inverval, and*

$$\hat{E}(\hat{f}) = \int_0^1 \hat{f}(\hat{x})\,\mathrm{d}\hat{x} - \frac{1}{2}[\hat{f}(0)+\hat{f}(1)]$$

*be the quadrature error on a reference interval $[0,1]$. Then $\hat{E}$ is the continuous linear form $\mathcal{L}$ in the Bramble-Hilbert Lemma on $\Omega = [0,1]$ and we have*

$$|E_i(f)| = h|\hat{E}(\hat{f})| \le Ch|\hat{f}|_2 = Ch\sqrt{\int_0^1 [\hat{f}''(\hat{x})]^2\,\mathrm{d}\hat{x}} = Ch^{2.5}\sqrt{\int_{x_i}^{x_{i+1}} [f''(x)]^2\,\mathrm{d}x}.$$

*With Cauchy-Schwarz inequality for vectors $\sum_i a_i b_i \le \sqrt{\sum_i a_i^2}\sqrt{\sum_i b_i^2}$, we get the total quadrature error as*

$$\sum_{i=0}^{N} |E_i(f)| \le Ch^2 \sum_{i=0}^{N} \sqrt{h}\sqrt{\int_{x_i}^{x_{i+1}} [f''(x)]^2\,\mathrm{d}x}$$

$$\le Ch^2\sqrt{\sum_{i=0}^{N} h}\sqrt{\sum_{i=0}^{N}\int_{x_i}^{x_{i+1}} [f''(x)]^2\,\mathrm{d}x} = Ch^2|f|_2.$$

*So we have just proven the quadrature error estimate:*

$$\left|\int_0^1 f(x)\,\mathrm{d}x - \sum_{i=0}^{N}\frac{1}{2}h[f(x_i)+f(x_{i+1})]\right| \le Ch^2|f|_2. \tag{3.19}$$

### 3.7.3 *Error estimates*

We only demonstrate the main idea why the error estimates for the abstract finite element method can still hold after quadrature error is involved. We focus on the simplest example. Consider the scheme (3.9) for $a(x) \equiv 1$, i.e., the scheme $K\mathbf{u} = \mathbf{f}$. The integrand in the bilinear $\mathcal{A}(u_h, v_h)$ is simply piecewise constant because $u_h$ and $v_h$ are piecewise linear. Thus we have $\mathcal{A}(u_h, v_h) = \mathcal{A}_h(u_h, v_h)$ and the first Strang Lemma reduces to

$$\|u - u_h\|_1 \leq C \inf_{v_h \in V_0^h} \|u - v_h\|_1 + C \sup_{w_h \in V_0^h} \frac{|\langle f, w_h \rangle_h - (f, w_h)|}{\|w_h\|_1}.$$

For a piecewise polynomial $w_h$, its second order derivative only exists on each interval, thus we abuse the notation by letting $|w_h|_2$ denote (this is usually called Broken Sobolev space, i.e., the Sobolev space on each small interval):

$$|w_h|_2^2 = \sum_i \int_{x_i}^{x_{i+1}} [w_h''(x)]^2 \, dx.$$

With this modification of seminorm (you can verify that (3.19) still holds if replacing $f$ by $fw_h$), by (3.19), we have

$$|\langle f, w_h \rangle_h - (f, w_h)| \leq Ch^2 |fw_h|_2.$$

Notice that in each interval $(fw_h)'' = f''w_h + f'w_h'$ because $w_h$ is a linear polynomial and $w_h'' \equiv 0$ in each interval. By triangle inequality, we have

$$|\langle f, w_h \rangle_h - (f, w_h)| \leq Ch^2 |fw_h|_2 = Ch^2 \|f''w_h + 2f'w_h'\|_0$$

$$\leq Ch^2 (\|f''\|_0 \|w_h\|_\infty + 2\|f'\|_0 \|w_h'\|_\infty) \leq Ch^2 \|f\|_2 \|w_h\|_1,$$

where the last inequality is not obvious but can be derived following the arguments in [Li and Zhang (2020b), Theorem 2.4] and we do not discuss the details for it here. Therefore we obtain the $H^1$ estimate as

$$\|u - u_h\|_1 \leq Ch|u|_2 + Ch^2 \|f\|_2.$$

Similarly, the Aubin-Nitsche Lemma also holds up to quadrature error.

The conclusion is very simple: the orders in the estimates in (3.14) and (3.16) still hold in the estimates for the scheme with quadrature (3.9).

## 3.8 Generalization: general domain in two dimensions

We will have a quick glance at how everything can be easily extended to a general setup. Consider solving a two-dimensional Poisson equation:

$$-\nabla \cdot (A(x,y)\nabla u(x,y)) = f(x,y), \quad (x,y) \in \Omega$$

with homogeneous Dirichlet boundary conditions $u(x,y) = 0$ along the domain boundary $\Omega$ for a bounded region $\Omega$, where $A(x,y)$ is a $2 \times 2$ matrix coefficient.

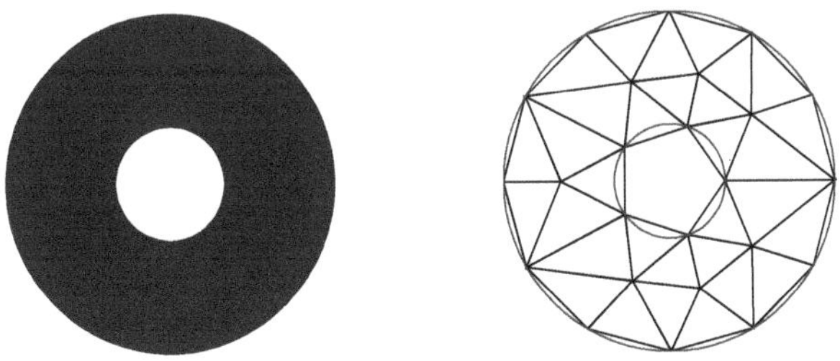

Fig. 3.3 Left: the domain $\Omega$. Right: the approximated domain $\Omega_h$ via a triangular mesh.

We just mention some key ingredients in the generalization to see how an easy extension is possible in the first place:

(1) Multiplying the test function and integration by parts, we get the equivalent variational formulation for the PDE:

$$\text{seek } u \in H_0^1(\Omega), \quad \iint_\Omega \nabla v^\top A \nabla u \, dxdy = \iint_\Omega f v \, dxdy, \quad \forall v \in H_0^1(\Omega),$$

which can be denoted as $\mathcal{A}(u,v) = (f,v)$.

(2) Construct an unstructured triangular mesh, which gives an approximated domain $\Omega_h$ as shown in Figure 3.3. Notice that the approximated boundary $\partial\Omega_h$ is a piecewise segment approximation to the curved boundary $\partial\Omega$, which induces a second order geometric error thus any finite element method defined on this $\Omega_h$ can be at most second order accurate even if using very high order polynomial basis. On the other hand, we can easily fix this issue by using curved triangle along the boundary, but quadrature on curved triangles are more expensive. For simplicity, we just consider the mesh shown in Figure 3.3.

(3) We define $V_0^h$ as the continuous piecewise linear polynomial space on the mesh shown in Figure 3.3, with the property of vanishing on the boundary $\partial\Omega_h$. An abstract finite element method is naturally given as

$$\text{seek } u_h \in V_0^h, \quad \iint_\Omega \nabla v_h^\top A \nabla u_h \, dxdy = \iint_\Omega f v_h \, dxdy, \quad \forall v_h \in V_0^h,$$

which can be denoted as $\mathcal{A}(u_h, v_h) = (f, v_h)$.

(4) Now let us consider what kind of coefficient $A(x, y)$ can ensure coercivity. For instance, if we assume that $A$ is real symmetric and its smallest eigenvalue has a uniform positive lower bound, i.e., $\lambda(A) \geq C > 0$ for any $(x, y)$, then by the Courant-Fisher-Weyl Min-Max principle,

$$\frac{\nabla v^\top A \nabla v}{\nabla v^\top \nabla v} \geq C \Rightarrow \mathcal{A}(v, v) \geq C|v|_1^2 \geq C\|v\|_1^2, \quad \forall v \in H_0^1(\Omega).$$

where we have used the fact that $H^1$-seminorm and $H^1$-norm are equivalent in $H_0^1(\Omega)$.

(5) Assume there are $N$ interior vertices in Figure 3.3, then $V_0^h$ is $N$-dimensional. We can define *Lagrange* basis (also called *nodal basis*) functions $\phi_i(x, y)$ on $\Omega_h$ just like the one-dimensional case. For instance, a linear polynomial is completely determined by its point values at three vertices on the triangle, and a quadratic polynomial is completely determined by its point values at three vertices and three edge centers on the triangle.

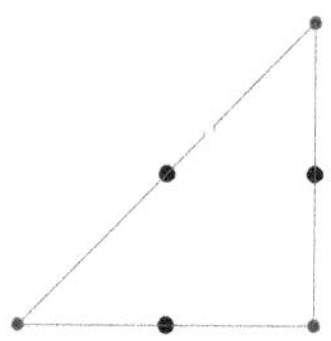

(6) Plugging in $u_h(x, y) = \sum_{j=1}^N u_j \phi_j(x)$ we get a linear system

$$\sum_{j=1}^N u_j \mathcal{A}(\phi_j, \phi_i) = (f, \phi_i), \quad i = 1, \cdots, N,$$

and the *stiffness* matrix $S$ has entries $S_{ij} = \mathcal{A}(\phi_j, \phi_i)$.

(7) It can be shown that weak partial derivatives of any $v_h \in V_0^h$ exist thus it is conforming: $V_0^h \subset H_0^1(\Omega)$. So the proof of Galerkin orthogonality holds. Coercivity and Galerkin orthogonality imply Céa's Lemma. Once we have Céa's Lemma, the $H^1$-norm error is controlled by the interpolation error, which can be given via Bramble-Hilbert Lemma. Similarly, the Aubin-Nitsche Lemma also holds.

(8) The quadrature using only three vertices is exact for linear polynomials on a triangle. The quadrature using only three edge centers is exact for quadratic polynomials on a triangle. With a suitable quadrature, the finite element method can be represented as

$$\text{seek } u_h \in V_0^h, \quad \mathcal{A}_h(u_h, v_h) = \langle f, v_h \rangle_h, \quad \forall v_h \in V_0^h.$$

(9) The scheme above can also be written in a matrix vector form

$$S\mathbf{u} = M\mathbf{f},$$

For simplicity, we just consider solving $-\Delta u = f$ as an example. With $P^1$ basis on an unstructured triangular mesh, $\mathbf{u}$ and $\mathbf{f}$ are column vectors of representing values at interior vertices in the mesh in Figure 3.3. If using the average of three vertex values for approximating integrals on a triangle as the quadrature in $\mathcal{A}_h$, then the matrices can be explicitly written as follows. Due to the quadrature, the mass matrix becomes a diagonal matrix with diagonal entries

$$M_{ii} = \mathcal{A}_h(\phi_i, \phi_i) = \frac{1}{3} \sum_{T \in \mathcal{T}_i} A_T, \tag{3.20a}$$

where $\mathcal{T}_i$ is the set of triangles containing the $i$-th vertex and $A_T$ denotes the area of the triangle $T$, see Figure 3.4. The stiffness matrix $S$ is sparse with $S_{ij} \neq 0$ only if $j \in \mathcal{R}_i$ where $\mathcal{R}_i$ is the set of indices for vertices connected to the vertex $p_i$ by edges, i.e., $j \in \mathcal{R}_i$ if vertices $p_i$ and $p_j$ are connected by an edge as shown in Figure 3.4. The nonzero entries of $S$ can be derived as

$$S_{ij} = -\frac{1}{2}\big(\cot \alpha_{ij} + \cot \beta_{ij}\big), \quad S_{ii} = \sum_{j \in \mathcal{R}_i} \frac{1}{2}\big(\cot \alpha_{ij} + \cot \beta_{ij}\big), \tag{3.20b}$$

where the two angles $\alpha_{ij}$ and $\beta_{ij}$ are defined in Figure 3.4.

(10) Finally, if you are curious whether this is still a finite difference scheme if using a structured triangular mesh, e.g., each rectangle is split into two triangles in a rectangular mesh, then the answer is yes! With no surprises, the $P^1$ finite element method will give the same 5-point discrete Laplacian scheme as in Chapter 2. The $P^2$ polynomial finite element method for $-\Delta u = f$ gives a fourth order accurate finite difference scheme (this is superconvergence, see Section 3.14) with the following stencil:

$$\text{edge center} \begin{matrix} & -1 & \\ -1 & 4 & -1 \\ & -1 & \end{matrix} \quad \text{vertex} \begin{matrix} & & 1 & & \\ & & -4 & & \\ 1 & -4 & 12 & -4 & 1 \\ & & -4 & & \\ & & 1 & & \end{matrix}.$$

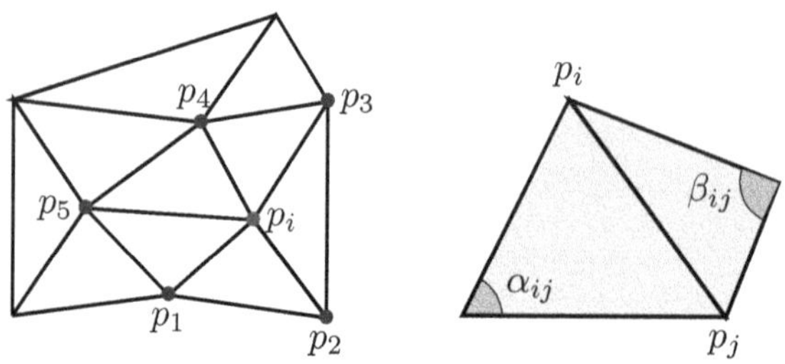


Fig. 3.4 Left: there are five points connecting the point $p_i$ and five triangles containing $p_i$ in the mesh, thus $\mathcal{R}_i = \{1, 2, 3, 4, 5\}$ and $\mathcal{T}_i$ contains five triangles. Right: the two angles for an interior edge connecting $p_i$ and $p_j$.

**Example 3.4.** *For a rectangular domain $\Omega$, consider a 2D variable coefficient problem*

$$-\nabla \cdot (a(\mathbf{x})\nabla u(\mathbf{x})) = f(\mathbf{x}), \quad \mathbf{x} \in \Omega$$

*with homogeneous Dirichlet boundary condition, where $a(\mathbf{x}) > 0$ is a scalar coefficient function. The equivalent variational form of the PDE can be written as*

$$\text{seek } u \in H_0^1(\Omega), \quad \iint_\Omega a\nabla v \cdot \nabla u \,\mathrm{d}x\mathrm{d}y = \iint_\Omega f v \,\mathrm{d}x\mathrm{d}y, \quad \forall v \in H_0^1(\Omega),$$

*which can be written as $A(u, v) = (f, v)$ with $\mathcal{A}(u, v) := \iint_\Omega a\nabla v \cdot \nabla u \,\mathrm{d}x\mathrm{d}y$.*

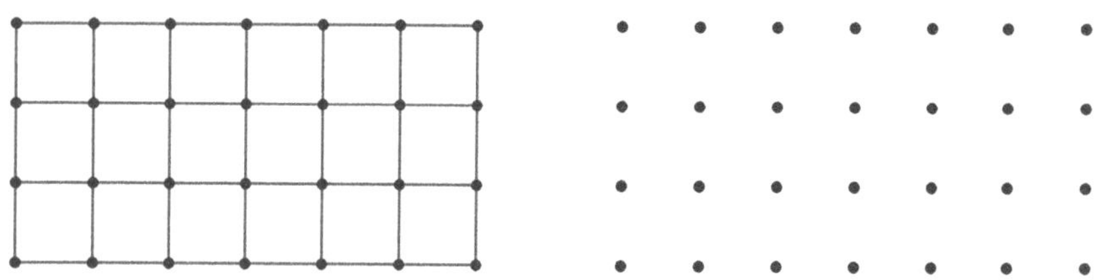

(a) The mesh and quadrature points (b) Corresponding Cartesian grid.

Fig. 3.5 An illustration of $Q^1$ element and the $2 \times 2$ Gauss-Lobatto quadrature.

*Let $\Omega_h$ be the uniform rectangular mesh of size $h$ for $\Omega$. On each cell $e$, we consider the $Q^1$ polynomial space spanned by $\{1, x, y, xy\}$. Define $V_0^h$ as the linear space of continuous piecewise $Q^1$ polynomial on $\Omega_h$ with zero values along the boundary of $\Omega_h$. Then the abstract finite element method is*

$$\text{seek } u_h \in V_0^h, \quad A(u_h, v_h) = (f, v_h), \quad \forall v_h \in V_0^h.$$

*The simplest quadrature is trapezoidal quadrature for both $x$ and $y$ variables, e.g., using the average of values at four vertices to approximate the integral over a square, see Figure 3.5. Then the finite element method is to seek $u_h \in V_0^h$ satisfying*

$$\mathcal{A}_h(u_h, v_h) = \langle f, v_h \rangle, \quad \forall v_h \in V_0^h.$$

*Using notation in Chapter 2, the scheme can be written as a finite difference scheme defined on interior grid points of Figure 3.5 (b).*

$$(D_x^\top \otimes I_y)A_1(D_x \otimes I_y) + (I_x \otimes D_y^\top)A_2(I_x \otimes D_y)]\, vec(U) = vec(F),$$

*where $D$ is defined in* (3.12)*, and $A_1, A_2$ are two diagonal matrices defined as follows.*

*Let $a_1$ be a 2D array of size $N_y \times (N_x + 1)$ satisfying $a_1(j, i) = \frac{1}{2}a(x_i, y_j) + \frac{1}{2}a(x_{i-1}, y_j)$ and $a_2$ be a 2D array of size $(N_y + 1) \times N_x$ satisfying $a_2(j, i) = \frac{1}{2}a(x_i, y_j) + \frac{1}{2}a(x_i, y_{j-1})$. Then $A_1$ and $A_2$ can be easily generated in MATLAB as sparse diagonal matrices:*

```
A1=sparse(diag(a1(:)));
A2=sparse(diag(a2(:)));
```

**Problem 3.8.** *Verify that the finite element method in the example above can be written as the linear system above with the matrix*

$$(D_x^\top \otimes I_y)A_1(D_x \otimes I_y) + (I_x \otimes D_y^\top)A_2(I_x \otimes D_y).$$

**Problem 3.9.** *In the scheme in Example 3.4, when $a(\mathbf{x}) \equiv 1$, $A_1$ and $A_2$ become the identity matrix, and the coefficient matrix in the system becomes $K2D$ matrix in Chapter 1. Implement this scheme and test the accuracy for a smooth solution with a smooth coefficient function. Use $K2D$ matrix as the preconditioner to implement Preconditioned Conjugate Gradient method in Appendix B solving this linear system. When inverting $K2D$ matrix, use the method described by* (2.9).

## 3.9 Generalization: purely Neumann b.c.

Consider a one-dimensional problem

$$-(a(x)u'(x))' = f(x), \quad x \in (0, 1)$$
$$u'(0) = \sigma_0, u'(1) = \sigma_1.$$

Recall that $f(x)$ must be compatible with the boundary conditions:

$$\int_0^1 f(x)\, \mathrm{d}x = -a_1\sigma_1 + a_0\sigma_0, \tag{3.21}$$

which is obtained by integrating the PDE.

### 3.9.1 Quotient space $H^1(\Omega)/P^0(\Omega)$

Recall that this boundary value problem does not have a unique solution: if $u(x)$ is a solution, then so is $u(x)+c$ for any constant $c$. This non-uniqueness issue must be addressed properly. To this end, it is natural to consider a quotient space in which two functions differing by only a constant are regarded as the same function.

Let $P^0(\Omega)$ be the linear space of all polynomials of degree zero, i.e., all constants. We first introduce an *equivalent* class by

$$\dot{v}(x) := \{w(x) = v(x) + c, \quad c \in P^0(\Omega)\}.$$

In other words, if two functions $v(x)$ and $w(x)$ are different only by a constant $c$, then we regard them to be in the same equivalent class, which is a set. Any element $w(x)$ in an equivalent class $\dot{v}(x)$ is called a representation of the equivalent class $\dot{v}(x)$. For instance, in this section, $v(x)$ means a representation of the equivalent class $\dot{v}(x)$ which $v(x)$ belongs to.

The quotient space $H^1(\Omega)/P^0(\Omega)$ is defined as

$$H^1(\Omega)/P^0(\Omega) = \{\dot{v}(x) : v(x) \in H^1(\Omega)\}.$$

The norm of the quotient space $H^1(\Omega)/P^0(\Omega)$ is defined as

$$\|\dot{v}\|_1 := \inf_{w\in\dot{v}} \|w\|_1,$$

where $\|w\|_1$ is the $H^1$-norm of the representation $w(x)$. This definition can be explicitly written as

$$\|\dot{v}\|_1 := \inf_{c\in P^0(\Omega)} \|v(x)+c\|_1 = \min_{c\in\mathbb{R}} \sqrt{\int_\Omega |v(x)+c|^2 \,\mathrm{d}x + \int_\Omega \left|\frac{\mathrm{d}}{\mathrm{d}x}(v(x)+c)\right|^2 \mathrm{d}x}.$$

So we get

$$\|\dot{v}\|_1^2 := \min_{c\in\mathbb{R}} \int_\Omega |v(x)+c|^2 \,\mathrm{d}x + \int_\Omega |v'(x)|^2 \,\mathrm{d}x,$$

which is nothing but a minimization with respect to $c$. Also, it is a simple quadratic function of the number $c$, so the minimizer is the average of $v(x)$, $c = \frac{1}{|\Omega|}\int_\Omega v(x)\,\mathrm{d}x$. For the domain $\Omega = (0,1)$, let $\bar{v} = \int_0^1 v(x)\,\mathrm{d}x$ be the average of the function $v(x)$ over $\Omega$. Then the norm of the quotient space $H^1(\Omega)/P^0(\Omega)$ can be equivalently written as

$$\|\dot{v}\|_1^2 = \int_\Omega |v(x)-\bar{v}|^2 \,\mathrm{d}x + \int_\Omega |v'(x)|^2 \,\mathrm{d}x.$$

This quotient space norm is also equivalent to the seminorm $|v|_1$:

$$C\|\dot{v}\|_1 \le |v|_1 \le \|\dot{v}\|_1, \quad C > 0. \tag{3.22}$$

The second inequality is trivial. The first inequality is due to the *Poincaré inequality* (Theorem 3.1):

$$\int_\Omega |v(x)-\bar{v}|^2 \,\mathrm{d}x \le C \int_\Omega |v'(x)|^2 \,\mathrm{d}x. \tag{3.23}$$

### 3.9.2 *Variational formulation and coercivity*

Multiplying a test function, after integration by parts, we can get a variational form:

$$\int_0^1 a(x)u'(x)v'(x)\,\mathrm{d}x = \int_0^1 f(x)v(x)\,\mathrm{d}x + a_1\sigma_1 v(1) - a_0\sigma_0 v(0).$$

Obviously, both side stay the same if we replace $u(x)$ by $u(x)+c$ for any constant $c$. Now if we replace $v(x)$ by $v(x)+c$, the left hand side stays the same, and the right hand side also stays the same because of the compatibility condition (3.21).

So the equivalent variational formulation is to seek $\dot{u} \in H^1(\Omega)/P^0(\Omega)$ such that any representation $u$ of the equivalent class $\dot{u}$ satisfy

$$\int_\Omega a(x)u'(x)v'(x)\,\mathrm{d}x = \int_\Omega f(x)v(x)\,\mathrm{d}x + a_1\sigma_1 v(1) - a_0\sigma_0 v(0), \quad \forall \dot{v} \in H^1(\Omega)/P^0(\Omega).$$

It can be denoted by the same short hand notation as:

$$\mathcal{A}(u,v) = (f,v) + a_1\sigma_1 v(1) - a_0\sigma_0 v(0), \quad \forall \dot{v} \in H^1(\Omega)/P^0(\Omega).$$

The Cauchy-Schwarz inequality implies the continuity of the bilinear form. Since quotient space norm is also equivalent to the $H^1$ seminorm, we also have the coercivity:

$$\mathcal{A}(v,v) \geq C\|\dot{v}\|_1^2, \quad \forall v \in H^1(\Omega)/P^0(\Omega).$$

**Remark 3.6.** *The quotient space is an elegant tool for analysis, e.g., coercivity is implied by* (3.22)*, which holds due to the Poincaré inequality. On the other hand, one can also use the Poincaré inequality* (3.23) *to do analysis, without invoking the quotient space, which will be demonstrated in Section 6.5.*

### 3.9.3 *The finite element method*

On a mesh with intervals $I_j$, we have defined the space $V^h$ as an approximation to $H^1(0,1)$. We can also define a quotient space $V^h/P^0$ similarly:

$$V^h/P^0 = \{\dot{v}_h(x) : v_h(x) \in V^h\}.$$

The finite element method is to seek $\dot{u}_h(x) \in V^h/P^0$ such that

$$\mathcal{A}(u_h, v_h) = (f, v_h) + a_1\sigma_1 v_h(1) - a_0\sigma_0 v_h(0), \quad \forall \dot{v}_h \in V^h/P^0.$$

Notice that we use representations $u_h$ and $v_h$ in the bilinear form $\mathcal{A}(u_h, v_h)$, instead of their equivalent classes $\dot{u}_h$ and $\dot{v}_h$. All the previous arguments for error estimates can be established similarly, and the only difference is that the underlying function space is the quotient space $H^1(\Omega)/P^0(\Omega)$, even though we just plug in functions into the variational form as before.

### 3.9.4 *Coercivity implies the stiffness matrix null space*

For simplicity, we assume homogeneous Neumann boundary condition $\sigma_0 = \sigma_1 = 0$, and constant coefficient $a(x) \equiv 1$. Then for the $P^1$ basis finite element method, the bilinear form with trapezoidal quadrature $\mathcal{A}_h(u_h, v_h)$ is the same as $\mathcal{A}(u_h, v_h)$.

Recall our uniform grid points are

$$0 = x_0 < x_1 < \cdots < x_N < x_{N+1} = 1.$$

Then $V^h$ is $(N+2)$-dimensional. As a comparison, $V_0^h$ for Dirichlet boundary is $N$-dimensional. Let $\phi_i(x), i = 0, 1, \cdots, N+1$ be the Lagrange basis or *nodal basis* of $V^h$. Then the stiffness matrix $S \in \mathbb{R}^{(N+2)\times(N+2)}$ has entries $S_{ij} = \mathcal{A}_h(\phi_j, \phi_i) = \mathcal{A}(\phi_j, \phi_i)$. Here we have abused notation by allowing indices $i, j$ to take value 0.

With similar notation as before, e.g., $\mathbf{v} \in \mathbb{R}^{N+2}$ denoting a vector of point values $v_h(x_j)$, we have

$$\mathbf{v}^\top S\mathbf{v} = \mathcal{A}(v_h, v_h) \geq C\|\dot{v}_h\|_1 \geq 0,$$

thus $S$ is still real symmetric and positive semi-definite.

Since the boundary value problem does not have a unique solution, the stiffness matrix $S$ must have a nontrival null space. As a matter of fact, the constant one vector $\mathbf{1}$ is in its null space. We first have

$$\forall \mathbf{v}, \quad \mathbf{v}^\top S\mathbf{1} = \mathcal{A}(1, v_h) = 0 \Rightarrow \mathbf{v} \perp S\mathbf{1}, \forall \mathbf{v} \Rightarrow S\mathbf{1} = \mathbf{0}.$$

Next, we want to show that the coercivity implies that the null space of $S$ is one-dimensional:

$$S\mathbf{v} = \mathbf{0} \Leftrightarrow \mathbf{v}^\top S\mathbf{v} = 0 \Leftrightarrow \mathcal{A}(v_h, v_h) = 0 \Rightarrow \|\dot{v}_h\|_1 = 0,$$

where the last step is due to the coercivity. Thus

$$S\mathbf{v} = \mathbf{0} \Leftrightarrow \|\dot{v}_h\|_1 = 0 \Leftrightarrow \dot{v}_h(x) = \dot{0} \Leftrightarrow v_h(x) \equiv c \Leftrightarrow \mathbf{v} = c\mathbf{1},$$

because a function in the quotient space has zero norm if and only if it is $\dot{0}$, which is the property of a norm.

### 3.9.5 *The finite difference form*

For simplicity, just consider the constant coefficient case $a(x) = 1$, for piecewise linear basis with trapezoidal quadrature on the uniform grid, the finite element method can be equivalently written as

$$\frac{1}{h}(u_1 - u_0) = \frac{h}{2}f_0 + a_0\sigma_0$$

$$\frac{1}{h}(-u_{j-1} + 2u_j - u_{j+1}) = hf_j, \quad j = 1, \cdots, N$$

$$\frac{1}{h}(u_{N+1} - u_N) = \frac{h}{2}f_{N+1} + a_1\sigma_1$$

which is exactly the same as the traditional finite difference scheme in Section 2.5.2.

Now the finite element theory can give error estimates like (3.14) and (3.16). On the other hand, it is straightforward to check the truncation error at $x_0$ or $x_{N+1}$ is only first order, even though the Neumann boundary condition was approximated by a second order centered difference in Section 2.5.2. It is quite difficult to show that this scheme is second order accurate following arguments in Chapter 2, especially for a variable coefficient problem in multiple dimensions. But we know this scheme is indeed second order accurate in the sense of (3.16), which demonstrates the superiority of the finite element method compared to traditional finite difference method, although both approaches give the same scheme.

For $a(x) = 1$, the stiffness matrix $S$ with entries $\mathcal{A}_h(\phi_j, \phi_i)$ can be written out as

$$S = \frac{1}{h}\begin{pmatrix} 1 & -1 & & & \\ -1 & 2 & -1 & & \\ & \ddots & \ddots & \ddots & \\ & & -1 & 2 & -1 \\ & & & -1 & 1 \end{pmatrix}_{(N+2)\times(N+2)}.$$

For the variable coefficient, we have

$$S = \frac{1}{h}\frac{1}{2}\begin{pmatrix} a_0+a_1 & -a_0-a_1 & & & \\ -a_0-a_1 & a_0+2a_1+a_2 & -a_1-a_2 & & \\ & -a_1-a_2 & a_1+2a_2+a_3 & -a_2-a_3 & \\ & & & & \\ & & \ddots & \ddots & \ddots \end{pmatrix}_{(N+2)\times(N+2)}.$$

### 3.9.6 *How to solve the singular linear system*

Now consider the finite element method with quadrature for a variable coefficient problem:

$$\mathcal{A}_h(u_h, v_h) = \langle f, v_h\rangle_h + a_1\sigma_1 v_h(1) - a_0\sigma_0 v_h(0), \quad \forall \dot{v}_h \in V^h/P^0.$$

For piecewise linear finite element method, if using a Lagrange basis or *nodal basis* $\phi_j(x)$, by plugging in $v_h = \phi_i(x)$, we get a singular linear system

$$S\begin{pmatrix} u_0 \\ u_1 \\ \vdots \\ u_N \\ u_{N+1} \end{pmatrix} = h\begin{pmatrix} \frac{1}{2}f_0 \\ f_1 \\ \vdots \\ f_N \\ \frac{1}{2}f_{N+1} \end{pmatrix} + \begin{pmatrix} a_0\sigma_0 \\ 0 \\ \vdots \\ 0 \\ a_1\sigma_1 \end{pmatrix}.$$

The finite element method is defined on a quotient space $V^h/P^0$, and we look for a solution in a quotient space $H^1(\Omega)/P^0(\Omega)$ in the original variational formulation. Corresponding to these facts, the null space of $S$ is Span$\{\mathbf{1}\}$, and we look for the least square solution of this singular linear system. We do not elaborate more about it here, but at least all discussions about quotient spaces and the null spaces are consistent.

Let $S\mathbf{u} = \tilde{f}$ denote this singular linear system. The first question we should ask is whether it has a solution since $S$ is not invertible. A linear system $S\mathbf{u} = \tilde{f}$ has a solution if and only if $\tilde{f}$ is in the column space of $S$, denoted by Col$(S)$. To this end, we need the orthogonal complement of Col$(S)$, denoted by Col$(S)^\perp$:

$$\mathrm{Col}(S)^\perp = \{\mathbf{y} \in \mathbb{R}^{N+2} : \mathbf{y}^\top S = \mathbf{0}.\}$$

Since $S$ is symmetric, $\mathbf{y}^\top S = \mathbf{0} \Leftrightarrow S\mathbf{y} = \mathbf{0}$, thus Col$(S)^\perp$ is the null space of $S$. In particular, we know that $\mathbf{1}$ is the basis of Col$(S)^\perp$. We have

$$\tilde{f} \in \mathrm{Col}(S) \Leftrightarrow \tilde{f} \perp \mathrm{Col}(S)^\perp \Leftrightarrow \tilde{f} \perp \mathbf{1}$$

$$\Leftrightarrow \frac{1}{2}hf_0 + h\sum_{j=1}^{N} f_j + \frac{1}{2}hf_{N+1} + a_0\sigma_0 + a_1\sigma_1 = 0$$

which is nothing but a discrete compatibility condition.

For a function $f(x)$ satisfying the compatibility condition, its point values may not necessarily satisfy the discrete compatibility condition. We can simply project $\tilde{f}$ to the column space of $S$. Let $\bar{f}$ be the projection vector, then $S\mathbf{u} = \bar{f}$ is ensured to have a solution, and we can use iterative solvers in Appendix B such as conjugate gradient method or its preconditioned version directly on $S\mathbf{u} = \bar{f}$ to find the least square solution to $S\mathbf{u} = \tilde{f}$. Since we know what Col$(S)^\perp$ is, the projection $\bar{f}$ is quite easy to find. We summarize it as follows:

(1) The projection $\bar{f}$ is computed as

$$\bar{f} = \tilde{f} - \frac{\langle \mathbf{1}, \tilde{f}\rangle}{\|\mathbf{1}\|^2}\mathbf{1}.$$

(2) Solve $S\mathbf{u} = \bar{f}$ by direct or iterative solvers. See Appendix B.

**Remark 3.7.** *Iterative solvers like conjugate gradient may not work well directly on $S\mathbf{u} = \tilde{f}$ especially if the discrete compatibility error is large.*

### 3.9.7 *Why symmetry of S is important*

To find the least square solution to $S\mathbf{u} = \tilde{f}$, it is mathematically equivalent to solve the normal equation $S^\top S\mathbf{u} = S^\top \tilde{f}$ which is ensured to have a solution for any $\tilde{f}$. However, $S^\top S\mathbf{u} = S^\top \tilde{f}$ is much harder to solve. For example, if $S$ is invertible, then the condition number of $S^\top S$ is about the square of the condition number of $S$. See Appendix B for the definition of condition number and its effects on solving linear systems. In other words, in the context of solving Poisson equation, it is absolutely a bad idea to solve the normal equation.

If $S$ is not symmetric, in order to find the projection $\bar{f}$ (as we have also seen in Section 2.11.3), we need to compute the left null vector $\mathbf{y}$ first: solving $S^\top \mathbf{y} = \mathbf{0} = 0 \cdot \mathbf{y}$ is an eigenvector problem, which is much more expensive than solving a linear system of the same size. For instance, for a nonsingular system $A\mathbf{x} = b$, iterative solvers are based on minimizing a function $f(\mathbf{x}) = \frac{1}{2}\mathbf{x}^\top A\mathbf{x} - b^\top \mathbf{x}$. For $A\mathbf{x} = 0$, if minimizing $f(\mathbf{x}) = \frac{1}{2}\mathbf{x}^\top A\mathbf{x}$, we simply get $\mathbf{x} = \mathbf{0}$, which is a solution that we do not want at all. For getting the nonzero solution to $A\mathbf{x} = 0$, roughly speaking, we would have to minimize $f(\mathbf{x}) = \frac{1}{2}\mathbf{x}^\top A\mathbf{x}$ over the sphere $\{\mathbf{x} : \|\mathbf{x}\| = 1\}$, which is a harder problem than solving linear systems.

For all remarks above, it is highly desired to have a symmetric $S$. The symmetry of the matrix $S$ with entries $S_{ij} = \mathcal{A}_h(\phi_j, \phi_i) = \mathcal{A}_h(\phi_i, \phi_j)$ holds trivially even for a two-dimensional or three-dimensional problem $-\nabla \cdot (A\nabla u) = f$ with a real symmetric matrix coefficient $A$. This is one of the key advantages of using finite element method for purely Neumann boundary conditions, which appear in many applications. It is in general quite difficult to construct a real symmetric matrix for variable coefficient problems with Neumann boundary in multiple dimensions by traditional finite difference method.

## 3.10 Generalization: nonhomogeneous Dirichlet b.c.

Consider solving

$$
\begin{aligned}
-u''(x) = f(x), \quad x \in (0,1),\\
u(0) = \sigma_0, u(1) = \sigma_1.
\end{aligned}
$$

One approach is to assume that there exists a smooth enough function $g(x)$ satisfying the same boundary condition. Then the function $\tilde{u} = u - g$ satisfies

$$
-\tilde{u}''(x) = f(x) + g''(x), \quad x \in (0,1),
$$

$$\tilde{u}(0) = \tilde{u}(1) = 0.$$

By multiplying a test function $v \in H_0^1(\Omega)$ and integration by parts, we get the equivalent variational form for seeking $\tilde{u} \in H_0^1(\Omega)$ satisfying

$$\int_0^1 \tilde{u}'(x)v'(x)\,\mathrm{d}x = \int_0^1 f(x)v(x)\,\mathrm{d}x - \int_0^1 g'(x)v'(x)\,\mathrm{d}x, \quad \forall v \in H_0^1(\Omega),$$

which can be denoted as

$$\mathcal{A}(\tilde{u}, v) = (f, v) - \mathcal{A}(g, v), \quad \forall v \in H_0^1(\Omega).$$

Obviously everything in Section 3.3 can be easily applied to construct and analyze a finite element method for $\tilde{u} \in H_0^1(\Omega)$, provided that we know what $g(x)$ is, which is easy to construct in one-dimension but not necessarily in multiple dimensions.

In this section, we will introduce a simple method in which we only need to assume the existence of the smooth function $g(x)$ and **we do not need to know what function $g(x)$ is to implement the scheme.**

### 3.10.1 *A scheme in theory*

An abstract finite element method that we should never implement is to find $\tilde{u}_h \in V_0^h$ satisfying

$$\mathcal{A}(\tilde{u}_h, v_h) = (f, v_h) - \mathcal{A}(g, v_h), \quad \forall v_h \in V_0^h.$$

Assume $g(x)$ is a nice function so that we can still derive the error estimates (3.14) and (3.16). For example, if $g''(x)$ exists, then after integration by parts for test function $v_h(x) \in V_0^h$, the abstract finite element is equivalent to seeking $\tilde{u}_h \in V_0^h$ satisfying

$$\mathcal{A}(\tilde{u}_h, v_h) = (f + g'', v_h), \quad \forall v_h \in V_0^h.$$

If we treat $f - g''$ as the right hand side function, then the error estimates (3.14) and (3.16) can still hold for $\tilde{u}_h - \tilde{u}$.

The numerical solution that we want is

$$u_h := \tilde{u}_h + g(x).$$

Be aware that we no longer have $u_h \in V^h$. By moving $\mathcal{A}(g, v_h)$ to the left hand side, we get

$$\mathcal{A}(u_h, v_h) = (f, v_h), \quad \forall v_h \in V_0^h.$$

Also $u_h - u$ satisfies the error estimates (3.14) and (3.16), because of the error estimates for $\tilde{u}_h - \tilde{u}$.

Next, assume we use quadrature, so we have

$$\mathcal{A}_h(\tilde{u}_h, v_h) = \langle f, v_h \rangle_h - \mathcal{A}_h(g, v_h), \quad \forall v_h \in V_0^h, \tag{3.24}$$

or equivalently

$$\mathcal{A}_h(u_h, v_h) = \langle f, v_h \rangle_h, \quad \forall v_h \in V_0^h. \tag{3.25}$$

The estimates (3.14) and (3.16) still hold after using quadrature for the scheme (3.24). **In this method, we need $g(x)$ to implement the scheme.**

### 3.10.2 *A simple and practical scheme*

Now we introduce a method without explicitly using $g(x)$. We consider the piecewise linear Lagrange interpolation polynomial for $g(x)$ at grid points $x_i$, denoted by $g_h(x) = \Pi_1 g(x) \in V^h$. For *nodal basis* $\{\phi_j(x)\}_{j=0}^{N+1}$ of $V^h$, we simply have

$$g_h(x) = \sum_{j=0}^{N+1} g_j \phi_j(x) \in V^h,$$

where $g_0 = \sigma_0, g_{N+1} = \sigma_1, x_0 = 0, x_{N+1} = 1$. Then we consider a new scheme seeking $\tilde{u}_h \in V_0^h$ satisfying

$$\mathcal{A}_h(\tilde{u}_h, v_h) = \langle f, v_h \rangle_h - \mathcal{A}_h(g_h, v_h), \quad \forall v_h \in V_0^h. \tag{3.26}$$

The difference between the scheme (3.26) and the scheme (3.24) is whether using $g(x)$ or its polynomial interpolation $g_h(x)$.

Let $u_h(x) = \tilde{u}_h(x) + g_h(x) \in V^h$, then we can rewrite (3.26) equivalently as

$$\mathcal{A}_h(u_h, v_h) = \langle f, v_h \rangle_h, \quad \forall v_h \in V_0^h. \tag{3.27}$$

Since $u_h(x) = \tilde{u}_h(x) + g_h(x) \in V^h$, we have

$$u_h(x) = \sum_{j=0}^{N+1} u_j \phi_j(x),$$

where $u_i = u_h(x_i)$. Here we should pay attention to the facts that $V^h$ is $(N+2)$-dimensional with basis $\{\phi_j(x)\}_{j=0}^{N+1}$, whereas the test function space $V_0^h$ is only $N$-dimensional with basis $\{\phi_j(x)\}_{j=1}^{N}$.

Obviously, plugging this representation into (3.27) and test function space basis $\phi_i(x)$ for $i = 1, \cdots, N$, we get a linear system

$$\sum_{j=0}^{N+1} \mathcal{A}_h(\phi_j, \phi_i) u_j = h f_i, \quad \forall i = 1, \cdots, N,$$

in which we do not need what $g(x)$ is. Of course, the linear system should have only $N$ unknowns because of Dirichlet boundary $u_0 = \sigma_0$ and $u_{N+1} = \sigma_1$. For $a(x) \equiv 1$, this scheme is precisely

$$\frac{1}{h}(-u_{j-1} + 2u_j - u_{j+1}) = hf_j, \quad j = 1, \cdots, N, \tag{3.28}$$

where $u_0 = \sigma_0, u_{N+1} = \sigma_1$.

**Remark 3.8.** *Notice that the scheme* (3.27) *is equivalent to the following scheme seeking* $u_h(x) \in V_0^h$ *satisfying*

$$A_h(u_h, v_h) = \langle f, v_h \rangle_h - A_h(\sigma_h, v_h), \quad \forall v_h \in V_0^h, \tag{3.29}$$

*where* $\sigma_h \in V^h$ *is the Lagrange interpolation of the trivial nonsmooth extension function:*

$$\sigma_h(x_0) = \sigma_0, \sigma_h(x_{N+1}) = \sigma_1, \quad \sigma_h(x_i) = 0, i = 1, \cdots, N.$$

**Remark 3.9.** *The scheme* (3.27) *or* (3.29) *has nothing to do with what* $g(x)$ *is. On the other hand, with the existence of smooth* $g(x)$*, the error estimates can be easily established via the analysis of the scheme* (3.24)*. To establish error estimates for* (3.27) *or* (3.29)*, notice that their only difference from* (3.24) *is the following*

$$\mathcal{A}_h(g, v_h) - \mathcal{A}_h(g_h, v_h),$$

*which can be analyzed through the interpolation error estimates on* $\|g - g_h\|_1$ *and* $\|g - g_h\|_0$*. For instance, for convergence in* $H^1$ *norm, similar to the First Strang Lemma in Theorem 3.6, we will have to deal with*

$$\sup_{w_h \in V_0^h} \frac{|\mathcal{A}_h(g, w_h) - \mathcal{A}_h(g_h, w_h)|}{\|w_h\|_1},$$

*which can be easily done by discrete continuity of the bilinear form:*

$$\frac{|\mathcal{A}_h(g, w_h) - \mathcal{A}_h(g_h, w_h)|}{\|w_h\|_1} \le C\|g - g_h\|_1.$$

The scheme (3.28) is exactly the same as taking the scheme for purely Neumann boundary at interior grid points $j = 1, \cdots, N$ in Section 3.9. This is not a coincidence at all. This fact remains true even for high order polynomial basis with variable coefficients, which means that we have a neat treatment of boundary condition in finite element method. In particular, for a variable coefficient problem, by taking the scheme at interior grid points $j = 1, \cdots, N$ in Section 3.9, we obtain the $P^1$ finite element method with trapezoidal quadrature for the nonhomogeneous Dirichlet b.c. as

$$\frac{-(a_{j-1} + a_j)u_{j-1} + (a_{j-1} + 2a_j + a_{j+1})u_j - (a_j + a_{j+1})u_{j+1}}{2h} = hf_j, \tag{3.30}$$

for $j = 1, \cdots, N$, where $u_0 = \sigma_0, u_{N+1} = \sigma_1$.

**Remark 3.10.** *Recall that in Section 2.4 we discussed the M-matrix structure of $K$ matrix, which is often used to construct maximum-principle-satisfying or bound-preserving numerical schemes. We can see that the scheme* (3.30) *also has an M-matrix structure, e.g., the coefficients on the left hand side of* (3.30) *satisfies Theorem 2.1. Such an M-matrix structure in general may not hold in a very high order scheme in multiple dimensions. On the other hand, usually the M-matrix structure can be achieved in finite element method using linear basis, see [Xu and Zikatanov (1999); Li and Zhang (2025)].*

### 3.10.3 *A scheme in theory for a general domain $\Omega$*

Consider solving a two-dimensional Poisson equation with nonhomogeneous Dirichlet boundary condition for a bounded region $\Omega$:

$$-\nabla \cdot (A(x,y)\nabla u(x,y)) = f(x,y), \quad (x,y) \in \Omega,$$

$$u(x,y) = \sigma(x,y), \quad (x,y) \in \partial\Omega,$$

where $A(x,y)$ is a $2 \times 2$ matrix coefficient.

Assume there exists a smooth extension function $g(x,y)$ satisfying that $g|_{\partial\Omega}(x,y) = \sigma(x,y)$, then $\tilde{u} = u - g \in H_0^1(\Omega)$ satisfying

$$\mathcal{A}(\tilde{u}, v) = (f, v) - \mathcal{A}(g, v), \quad \forall v \in H_0^1(\Omega)$$

where the bilinear form is $\mathcal{A}(u,v) = \iint_\Omega \nabla v^\top A \nabla u \, dxdy$.

Given a triangulation of the domain $\Omega_h$ as shown in Figure 3.3, assume either $\Omega$ is polygonal or we use curved triangles, so that $\partial\Omega_h = \partial\Omega$. Define a continuous piecewise polynomial space $V_0^h \subset H_0^1(\Omega)$, then an abstract finite element method that can be easily analyzed is to find $\tilde{u}_h \in V_0^h$ satisfying

$$\mathcal{A}(\tilde{u}_h, v_h) = (f, v_h) - \mathcal{A}(g, v_h), \quad \forall v_h \in V_0^h.$$

The scheme with quadrature is written as

$$\mathcal{A}_h(\tilde{u}_h, v_h) = \langle f, v_h \rangle_h - \mathcal{A}_h(g, v_h), \quad \forall v_h \in V_0^h,$$

or equivalently

$$u_h = \tilde{u}_h + g, \quad \mathcal{A}_h(u_h, v_h) = \langle f, v_h \rangle_h, \quad \forall v_h \in V_0^h. \tag{3.31}$$

The error estimates of (3.31) can be similarly established as in the 1D case.

### 3.10.4 A simple and practical scheme in two dimensions

For the ease of implementation, we define $g_h(x)$ as the Lagrange interpolation of $g(x)$ over nodal points in the mesh, which will be explained below. Then we implement a different scheme

$$\tilde{u}_h \in V_0^h, \quad \mathcal{A}_h(\tilde{u}_h, v_h) = \langle f, v_h \rangle_h - \mathcal{A}_h(g_h, v_h), \quad \forall v_h \in V_0^h,$$

or equivalently

$$u_h = \tilde{u}_h + g_h \in V^h, \quad \mathcal{A}_h(u_h, v_h) = \langle f, v_h \rangle_h, \quad \forall v_h \in V_0^h. \tag{3.32}$$

For convenience, let $\mathbf{x}$ denote $(x, y)$. Now we need to make some assumptions which are quite practical at least for $P^1$ and $P^2$ bases:

I. $V_0^h$ is $N$-dimensional and $V^h$ is $(N+n)$-dimensional.

II. $V^h$ has a Lagrange basis (*nodal basis*) $\{\phi_j(\mathbf{x})\}_{j=1}^{N+n}$ satisfying

$$\phi_j(\mathbf{x}_i) = \delta_{ij} = \begin{cases} 1, & i = j \\ 0, & i \neq j \end{cases},$$

for the points $\mathbf{x}_i : i = 1, \cdots, N+n$.

III. $V_0^h$ has a Lagrange basis $\{\phi_j(\mathbf{x})\}_{j=1}^{N}$ satisfying

$$\phi_j(\mathbf{x}_i) = \delta_{ij} = \begin{cases} 1, & i = j \\ 0, & i \neq j \end{cases},$$

for the points $\mathbf{x}_i, i = 1, \cdots, N$. For example, $\mathbf{x}_i$ are three vertices of all triangles for a continuous piecewise linear polynomial on a triangular mesh. For a continuous piecewise quadratic polynomial on a triangular mesh, $\mathbf{x}_i$ are three vertices and three edge centers of all triangles.

IV. The quadrature points used in $\mathcal{A}_h(\cdot, \cdot)$ is a subset of $\{\mathbf{x}_i, i = 1, \cdots, N\}$. For instance, the quadrature using three vertices with equal weight is exact for integrating a linear polynomial on a triangle thus second order accurate by Bramble-Hilbert Lemma, and the quadrature using only three edge centers with equal weight is exact for integrating a quadratic polynomial on a triangle thus third order accurate by Bramble-Hilbert Lemma.

So the points $\{\mathbf{x}_i\}_{i=1}^N$ are interior points inside the domain $\Omega$ and the points $\{\mathbf{x}_i\}_{i=N+1}^{N+n}$ are boundary points, along the boundary $\partial\Omega_h = \Omega$ (not true in general but we have assumed such a geometry-conforming mesh).

Let $u_j = u_h(\mathbf{x}_j)$ and $\sigma_j = \sigma(\mathbf{x}_j)$, then

$$u_h(\mathbf{x}) = \sum_{j=1}^{N+n} u_j \phi_j(\mathbf{x}) = \sum_{j=1}^{N} u_j \phi_j(\mathbf{x}) + \sum_{j=N+1}^{N+n} \sigma_j \phi_j(\mathbf{x}).$$

Under these assumptions, the scheme (3.32) is exactly the same as

$$\mathcal{A}_h(\sum_{j=1}^{N+n} u_j\phi_j, v_h) = \langle f, v_h\rangle_h, \quad \forall v_h \in V_0^h, \tag{3.33}$$

or equivalently

$$\sum_{j=1}^{N} \mathcal{A}_h(\phi_j, \phi_i)u_j = \langle f, \phi_i\rangle_h - \sum_{j=N+1}^{N+n} \mathcal{A}_h(\phi_j, \phi_i)\sigma_j, \quad i = 1, \cdots, N.$$

If you ever wonder what the simplest boundary treatment for a high order accurate scheme should be, (3.32) gives a perfect answer.

To establish the convergence in $H^1$-norm and $L^2$-norm for the scheme (3.32) or (3.33), we first can have the error estimates for (3.31), then analyze the only difference between (3.32) and (3.31):

$$\mathcal{A}_h(g, v_h) - \mathcal{A}_h(g_h, v_h),$$

which is related to the interpolation error estimates on $\|g - g_h\|_1$ and $\|g - g_h\|_0$.

Finally, we should emphasize that the Dirichlet boundary treatment in this method is as simple as in Chapter 2 but with a price to pay: we have to assume existence of the extension function $g(\mathbf{x})$, and assume continuity of the boundary data function $\sigma(\mathbf{x})$ so that its nodal values $\sigma_j$ are well defined. To treat the Dirichlet boundary condition with weaker assumptions, there are other more sophisticated methods, which we do not consider in this book.

### 3.10.5 *The error in the 2-norm over grid point values*

Obviously, the implementation in previous subsection has absolutely nothing to do with what $g(x)$ is. As a matter of fact, the implementation (3.28) is our classical finite difference scheme. But there is still one catch that we have not mentioned, for implementing the finite element method as a finite difference scheme.

To be specific, in (3.28), we can only get point values of $u_h(x)$ at $x_j$, even though in practice we are quite happy with that already. On the other hand, if we have $g(x)$ and we solve (3.24), then we get $u_h(x) = \tilde{u}_h(x) + g(x)$ for any $x \in (0, 1)$.

In terms of the error estimates, the $L^2$-norm (3.16) measures the error for all $x$ in the interval $(0, 1)$. In the scheme (3.28), since we only have $u_h(x_j)$, the errors can be measured only at these grid points. For $P^1$ finite element method, it is straightforward to show that (3.16) for $u_h(x)$ implies the scheme (3.28) is second order accurate in the 2-norm:

$$\|\mathbf{e}\|_2 = \sqrt{h\sum_{j=1}^{N} e_j^2} = \sqrt{h\sum_{j=1}^{N} |u_j - u(x_j)|^2}.$$

The 2-norm above is an approximation to $L^2$-norm error $\|e_h\|_0$ by the trapezoidal quadrature for the error $e_h = u_h - u$:

$$\|e_h\|_0 = \sqrt{\int_0^1 |e_h(x)|^2 \, dx},$$

where $e_h(0) = e_h(1) = 0$ because $u_h(x)$ satisfies the boundary condition.

**Remark 3.11.** *For $P^k$ basis finite element method with $k \geq 2$, the error order for function values at (k+1)-point Gauss-Lobatto quadrature points are $(k+2)$-th order in the 2-norm. This one order higher phenomenon is called superconvergence of function values. We can use finite element method with quadratic polynomial to get a fourth order accurate finite difference scheme! Of course it can no longer be derived from $L^2$-norm estimate* (3.16)*, which is only third order accurate for $P^2$. This will be explained in Section 3.14.*

## 3.11 Generalization: a general elliptic operator

Next, we consider an elliptic equation in the following form

$$-(a(x)u'(x))' + b(x)u'(x) + c(x)u(x) = f(x), \quad x \in (0,1), \quad u(0) = u(1) = 0,$$

where $a(x) \geq \min_x a(x) > 0$ and $c(x) \geq 0$.

The variational form is still $\mathcal{A}(u,v) = (f,v)$ where

$$\mathcal{A}(u,v) = \int_0^1 a(x)u'(x)v'(x) + b(x)u'(x)v(x) + c(x)u(x)v(x) \, dx.$$

First of all, unless $b(x) = 0$, we lose the symmetry of the bilinear form, and $\mathcal{A}(u,v) = \mathcal{A}(v,u)$ is not true in general. Thus the stiffness matrix will no longer be symmetric. But other than this, almost everything above can be extended, under suitable assumptions.

For simplicity, we will just focus on how to establish the coercivity. Since $c(x) \geq 0$, we have

$$\mathcal{A}(v,v) \geq \int_0^1 a(x)|v'(x)|^2 \, dx + \int_0^1 b(x)v'(x)v(x) \, dx.$$

For the second order derivative term, recall that *Poincaré inequality* gives

$$\int_0^1 a(x)|v'(x)|^2 \, \mathrm{d}x \geq \min_x a(x)|v|_1^2 \geq C \min_x a(x)\|v\|_1^2, \quad \forall v \in H_0^1(\Omega),$$

where the constant $C$ depends only on $\Omega$.

For the first order derivative, after integration by parts, $\forall v \in H_0^1(\Omega)$,

$$\int_0^1 b(x)v'(x)v(x) \, \mathrm{d}x = \int_0^1 b(x)\frac{\mathrm{d}}{\mathrm{d}x}\frac{v^2(x)}{2} \, \mathrm{d}x = -\int_0^1 b'(x)\frac{v^2(x)}{2} \, \mathrm{d}x. \quad (3.34)$$

In two dimensions, for a first order derivative term like $\mathbf{b} \cdot \nabla u$, after integration by parts, we have

$$\iint_\Omega (\mathbf{b} \cdot \nabla v) v \mathrm{d}\mathbf{x} = -\iint_\Omega \frac{v^2}{2}(\nabla \cdot \mathbf{b}) \mathrm{d}\mathbf{x}, \quad \forall v \in H_0^1(\Omega). \quad (3.35)$$

To get the *coercivity* $\mathcal{A}(v, v) \geq C\|v\|_1^2$, we need more assumptions. Below are a few quick examples:

(1) If $b'(x) \equiv 0$, then the term in (3.34) is gone. In two dimensions, if $\nabla \cdot \mathbf{b} \equiv 0$, i.e., $\mathbf{b}$ is *incompressible*, then the term in (3.35) is gone in two dimensions.

(2) If $b'(x) \leq 0$ in one dimension or $\nabla \cdot \mathbf{b} \leq 0$ in two dimensions, then we have

$$\mathcal{A}(v, v) \geq \int_0^1 a(x)|v'(x)|^2 \, \mathrm{d}x \geq C\|v\|_1^2.$$

(3) If $b'(x) \geq 0$, and assume $C \min_x a(x) - \frac{1}{2} \max_x b'(x) \geq C_1$ where $C$ is the constant in the *Poincaré inequality*, then

$$\mathcal{A}(v, v) \geq C \min_x a(x)\|v\|_1^2 - \max_x b'(x)\frac{1}{2}\|v\|_0^2 \geq C_1\|v\|_1^0.$$

**Remark 3.12.** *For the case $b'(x) \geq 0$, obviously we need the diffusion term $-(au')'$ to be strong enough to dominate the convection term $bu'$. However, if the diffusion coefficient is very small compared to $b'(x)$, then the coercivity will be lost, thus all arguments in finite element theory based on coercivity will also break down. In practice, this reflects on the difficulties of using conventional finite element theory to construct a scheme for convection dominated problems, e.g.,* $\max_x b'(x) \gg \max_x a(x)$ *or* $a(x)$ *is nearly zero.*

## 3.12 Generalization: higher order accuracy via $P^2$

We only discuss the constant coefficient case. If you are interested, you can find the variable coefficient case in [Li and Zhang (2020c)].

### 3.12.1 *Dirichlet b.c.*

Let $V^h$ and $V_0^h$ denote the corresponding spaces of continuous piecewise quadratic polynomial was shown in Figure 3.2. The difference between $V^h$ and $V_0^h$ is that elements in $V_0^h$ are always zero on the boundary.

The scheme (3.9) with piecewise quadratic basis and Simpson's quadrature (3-point Gauss-Lobatto quadrature) has a matrix form $S\mathbf{u} = M\mathbf{f}$ where $S$ is stiffness matrix, and a diagonal mass matrix $M$ with entries $\langle \phi_i, \phi_j \rangle_h = \delta_{ij}$, because of the choice of the Lagrange basis and the special quadrature. The mass matrix in the original finite element method has entries $M_{ij} = \int_0^1 \phi_i(x)\phi_j(x)\,\mathrm{d}x$, which in general cannot be $M_{ij} = \delta_{ij}$ unless we use Lagrangian basis and some special quadrature rule such that $\langle \phi_i, \phi_j \rangle_h = \delta_{ij}$. On a uniform mesh, it can be written as $H\mathbf{u} = \mathbf{f}$ with $H = M^{-1}S$, where

$$H = \frac{1}{h^2}\begin{pmatrix} 2 & -1 & & & & & & \\ -2 & \frac{7}{2} & -2 & \frac{1}{4} & & & & \\ & -1 & 2 & -1 & & & & \\ & \frac{1}{4} & -2 & \frac{7}{2} & -2 & \frac{1}{4} & & \\ & & & -1 & 2 & -1 & & \\ & & & & \ddots & \ddots & & \\ & & & & \frac{1}{4} & -2 & \frac{7}{2} & -2 \\ & & & & & & -1 & 2 \end{pmatrix}, \quad S = \frac{1}{h}\frac{1}{3}\begin{pmatrix} 4 & -2 & & & & & & \\ -2 & \frac{7}{2} & -2 & \frac{1}{4} & & & & \\ & -2 & 4 & -2 & & & & \\ & \frac{1}{4} & -2 & \frac{7}{2} & -2 & \frac{1}{4} & & \\ & & & -2 & 4 & -2 & & \\ & & & & \ddots & \ddots & & \\ & & & & \frac{1}{4} & -2 & \frac{7}{2} & -2 \\ & & & & & & -2 & 4 \end{pmatrix},$$

and

$$M = h \cdot \operatorname{diag}\{\frac{2}{3}, \frac{1}{3}, \frac{2}{3}, \frac{1}{3}, \cdots, \frac{1}{3}, \frac{2}{3}\}.$$

For the nonhomogeneous Dirichlet boundary value problem, the scheme (3.31) can be written as

$$\frac{-u_{i-1} + 2u_i - u_{i+1}}{h^2} = f_i, \quad \text{if } x_i \text{ is a mid point}, \tag{3.36a}$$

$$\frac{u_{i-2} - 8u_{i-1} + 14u_i - 8u_{i+1} + u_{i+2}}{4h^2} = f_i, \quad \text{if } x_i \text{ is a cell end}, \tag{3.36b}$$

where $u_0 = \sigma_0$ and $u_{N+1} = \sigma_1$

**Remark 3.13.** *The finite difference scheme* (3.36) *can only be defined on a grid with odd number of grid points, because it is obtained by taking end points and midpoints in a finite element mesh as shown in Figure 3.2. This is the only drawback when using* (3.36) *as a finite difference scheme.*

Now consider solving $-\Delta u = f$ on a rectangular domain with the homogeneous Dirichlet boundary condition. We can use (3.9) with piecewise $Q^2$ (tensor product of quadratic polynomial) basis and Simpson's rule or $3 \times 3$

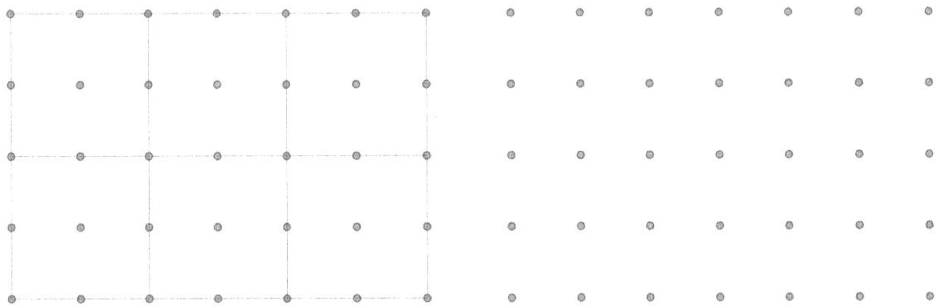

(a) The quadrature points (b) Finite difference grid

Fig. 3.6 An illustration of $Q^2$ element and the $3 \times 3$ Gauss-Lobatto quadrature.

Gauss-Lobatto quadrature rule on a uniform rectangular mesh as shown in Figure 3.6. With the same array notation in Chapter 2, the equivalent matrix-vector form of the scheme (3.9) is

$$(H_x \otimes I_y + I_x \otimes H_y) vec(U) = vec(F), \tag{3.37}$$

or equivalently

$$(S_x \otimes M_y + M_x \otimes S_y) vec(U) = (M \otimes M) vec(F).$$

**Remark 3.14.** *The linear system in (3.37) can be easily solved by first computing eigenvalue decomposition of $H$ then the eigenvector method as in Chapter 2. The eigenvalue decomposition of $H$ can be computed in MATLAB, which is affordable since $H$ is a small matrix compared to the big linear system to solve.*

**Remark 3.15.** *The stiffness matrix $S$ is always symmetric and the lumped mass matrix $M$ is diagonal. The matrix $H$ or $H_x \otimes I_y + I_x \otimes H_y$ is not symmetric, but $S_x \otimes M_y + M_x \otimes S_y$ is real symmetric. If a symmetric linear system is preferred, then the symmetric form can be used.*

**Remark 3.16.** *The convergence of finite difference scheme $KU = F$ in the maximum norm was discussed in Section 2.4 by using the fact that $K$ is an M-matrix thus a monotone matrix, i.e., $K^{-1}$ has positive entries. In general, high order finite element methods in multiple dimensions do not have monotonicity on unstructured meshes. On the other hand, the matrix $H$ or $S$ is not an M-matrix, but they are still monotone matrices, i.e., $H^{-1}$ and $S^{-1}$ have positive entries, and it holds on quasi-uniform rectangular meshes in multiple dimensions [Li and Zhang (2020a); Cross and Zhang (2024a)]. The monotonicity of discrete Laplacian of $Q^3$ element can also be proven [Cross and Zhang (2024b)]. See [Zhang (2025)] and references* *therein for high order accurate monotone discrete Laplacian schemes. The monotonicity of these high order methods can be used to construct positivity or bound preserving schemes for many interesting applications [Shen and Zhang (2022); Hu and Zhang (2023); Liu and Zhang (2023); Liu* et al. *(2024a)].*

**Problem 3.10.** *Show that the $P^2$ finite element method with Simpson's quadrature rule can be equivalently written as* (3.36).

**Problem 3.11.** *For solving* $-u''(x) = f(x), \quad x \in (0,1)$ *with homogeneous Dirichlet boundary condition* $u(0) = u(1) = 0$, *consider the scheme* (3.36). *In other words, use a uniform interior grid* $x_j = jh$ *with spacing* $h = \frac{1}{n+1}$ **where $n$ must be odd**, *then consider the following finite difference scheme:*

$$\frac{-u_{j-1} + 2u_j - u_{j+1}}{h^2} = f_j, \quad \textit{if } j \textit{ is odd;}$$

$$\frac{u_{j-2} - 8u_{j-1} + 14u_j - 8u_{j+1} + u_{j+2}}{4h^2} = f_j, \quad \textit{if } j \textit{ is even.}$$

*(1) Find the local truncation error (see Chapter 2 for the definition) of the scheme.*

*(2) Implement this scheme for an exact solution* $u(x) = sin(\pi x) + 1.1sin(2\pi x) - sin(3\pi x)$ *with* $f(x) = -u''(x)$ *and compute the error in 2-norm (defined in* (2.3)*), by modifying the MATLAB code in Section 2.2.3. What is the order of accuracy that you observe?*

### 3.12.2 *Neumann b.c.*

For one-dimensional homogeneous Neumann boundary, the scheme can be written as

$$\begin{aligned}
\frac{7u_0 - 8u_1 + u_2}{2h^2} &= f_0, \\
\frac{-u_{i-1} + 2u_i - u_{i+1}}{h^2} &= f_i, \quad \text{if } x_i \text{ is a mid point} \\
\frac{u_{i-2} - 8u_{i-1} + 14u_i - 8u_{i+1} + u_{i+2}}{4h^2} &= f_i, \quad \text{if } x_i \text{ is a cell end but not on } \partial\Omega, \\
\frac{u_{N-1} - 8u_N + 7u_{N+1}}{2h^2} &= f_{N+1}.
\end{aligned}$$

### 3.12.3 The fourth order accuracy of $P^2$ FEM as a finite difference scheme

The standard finite element error estimate for schemes in this section is third order in $L^2$-norm. But it can be proven that (3.36) is actually fourth order accurate in the 2-norm over grid points.

First of all, we can check that the finite difference approximation to the second order derivative in (3.36) is only second order accurate, even for the one in (3.36b). Second, if we use this second order approximation to solve a second order PDE such as $-u''(x) = f$, we get a fourth order accurate scheme (in $\ell^2$-norm or 2-norm)! As a matter of fact, it can be rigorously proven that this scheme is fourth order accurate for commonly used linear second order PDEs for

- Elliptic equation $-\Delta u = f$.
- Parabolic equation $u_t = \Delta u$.
- Wave equation $u_{tt} = \Delta u$.
- Schrödinger equation $\mathbb{i} u_t = \Delta u$.
- Helmholtz equation $-\Delta u - k^2 u = f$.
- Variable coefficient version of the equations above.

**Remark 3.17.** *In general, by the standard superconvergence theory of $P^k$ ($k \geq 2$) finite element method* (3.41) *(even for a variable coefficient problem in multiple dimensions), function values of $u_h(x)$ are $(k+2)$-th order accurate at Gauss-Lobatto points for each small interval in 2-norm, as opposed to $(k+1)$-th order in the $L^2$-norm error estimate, and derivatives of $u_h(x)$ are $(k+1)$-th order accurate at Gauss points, as opposed to $k$-th order in the $H^1$-norm error estimate. For a discrete Laplacian scheme like* (3.36), *i.e., the finite difference implementation of* (3.41), *the proof of the $(k+2)$-th order accuracy in 2-norm in 2D and 3D can be found in [Li and Zhang (2020c)] for elliptic equations, in [Li* et al. *(2022)] for parabolic, wave and Schrödinger equations, and in [Zhang (2025)] for the Helmholtz equation.*

All error estimates here are *a priori* error estimates, which means that the order holds if the exact solution $u(x)$ is smooth enough. For instance, the fourth order accuracy of (3.37) can be proven only if assuming $u \in H^4(\Omega)$. In practice, we often use high order accurate schemes for nonsmooth solutions, for which high order *a priori* error estimates can no longer hold. So a natural question is whether it still makes sense to use a high order accurate scheme like (3.36) on uniform meshes, which is nonetheless often used in applications. Figure 3.7 shows a comparison between the second order finite difference (3.28) and the fourth order finite difference (3.36) for solving the following generalized Allen-Cahn equation

$$\phi_t + u\phi_x + v\phi_y = \mu\Delta\phi - \frac{F'(\phi)}{\varepsilon}, \quad (x, y) \in \Omega, \tag{3.38}$$

where $u, v$ are given incompressible velocity field, and $F'(\cdot)$ is some fixed energy potential term. With the first order accurate implicit explicit (IMEX) time discretization, it becomes

$$\frac{\phi^{n+1} - \phi^n}{\Delta t} + u^{n+1}\phi_x^{n+1} + v^{n+1}\phi_y^{n+1} = \mu\Delta\phi^{n+1} - \frac{F'(\phi^n)}{\varepsilon}. \tag{3.39}$$

For the differential operators in (3.39), we can used two finite difference schemes derived from $P^1$ and $P^2$ finite element method with quadrature. For the second order derivative, they are (3.28) and (3.36). In Figure 3.7, we can see that the solution $\phi(x, y)$ has a sharp interface with large gradient, thus the smoothness assumption we need for proving that (3.36) is a fourth order scheme may no longer hold. Nonetheless, the fourth order spatial discretization is still superior because the second order spatial discretization gives a wrong solution on the same coarse $239 \times 239$ grid. Higher order time accuracy here does not help the second order spatial discretization on the same coarse $239 \times 239$ grid. This is somehow intuitive since usually time evolution is a lot smoother, thus spatial error is dominant in these problems. In other words, when spatial error dominates, higher order accurate spatial discretization often performs better.

## 3.13 Efficient inversion of Laplacian in three dimensions

For solving the Poisson equation on a rectangular domain, the same simple and efficient implementation in Section 2.9 can also be extended to arbitrary high order finite element method with $Q^k$ polynomial basis on rectangular meshes. In two dimensions, the $Q^k$ polynomial space has dimension $(k+1)^2$, spanned by $x^i y^j$ for $i, j = 0, 1, \cdots, k$. In three dimensions, $Q^k$ polynomial space is similarly defined with dimension $(k+1)^3$, while $P^k$ polynomial space has dimension $\frac{1}{6}(k+1)(k+2)(k+3)$. The $Q^k$ basis is used for finite element methods on rectangular meshes, while $P^k$ is used for finite element methods on simplicial meshes consisting of triangles and pyramids.

Now consider solving $-u_{xx} - u_{yy} - u_{zz} = f$ on a cube $\Omega = [0, 1]^3$ with homogeneous Dirichlet boundary conditions. Let us use $Q^2$ element as

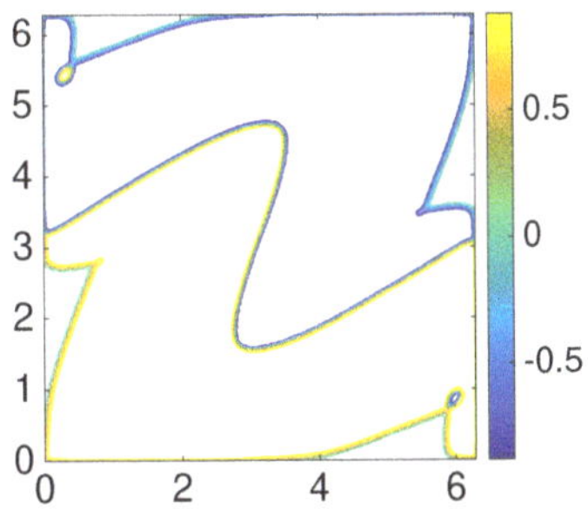


(a) Second order scheme with first order IMEX on a 239 × 239 grid

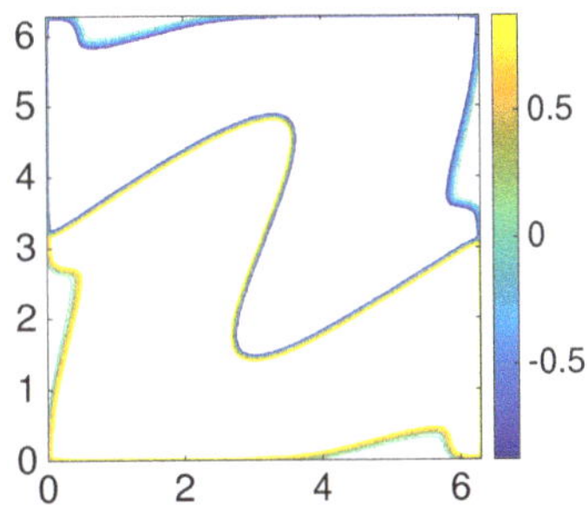


(b) Fourth order scheme with first order IMEX on a 239 × 239 grid

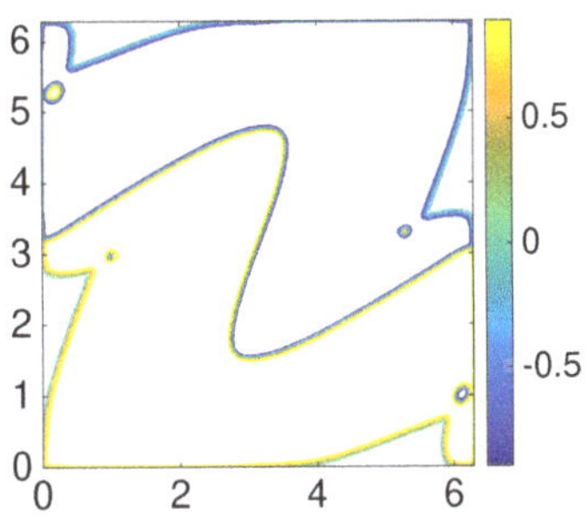


(c) Second order scheme with third order IMEX BDF on a 239 × 239 grid

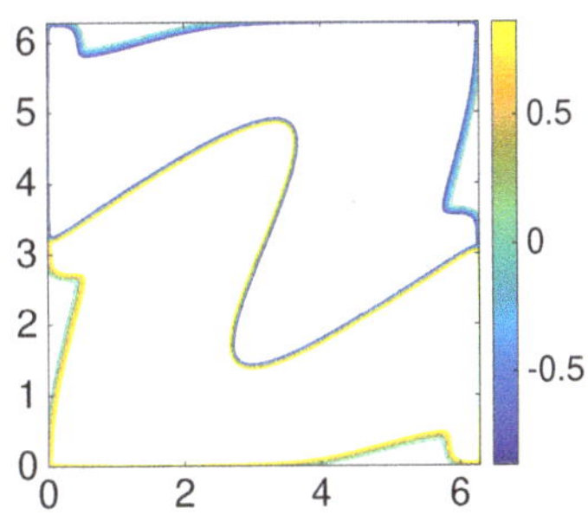


(d) Reference Solution

Fig. 3.7 A generalized Allen-Cahn equation coupled with a passive convection and log energy at $T = 1.8$.

an example, and consider a 3D rectangular mesh shown in Figure 3.8 (a). Consider the quadrature of using Simpson's rule for $x, y, z$ variables, i.e., $3 \times 3 \times 3$ Gauss-Lobatto quadrature. There are 27 quadrature points in each cell, see Figure 3.8 (b).

Consider a uniform rectangular mesh for $\Omega$. Let $V_0^h$ be the linear space of continuous piecewise $Q^2$ polynomial on this mesh with zero values on the domain boundary. Since a $Q^2$ polynomial has 27 coefficients, on a cubic cell it can be uniquely represented by the 27 quadrature points shown in Figure 3.8. All such quadrature points in a rectangular mesh form a uniform Cartesian in 3D with grid points $(x_i, y_j, z_k)$ $(i = 1, \cdots, N_x; j = 1, \cdots, N_y; k = 1, \cdots, N_z)$.

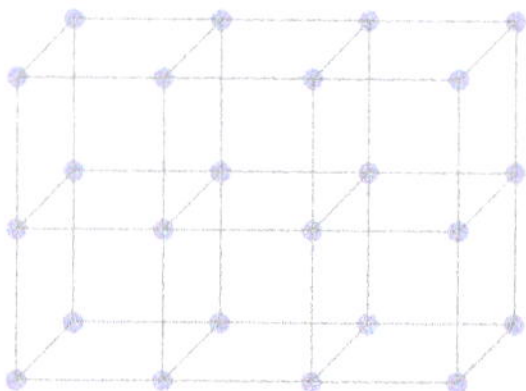

(a) A rectangular mesh of six cells.

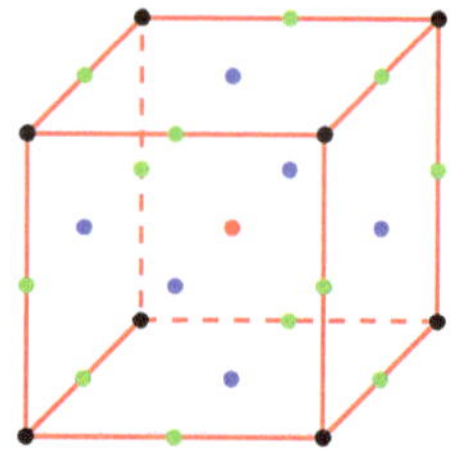

(b) Quadrature points on one cell.

Fig. 3.8 An illustration of the $3 \times 3 \times 3$ Gauss-Lobatto quadrature on one rectangular cell.

By the same notation in Section 2.9, the piecewise $Q^2$ finite element solution can be represented by a 3D array $U$ of size $N_z \times N_y \times N_x$ with $(k, j, i)$-th entry $U(k, j, i)$ denoting the point value at $(x_i, y_j, z_k)$. Let $F$ be a 3D array of size $N_z \times N_y \times N_x$ with $(k, j, i)$-th entry $F(k, j, i)$ equal to the point value $f(x_i, y_j, z_k)$.

By the same notation and operations defined in Section 2.9, the finite element method $\mathcal{A}_h(u_h, v_h) = \langle f, v_h \rangle_h$ implemented by Simpson's quadrature rule can be written as

$$(S_x \otimes M_y \otimes M_z + M_x \otimes S_y \otimes M_z + M_x \otimes M_y \otimes S_z)\, vec(U) = (M_x \otimes M_y \otimes M_z)\, vec(F),$$

or equivalently,

$$(H_x \otimes I_y \otimes I_z + I_x \otimes H_y \otimes I_z + I_x \otimes I_y \otimes H_z)\, vec(U) = vec(F), \quad (3.40)$$

where the matrices $S, H, M$ are the same as those in Section 3.12.1.

Suppose we have eigenvalue decomposition $H = S\Lambda S^{-1}$, then (3.40) can be implemented in exactly the same way described in Section 2.9.

If using the $Q^k$ element for $k \geq 3$ with $(k+1)^3$-point Gauss-Lobatto quadrature, the quadrature points within each cell no longer form a uniform grid, but the finite element method can still be written as (3.40) with $U$ and $F$ denoting point values at all these quadrature points.

In general, it is not possible to have an explicit formula for eigenvalue decomposition of $H$ matrix for $Q^k$ element. Since $M$ is a diagonal matrix, the eigenvalue decomposition for the nonsymmetric matrix $H = S\Lambda S^{-1}$ can be approximated by numerically solving the following symmetric eigenvalue problem, especially for a very high order polynomial basis. Let

$$H = M^{-1}S = M^{-1/2}(M^{-1/2}SM^{-1/2})M^{1/2}.$$

Since $S_1 = M^{-1/2}SM^{-1/2}$ is real symmetric, we can first find its eigenvalue decomposition as $S_1 = Q\Lambda Q^\top$ where $\Lambda$ is a diagonal matrix and $Q$ is an orthogonal matrix. Then, we have

$$H = M^{-1/2}(Q\Lambda Q^\top)M^{1/2} = T\Lambda T^{-1},$$

with $T = M^{-1/2}Q$ and $T^{-1} = Q^\top M^{1/2}$.

The MATLAB code for forming the matrices $S, M, H$ and computing the composition $H = S\Lambda S^{-1}$ for $Q^k$ element with Dirichlet, or Neumann, or periodic boundary conditions is given in [Liu *et al.* (2024b), Appendix].

**Problem 3.12.** *For solving $-u_{xx} - u_{yy} = f$ with homogeneous Dirichlet b.c. on $[0,1]^2$ by $Q^2$ element and Simpson quadrature rule, show that the scheme can be written as*

$$(S_x \otimes M_y + M_x \otimes S_y)\, vec(U) = (M_x \otimes M_y)\, vec(F),$$

*where the matrices $S, M$ are the same as those in Section 3.12.1.*

**Problem 3.13.** *Use the code in [Liu et al. (2024b), Appendix] to write a MATLAB code for $Q^k$ element of your chosen $k$ solving $-u_{xx} - u_{yy} - u_{zz} = 14\pi^2 \sin(\pi x)\sin(2\pi y)\sin(3\pi z)$ with homogeneous Dirichlet b.c. on the domain $[0,1]^3$. Test your code for the exact solution $u(x,y,z) = \sin(\pi x)\sin(2\pi y)\sin(3\pi z)$. Test the convergence rate on several different grids. On a modern desktop computer, the implementation in this section should be quite efficient for a grid of size $200^3$ or even $300^3$. If your computer has large enough memory, try to push the grid size to as large as it allows. For $k \geq 2$, you should observe $(k+2)$-th order accuracy in 2-norm, which is the superconvergence in Section 3.14.*

**Problem 3.14.** *Modify your code in the previous problem to solve $-u_{xx} - u_{yy} - u_{zz} = 14\pi^2 \cos(\pi x)\cos(2\pi y)\cos(3\pi z)$ with homogeneous Neumann b.c. on the domain $[0,1]^3$. Notice that $\Lambda(1,1,1) = 0$ so you need to use the trick in Chapter 2 to redefine $\frac{1}{\Lambda(1,1,1)} = 0$.*

## 3.14 *Superconvergence

Beginners may skip this section. For the scheme (3.36), the error order at quadrature points (two cell ends and the middle point) is one order higher than $L^2$ error, which is computed for all $x$ in the domain. Such a phenomenon that error at certain points is smaller is called *superconvergence*. On the other hand, it is straightforward to verify that the local truncation error of (3.36a) and (3.36b) is only second order. Recall that the local truncation error is not the true error. Sometimes, the local truncation error at particular locations can be lower order than the true error order, and such a phenomenon is called *supraconvergence* [Ferreira and Grigorieff (1998)], which is different from superconvergence. The full proof of why the scheme (3.36) in 2D is fourth order accurate in 2-norm over all grid points is quite complicated, see [Li and Zhang (2020c)]. In this section, we will only learn some quick reasons why superconvergence is even possible in one dimension.

### 3.14.1 *The delta function*

Our heuristic understanding of the delta function is $\delta(x) = \begin{cases} +\infty, & x = 0 \\ 0, & x \neq 0 \end{cases}$, which is however not a conventional function at all. One rigorous understanding of it is to define it as a *functional*, which maps any continuous function with compact support $f(x)$ linearly to a number $f(0)$. It is often denoted by an integral, i.e., the symbol $\delta(x)$ is defined to satisfy

$$\int_{-\infty}^{+\infty} f(x)\delta(x)\,\mathrm{d}x = f(0), \quad \forall f(x) \in C_0(\mathbb{R}).$$

In other words, the mapping $\delta(x)$ maps $f(x)$ to $f(0)$, and we write the action of $\delta$ onto $f$ by an integral $\int_{-\infty}^{+\infty} f(x)\delta(x)\,\mathrm{d}x$. By writing a mapping in such an integral form, it allows us to be able to pretend that $\delta(x)$ is some function.

Recall that the function $f(x) = |x|$ is not differentiable but we can define its *weak or generalized derivative* as the step function $g(x) = \begin{cases} 1 & x \geq 0 \\ -1 & x < 0 \end{cases}$.

Now let us compute the weak derivative of the step function by integration by parts:

$$\forall v(x) \in C_0^\infty(\mathbb{R}), \quad \int_{-\infty}^{+\infty} g(x)v'(x)\,\mathrm{d}x = \int_0^{+\infty} v'(x)\,\mathrm{d}x - \int_{-\infty}^{0} v'(x)\,\mathrm{d}x = -2v(0).$$

With the definition of $\delta(x)$ above, we have

$$\forall v(x) \in C_0^\infty(\mathbb{R}) \subset C_0(\mathbb{R}), \quad \int_{-\infty}^{+\infty} v(x)\delta(x)\,\mathrm{d}x = v(0),$$

thus

$$\forall v(x) \in C_0^\infty(\mathbb{R}), \quad \int_{-\infty}^{+\infty} \frac{1}{2}g(x)v'(x)\,\mathrm{d}x = -\int_{-\infty}^{+\infty} v(x)\delta(x)\,\mathrm{d}x.$$

Therefore, we have obtained $\frac{\mathrm{d}^2}{\mathrm{d}x^2}|x| = 2\delta(x)$, in the weak derivative sense. The symbol $\delta_a(x) := \delta(x-a)$ is defined to satisfy

$$\int_{-\infty}^{+\infty} f(x)\delta_a(x)\,\mathrm{d}x = \int_{-\infty}^{+\infty} f(x)\delta(x-a)\,\mathrm{d}x = f(a), \quad \forall f(x) \in C_0(\mathbb{R}).$$

Thus we also have $\frac{\mathrm{d}^2}{\mathrm{d}x^2}\frac{1}{2}|x-a| = \delta(x-a)$.

### 3.14.2 *The one-dimensional Green's function*

For the boundary value problem $-u''(x) = f(x), \quad x \in (0,1), \quad u(0) = u(1) = 0$, its Green's function $G_a(x)$ is defined to satisfy

$$-\frac{\mathrm{d}^2}{\mathrm{d}x^2}G_a(x) = \delta_a(x), \quad G_a(0) = G_a(1) = 0,$$

where $a \in (0,1)$ is a fixed number.

Following the discussion in the previous subsection, it is straightforward to verify that

$$G_a(x) = \begin{cases} (1-a)x, & x \in [0,a] \\ (1-x)a, & x \in (a,1] \end{cases},$$

thus

$$\frac{\mathrm{d}}{\mathrm{d}x}G_a(x) = \begin{cases} (1-a), & x \le a \\ -a, & x > a \end{cases}, \quad -\frac{\mathrm{d}^2}{\mathrm{d}x^2}G_a(x) = \delta_a(x).$$

Notice that $G_a(x)$ is a continuous piecewise linear function, but this is true only for one-dimensional problems.

### 3.14.3 *Superconvergence at knots in one dimension*

For the one-dimensional problem $-u''(x) = f(x), \quad x \in (0,1), \quad u(0) = u(1) = 0$, assume we have a mesh of intervals $I_j$, on which we define continuous piecewise polynomial spaces $V^h$ and $V_0^h$.

The abstract finite element method is to seek $u_h \in V_0^h$ satisfying

$$(u_h', v_h') = (f, v_h), \quad \forall v_h \in V_0^h. \tag{3.41}$$

Recall that the solution $u_h$ has Galerkin orthogonality:

$$(u' - u_h', v_h') = 0, \quad \forall v_h \in V_0^h.$$

Let $e(x) = u(x) - u_h(x) \in H_0^1(0,1) \subset C_0(0,1)$, then Galerkin orthogonality can be written as

$$(e', v_h') = 0, \quad \forall v_h \in V_0^h.$$

Let $x_i$ be the cell end of some interval $I_j$ and we call $x_i$ a knot. Then we consider the Green's function at $a = x_i$, e.g, $G_{x_i}(x)$, which is a piecewise linear polynomial defined on the same mesh, thus $G_{x_i}(x) \in V_0^h$. Now let us take a special test function $v_h = G_{x_i}(x)$ in the Galerkin orthogonality:

$$(e', G_{x_i}(x)') = 0 \Rightarrow \int_0^1 e'(x) G_{x_i}(x)' \,\mathrm{d}x = 0 \Rightarrow -\int_0^1 e(x) \frac{\mathrm{d}^2}{\mathrm{d}x^2} G_{x_i}(x) \mathrm{d}x = 0$$

$$\Rightarrow -\int_0^1 e(x)\delta_{x_i}(x)\,\mathrm{d}x = 0 \Rightarrow -\int_{-\infty}^{+\infty} e(x)\delta_{x_i}(x)\,\mathrm{d}x = 0 \Rightarrow e(x_i) = 0,$$

where we have extended $e(x)$ to the whole real line by zero extension.

This means that the error at knots $x_i$ must be zero! Notice that this is the property to the abstract scheme (3.41) for any $P^k$ basis, which we however do not implement. For instance, for $P^1$ basis on a uniform mesh, the scheme (3.41) is the same as

$$\frac{1}{h}\begin{pmatrix} 2 & -1 & & & & \\ -1 & 2 & -1 & & & \\ & -1 & 2 & -1 & & \\ & & \ddots & \ddots & \ddots & \\ & & & -1 & 2 & -1 \\ & & & & -1 & 2 \end{pmatrix} \begin{bmatrix} u_1 \\ u_2 \\ u_3 \\ \vdots \\ u_{N-1} \\ u_N \end{bmatrix} = \begin{bmatrix} (f,\phi_1) \\ (f,\phi_2) \\ (f,\phi_3) \\ \vdots \\ (f,\phi_{N-1}) \\ (f,\phi_N) \end{bmatrix}. \tag{3.42}$$

But usually we implement it by approximating the integral $(f, \phi_i)$ by the trapezoidal rule, which is second order accurate. If we do compute the integrals $(f, \phi_i)$ exactly, then the scheme (3.42) has zero error!

For the $P^2$ basis on uniform mesh, the Simpson's rule is exact for the left hand integral $(u_h', v_h')$, thus (3.41) can be written as

$$\frac{2h}{3}\frac{-u_{i-1} + 2u_i - u_{i+1}}{h^2} = (f, \phi_i), \quad \text{if } x_i \text{ is a mid point,} \tag{3.43a}$$

$$\frac{h}{3}\frac{u_{i-2} - 8u_{i-1} + 14u_i - 8u_{i+1} + u_{i+2}}{4h} = (f, \phi_i), \quad \text{if } x_i \text{ is a cell end.} \tag{3.43b}$$

The error of the scheme (3.43) is zero at the cell end $x_i$ (knots). Of course, in the scheme (3.36), we use Simpson's rule for approximating the integrals $(f, \phi_i)$, which is fourth order accurate. So intuitively it is not a surprise that the scheme (3.36) should be fourth order accurate at the knots. For the fourth order accurate at the midpoint, we need some more advanced tools, which will not be explained here.

## 3.15 Comparison with traditional finite difference method

### 3.15.1 *Advantages of the finite element method*

Throughout this chapter, we have seen many things that cannot happen or cannot be explained in the traditional finite difference method. Even on uniform meshes for a rectangular domain, the finite element method is still superior from any perspective, because it gives us a finite difference scheme with all desired properties. We summarize some comparisons in Table 3.1.

Table 3.1 Comparison of traditional FD and finite element method for solving $-\nabla\cdot(A\nabla u)=f$.

| | traditional FD method | finite element method (FEM) |
|---|---|---|
| Tools needed | Calculus & Linear Algebra | functional analysis, PDE theory, etc |
| Approximation of what | approximates PDE | approximates variational form |
| Boundary conditions | direct approximation | absorbed in variational form |
| General domains | N/A | $\Omega$ is approximated by $\Omega_h$ |
| Rectangular $\Omega$ | a linear system | can be written as finite difference |
| Linear system | nonsymmetric in general | always symmetric |
| Implementation | form a matrix directly | approximate integrals |
| Purely Neumann b.c. | non unique solutions | unique solution in quotient space |
| Consistency | Taylor expansion | Galerkin orthogonality$\Leftarrow$ Conforming FEM |
| Stability | singular values | coercivity $\Leftarrow$ Poincaré inequality |
| Convergence | in 2-norm | $H^1$ and $L^2$ error estimates |
| Error order | truncation error order | interpolation error order |
| Higher order schemes | complicated near boundary | no difficulty near the boundary |
| Variable coefficient | difficult to design | easy to design |
| Superconvergence | N/A | $P^2$ gives a 4th order FD |

### 3.15.2 *Limitations of the finite element method*

In general, the finite element method is quite successful, for solving an elliptic equation $-\Delta u = f$ or some other types of equations including parabolic equations $u_t = \Delta u$, wave equations $u_{tt} = \Delta u$, Schrödinger equation $\mathrm{i}u_t = \Delta u$, etc. These equations all contain the Laplacian operator $-\Delta u$, for which a *coercive* bilinear form $\mathcal{A}(u,v) = (u',v')$ can be defined. Another different example is the biharmonic equation $u''''(x) = f$, for which we can also define a similar variational formulation with coercivity, thus the finite element method for this kind of fourth order PDE is also quite successful.

The foundation of the success for the finite element method, when missing, is also source of the limitations of the finite element method in applications. It could be extremely difficult to extend the same theory to equations that lack coercive operators, if not impossible. One simple example of such equations is the simple convection $u_t+u_x=0$ or its nonlinear version *nonlinear conservation laws* $u_t+f(u)_x=0$. Another example is the *Hamilton-Jacobi* equation $u_t+f(u_x)=0$, e.g., $u_t+|\nabla u|=0$, which is also closely related to *nonlinear conservation laws.* A formal application of the finite element method to these equations, with certain modifications to achieve stability or even convergence, is always possible, but many provable properties in this chapter will be no longer true.

# Chapter 4

# Applications and Examples on Cartesian Grids

Fast Poisson equation solvers play a central role in numerous applications, as the discrete Laplacian serves as a fundamental modeling tool not only in classical science and engineering but also in contemporary data science. In this chapter, we will examine several illustrative examples of these applications, which may serve as homework and exam problems in a computational math class, or even potential building blocks for designing research projects suitable for both undergraduate and graduate students.

In this chapter, the description of the applications and projects involves some background knowledge, which significantly exceeds the scope of the previous chapters and the main focus of this book, such as fluid dynamics equations in Section 4.1 and optimization algorithms in Section 4.2. For readers to be able to understand how Poisson equations emerge in these applications and finish computational projects in this chapter, it is not necessary to be familiar with these advanced materials. On the other hand, Appendix C lists some useful knowledge about convex optimization algorithms.

## 4.1 Poisson solver in computational fluid dynamics

In computational fluid dynamics, many PDEs involve the Poisson equation. One quick example is *implicit time stepping for solving parabolic equations.* Consider solving the heat equation $u_t = u_{xx}$ with an initial condition $u(x,0) = f(x)$ and homogeneous Dirichlet b.c. for $x \in [0,1]$. With the centered difference for the spatial derivative, and the explicit forward Euler time discretization, we have

$$U_j^{n+1} = U_j^n + \frac{\Delta t}{\Delta x^2}(U_{j-1}^n - 2U_j^n + U_{j+1}^n),$$

where $U_j^n$ denotes the value at location $x_j$ and time step $n$. It can be shown that the Lax-Richtmyer linear stability of this scheme holds if and only if $\frac{\Delta t}{\Delta x^2} \leq \frac{1}{2}$. In practice, the time step $\Delta t = \frac{1}{2}\Delta x^2$ is impractically too small. A practical time step like $\Delta t = \mathcal{O}(\Delta x)$ can be achieved if using the backward Euler *implicit* time discretization,

$$U_j^{n+1} = U_j^n + \frac{\Delta t}{\Delta x^2}(U_{j-1}^{n+1} - 2U_j^{n+1} + U_{j+1}^{n+1}).$$

For implementing this implicit scheme, a linear system $BU^{n+1} = U^n$ must be solved in each time step. Here the matrix is $B = I - \Delta t K$ with the same $K$ matrix in Chapter 2 thus a fast Poisson solver can be used to compute or approximate $B^{-1}$, e.g., the same inversion of discrete Laplacian in Section 2.9 can be used for solving the 3D heat equation. In other words, for solving a parabolic equation like $u_t = \nabla \cdot (A\nabla u)$ where $A$ is matrix coefficient, with an implicit time stepping using a time step $\tau = \Delta t$, we need to solve an elliptic equation $U^{n+1} - \tau\nabla \cdot (A\nabla U^{n+1}) = U^n$, for which methods in previous chapters may be used.

### 4.1.1 *Navier-Stokes equations in fluid dynamics*

The second example is the *variable density incompressible Navier-Stokes equations* for modeling a mixture of two fluids such as water and air,

$$\begin{cases} \rho_t + \nabla \cdot (\rho \mathbf{u}) = 0 \\ \rho(\mathbf{u}_t + (\mathbf{u} \cdot \nabla)\mathbf{u}) + \nabla p - \nu\Delta\mathbf{u} = \mathbf{f} \end{cases}, \tag{4.1}$$

where $\rho, \mathbf{u}, p$ are unknown density, velocity and pressure, $\mathbf{f}$ is given external force function such as gravity, and $\nu$ is the viscosity function which may depend on $\rho$. To recover the pressure from the evolving $\mathbf{u}$ and $\rho$, we can take the divergence of the second equation divided by $\rho$, then we get an elliptic equation for the pressure

$$\nabla \cdot (\frac{1}{\rho}\nabla p) = \nabla \cdot (\frac{1}{\rho}\mathbf{f} + \frac{1}{\rho}\nu\Delta\mathbf{u} - (\mathbf{u} \cdot \nabla)\mathbf{u}).$$

In other words, the Poisson equation naturally emerges here for recovering pressure.

If there is no external force, and only one fluid to model, then $\rho \equiv 1$, and (4.1) reduces to *constant density incompressible Navier-Stokes* equation

$$\mathbf{u}_t + (\mathbf{u} \cdot \nabla)\mathbf{u} + \nabla p = \nu\Delta\mathbf{u}, \quad \nabla \cdot \mathbf{u} = 0, \tag{4.2}$$

where $\nu = \frac{1}{\text{Re}}$ is a constant with Re being a given *Reynolds* number.

For two dimensional problems, if we take the curl of (4.2) in 2D, we get the *vorticity stream-function formulation of the 2D incompressible Navier-Stokes equations*:

$$\omega_t + u\omega_x + v\omega_y = \frac{1}{\text{Re}}\Delta\omega, \tag{4.3a}$$

$$\Delta\psi = \omega, \qquad \langle u, v\rangle = \langle -\psi_y, \psi_x\rangle, \tag{4.3b}$$

where $\psi(x, y)$ is the stream function and $\omega(x, y) = \nabla \times \mathbf{u} = v_x - u_y$ is the vorticity. Here $\nabla \times \mathbf{u}$ is a scalar because $\mathbf{u} = (u, v)$ is a 2D vector field. Suppose we use forward Euler time discretization with time step $\tau = \Delta t$ to approximate (4.3):

$$\omega^{n+1} = \omega^n - \tau u^n \omega_x^n - \tau v^n \omega_y^n + \tau \frac{1}{\text{Re}}\Delta\omega^n,$$

then in each time step we need to recover $\psi^n$ by solving $\Delta\psi^n = \omega^n$ to obtain the velocity by computing $u^n = -\psi_y^n, v^n = \psi_x^n$.

### 4.1.2 *Why Laplacian is hidden in incompressibility*

For solving the incompressible Navier-Stokes equation (4.2), there are quite a few different methods, including solving its equivalent form (4.3). Notice that (4.3) explicitly contains a Poisson equation $\Delta\psi = \omega$, but (4.2) does not contain a Poisson equation. Nonetheless, all methods solving (4.2) involve solving a Poisson equation due to the incompressibility condition $\nabla \cdot \mathbf{u} = 0$.

One quick explanation why incompressibility is connected to Laplacian, is to consider the projection of a velocity field $\mathbf{u}$ into an incompressible velocity field, then the projection operator naturally involves the Laplacian operator. To quickly see why this is true, we consider a discrete set up in one dimension. Consider a function $u(x)$ defined on $[0, 1]$, the incompressibility condition becomes $u'(x) = 0$. Consider a uniform grid $x_j = j\Delta x$ and $\Delta = \frac{1}{N+1}$. By forward difference approximating the derivative, we get the following matrix vector form for $u'(x) = 0$:

$$DU = 0, \quad U = \begin{pmatrix} U_1 \\ U_2 \\ \vdots \\ U_N \\ U_{N+1} \end{pmatrix}, \quad D = \frac{1}{\Delta x}\begin{pmatrix} -1 & 1 & & & \\ & -1 & 1 & & \\ & & \ddots & \ddots & \\ & & & -1 & 1 \end{pmatrix}_{N\times(N+1)}.$$

Given a vector $V$ that does not satisfy the equation $DV = 0$, the projection onto the set $\{U : DU = 0\}$ is given by

$$\text{Projection}(V) = V - D^\top(DD^\top)^{-1}DV.$$

We can check that $DD^\top = K$ with $K$ being the tridiagonal (-1, 2, -1) matrix, which is the negative discrete Laplacian matrix in Chapter 2.

This means that the Laplacian operator is related to enforcing the incompressible condition, or the projection onto the set of divergence free vector fields. The divergence free condition appears naturally, not only in the incompressible fluid, but also in optimal transport, as will be shown in Section 4.4.2.

### 4.1.3 *Project: incompressible flows with periodic boundary*

Consider the vorticity stream-function formulation of the *2D incompressible Navier-Stokes* given by (4.3) on a square domain with initial condition $\omega(x, y, 0) = \omega_0(x, y)$ and periodic boundary conditions.

Write a MATLAB code for solving this equation by following the instructions below step by step.

*Step I: 2D Poisson Solver.* We need to solve a Poisson equation $\Delta\psi = \omega$. Implement the eigenvector method in Chapter 2 for solving $u_{xx}+u_{yy} = f$ on $[0, 2\pi] \times [0, 2\pi]$ with periodic boundary conditions. Test your code with the following solution: $f = 2\cos(2x) + 2\cos(2y)$ and $u = \sin^2 x + \sin^2 y - 1$. Recall that the matrix is singular thus we usually redefine the division by zero eigenvalue as multiplying zero, which gives a zero sum numerical solution. For computing errors, we should compare the zero sum numerical solution with a zero sum exact solution, which can be obtained by shifting any exact solution by the average of its point values. Use *fft2* and *ifft2* functions for the eigenvector multiplication. Use uniform $N \times N$ meshes. Show loglog plot of the errors in max norm and compare it with the second order slope line for $N = 20, 40, 80, 160, 320$. Show your CPU time (use *tic, toc* functions in MATLAB to track CPU time) for $N = 1280, 2560, 5120$.

*Step II: Linear Convection Diffusion.* To solve the nonlinear convection diffusion like Navier-Stokes equation, we should first test our numerical scheme on a linear equation. Consider the following equation with periodic b.c. on $[0, 2\pi] \times [0, 2\pi]$:

$$u_t + u_x + u_y = d(u_{xx} + u_{yy}), \quad d > 0.$$

We can use centered difference for all spatial derivatives to achieve second order accuracy in space. Let $h = \Delta x = \Delta y$ denote mesh size of a uniform mesh and $\Delta_h$ denote the discrete Laplacian:

$$\frac{\mathrm{d}}{\mathrm{d}t}u_{i,j} = -\frac{u_{i+1,j} - u_{i-1,j}}{2h} - \frac{u_{i,j+1} - u_{i,j-1}}{2h} + d\Delta_h u_{i,j}. \tag{4.4}$$

If we use explicit methods like forward Euler or second order Runge Kutta (RK) method for (4.4) with $d = 0$, it is unstable for any time step because their stability regions do not contain any imaginary axis. Instead, higher order accurate RK methods will work well with a reasonable stable time step for small $d$. If we write the ODE system (4.4) as $\frac{\mathrm{d}}{\mathrm{d}t}U = \mathrm{RHS}(U)$, use the following Strong Stability Preserving (SSP) RK3 (third order RK) to solve $\frac{\mathrm{d}}{\mathrm{d}t}U = \mathrm{RHS}(U)$ with the time step $\Delta t = \min\{0.3\frac{h^2}{d}, 0.3h\}$ $(h = \Delta x = \Delta y)$:

```
U1=U+dt*RHS(U);
U2=0.75*U+0.25*(U1+dt*RHS(U1));
U=U/3+2/3*(U2+dt*RHS(U2));
```

Test your code using the exact solution $u(x,t) = exp(-2dt)\sin(x+y-2t)$ with $d = 0.01$ on $[0, 2\pi] \times [0, 2\pi]$. Run it till $T = 0.2$ with $N = 16, 32, 64, 128, 256, 512$. If the error has the order $e(h) = Ch^k$ on a mesh with size $h$, then on a refined mesh we have $e(h/2) = C(h/2)^k$. Therefore, we have

$$\log e(h/2) = \log C + k\log h - k\log 2, \quad \log e(h) = \log C + k\log h$$

thus we can compute the convergence order using the errors $e(h), e(h/2)$:

$$k = \frac{\log e(h) - \log e(h/2)}{\log 2} = \frac{1}{\log 2}\log\frac{e(h)}{e(h/2)}. \tag{4.5}$$

Show a table of errors and orders computed by refining the mesh: we have six errors on six meshes so we can compute 5 orders by (4.5).

*Step III: Accuracy Test for 2D Incompressible Flow.* Plugging the second equation of (4.3b) into (4.3a), we get

$$\omega_t - \psi_y\omega_x + \psi_x\omega_y = \frac{1}{\mathrm{Re}}\Delta\omega.$$

Let $D_x$ and $D_y$ denote the central difference operator for the first order partial derivatives, and let $\Delta_h$ be the discrete Laplacian using second order finite difference, we get

$$\omega_t = D_y\psi D_x\omega - D_x\psi D_y\omega + \frac{1}{\mathrm{Re}}\Delta_h\omega.$$

Use the RK3 above for the time derivative. At each time step $t^n$, first compute the maximum of velocity by $U^n = \max\{|u^n|, |v^n|\}$, then time step can be taken as $\Delta t = \min\{0.3\,\mathrm{Re}\,h^2, 0.3\frac{1}{U^n}h\}$ $(h = \Delta x = \Delta y)$. For each time stage, you need to compute/update the velocity by solving the Poisson equation $\psi$. Test the accuracy with the exact solution:

$\omega(x, y, t) = -2\exp\{(-2t/\operatorname{Re})\}\sin x \sin y$ on domain $[0, 2\pi] \times [0, 2\pi]$ with $\operatorname{Re} = 100$. Notice that the 2D Poisson equation needs to be solved for each of the three time stages in one RK step. Run your code till $T = 0.2$ with $N = 16, 32, 64, 128, 256$. List the error and order as a table.

*Step IV: Double Shear Layer Benchmark Test.* Once you have finished the three steps above, you can change your initial condition to further test your code on some benchmark tests. Take $\operatorname{Re} = 1000$. The initial condition is

$$\omega(x, y, 0) = \begin{cases} \delta\cos(x) - \frac{1}{\rho}\operatorname{sech}^2((y-\pi/2)/\rho) & y \le \pi \\ \delta\cos(x) + \frac{1}{\rho}\operatorname{sech}^2((3\pi/2-y)/\rho) & y > \pi \end{cases}$$

on domain $[0, 2\pi] \times [0, 2\pi]$, where we take $\rho = \pi/15$ and $\delta = 0.05$. Show your vorticity at $T = 6$ and $T = 8$ with $N = 256$ using 30 contour lines. Make sure your x-axis and y-axis are correctly shown.

For example, the initial value can be visualized using 30 contour lines on a $256 \times 256$ mesh as follows. See also Figure 4.1 for the solution at $T = 8$.

```
n=256; L  = 2*pi;
x  = linspace(0,L,n+1)'; x = x(2:end); y=x;
yy=y*ones(1,n); xx=ones(n,1)*x';
rho=pi/15; Delta=0.05;
omega1=Delta*cos(xx)-1/rho*sech((yy-pi/2)/rho).^2;
omega2=Delta*cos(xx)+1/rho*sech((3*pi/2-yy)/rho).^2;
indicator1=[ones(n/2,n); zeros(n/2,n)];
indicator2=[zeros(n/2,n); ones(n/2,n)];
omega=omega1.*indicator1+omega2.*indicator2;
contour(x,y,omega,30);colorbar; xlabel('X');ylabel('Y');
```

*Step V: Implicit diffusion treatment.* For small Re, the time step $\Delta t = 0.3\operatorname{Re} h^2$ is too small to use. Instead, we can consider using the implicit-explicit (IMEX) method, i.e., consider the following scheme which is first order accurate in time:

$$\omega^{n+1} = \omega^n - \Delta t u^n D_x \omega^n - \Delta t v^n D_y \omega^n + \frac{\Delta t}{\operatorname{Re}} \Delta_h \omega^{n+1},$$

$$u^n = -D_y \psi^n, \quad v^n = D_x \psi^n, \quad \Delta_h \psi^n = \omega^n.$$

Such an IMEX method allows a larger time step. At each time step, define $\|\mathbf{u}^n\|_\infty = \max\{\max_{i,j} |u^n_{i,j}|, \max_{i,j} |v^n_{i,j}|\}$. Then the linear stability analysis,

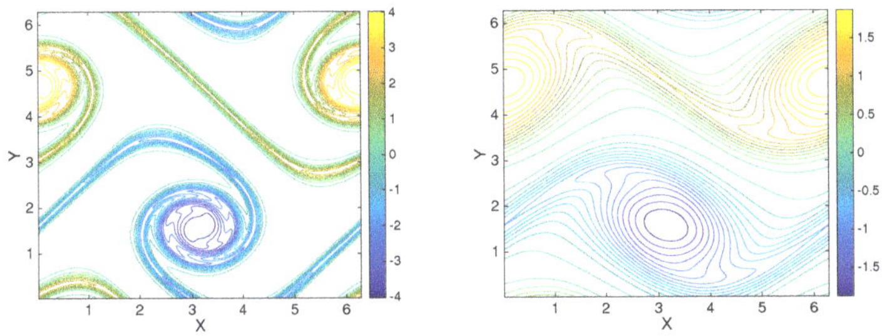


(a) Re $= 1000$. $T = 8$. $256 \times 256$ grid. (b) Re $= 70$. $T = 8$. $512 \times 512$ grid.

Fig. 4.1 Incompressible Navier Stokes equations: Double Shear Layer.

which is not covered in this book, suggests that the IMEX scheme is linearly stable if

$$\Delta t \leq \frac{1}{\|\mathbf{u}^n\|_\infty^2} \frac{1}{\mathrm{Re}}. \tag{4.6}$$

Implement this scheme for the Double Shear Layer problem using a time step as $\Delta t = \Delta x$ for Re $= 70$. Use the eigenvector method to invert the matrix and use FFT for the eigenvectors. Show your vorticity at $T = 6$ and $T = 8$ with $N = 512$ using 30 contour lines. You can also try a larger $\Delta t$ or larger Re so that (4.6) is violated on a coarse mesh, and see what happens.

*Bonus: high order accuracy by $Q^2$ finite element method.* Follow [Shen and Zhang (2022)] to use the fourth order finite difference therein, which is the $Q^2$ finite element method with Simpson quadrature (e.g., read Section 3.12). Implement it for the double shear layer problem and compare your results to those in [Shen and Zhang (2022)]. Notice that Section 3.12 is for Dirichlet b.c., and periodic boundary conditions are needed here.

### 4.1.4 *Project: Lid Driven Cavity flow*

The Lid Driven Cavity flow is a classical test problem for numerical methods solving incompressible Navier-Stokes equations. The domain $\Omega$ is a square $[0, 1] \times [0, 1]$ and the boundary conditions of the velocity is specified as in Figure 4.2. In other words, the velocity along the boundary is zero, except that $u$ is equal to some given velocity at the top boundary.

Consider the vorticity stream-function formulation of the *2D incompressible Navier-Stokes* given by (4.3) on a square domain $[0, 1]^2$ with the

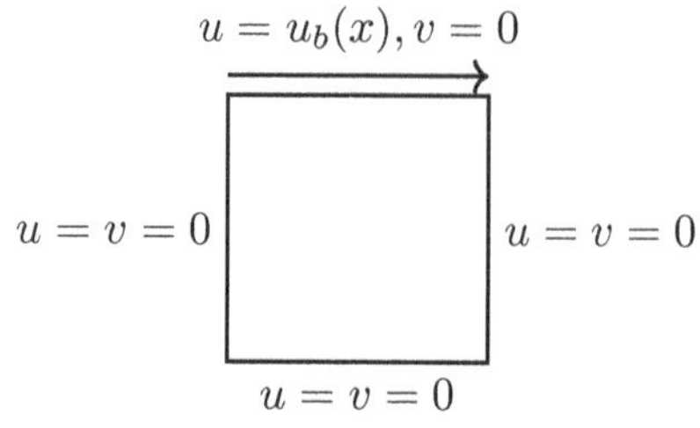


Fig. 4.2 Lid driven cavity flow. The top boundary condition is $u = u_b(x)$.

initial stream function $\psi_0(x, y) = 16(y^2 - y^3)x^2(1-x)^2$ and the initial vorticity

$$\omega_0(x, y) = \frac{\partial^2}{\partial x^2}\psi_0 + \frac{\partial^2}{\partial y^2}\psi_0.$$

We use the boundary condition shown in Figure 4.2 with the top horizontal velocity as $u = u_b(x) = 16x^2(1-x)^2$.

Consider a uniform $(N+2) \times (N+2)$ mesh for the domain $[0,1] \times [0,1]$ with $\Delta x = \Delta y = \frac{1}{N+1}$. For instance, $0 = x_0 < x_1 < \cdots < x_{N+1} = 1$ where $x_j = j\Delta x$. Use the centered difference for all spatial derivatives. You may assemble your solver for this problem following the steps below.

*Step I: Poisson solver.* Given $\omega^n$, we solve $\Delta\psi^n = \omega^n$ on the interior $N \times N$ grid points with homogeneous Dirichlet boundary conditions for $\psi^n$. The boundary values of $\psi^n$ are simply zero. Then on the interior $N \times N$ grid points, compute the velocity $u^n = -D_y\psi^n$ and $v^n = D_x\psi^n$ where $D_x$ and $D_y$ denote the central difference operator for the first order partial derivatives. By enforcing homogeneous Dirichlet boundary conditions for $\psi^n$ in the Poisson solver, we have explicitly enforced the no-penetration boundary condition, i.e., the normal velocity along the boundary is zero. The tangential velocity boundary condition is not enforced in the Poisson solver here but will be used later. The boundary values of $u^n$ and $v^n$ are simply those specified by the boundary condition as shown in Figure 4.2.

*Step II: time discretization.* Use explicit time discretization for both convection and diffusion terms. Use the classical fourth order Runge-Kutta for solving $u' = f(u,t)$ because it has larger stability region than RK3:

$$\begin{aligned}
U^{(1)} &= U^n \\
U^{(2)} &= U^n + \frac{1}{2}\Delta t f(U^{(1)}, t_n) \\
U^{(3)} &= U^n + \frac{1}{2}\Delta t f(U^{(2)}, t_{n+\frac{1}{2}}) \\
U^{(4)} &= U^n + \Delta t f(U^{(3)}, t_{n+\frac{1}{2}}) \\
U^{n+1} &= U^n + \frac{1}{6}\Delta t \left[ f(U^{(1)}, t_n) + 2f(U^{(2)}, t_{n+\frac{1}{2}}) + 2f(U^{(3)}, t_{n+\frac{1}{2}}) + f(U^{(4)}, t_{n+1}) \right].
\end{aligned}$$

Think of the forward Euler time discretization as an example of each stage in RK4:

$$\frac{\omega^{n+1} - \omega^n}{\Delta t} + u^n D_x \omega^n + v^n D_y \omega^n = \frac{1}{\text{Re}}(D_x^2 + D_y^2)\omega^n,$$

in which we still need the boundary values of $\omega^n$ to get the interior $N \times N$ values of $\omega^{n+1}$.

*Step III: numerical boundary conditions.* We will use the following *Thom's formula.* Let $x_{-1} = -x_1$, $y_{-1} = -y_1$, $x_{N+2} = 1 + \Delta x$ and $y_{N+2} = 1 + \Delta y$ be the ghost grid points. On the left boundary, the tangential velocity boundary condition $v = 0$ implies $\frac{\partial \psi}{\partial x} = v = 0$ can be approximated by $\frac{\psi_{1,j} - \psi_{-1,j}}{2\Delta x} = 0$. The normal velocity boundary condition $u = 0$ implies $\frac{\partial \psi}{\partial y} = -u = 0$ at $x = 0$, thus $\psi_{yy} = 0$ at the left boundary $x = 0$. Therefore, along the left boundary, the equation $\omega = \Delta\psi$ reduces to $\omega = \psi_{xx}$ thus we get

$$\omega_{0,j} = D_x^2 \omega_{1,j} = \frac{\psi_{-1,j} - 2\psi_{0,j} + \psi_{1,j}}{\Delta x^2} = 2\frac{\psi_{1,j}}{\Delta x^2},$$

where we have used $\psi_{0,j} \equiv 0$ because we have assumed homogeneous Dirichlet boundary for $\psi$ in Step I. Similarly, on the right and bottom boundaries, we have

$$\omega_{N+1,j} = 2\frac{\psi_{N,j}}{\Delta x^2}, \quad \omega_{i,0} = 2\frac{\psi_{i,1}}{\Delta y^2}.$$

At the top boundary, we have $\frac{\partial \psi}{\partial y} = -u = -u_b(x)$ thus $\frac{\psi_{i,N+2} - \psi_{i,N}}{2\Delta y} = -u_b(x_i)$, and $\omega = \Delta\psi$ reduces to $\omega = \psi_{yy}$, which gives

$$\omega_{i,N+1} = \frac{\psi_{i,N} - 2\psi_{i,N+1} + \psi_{i,N+2}}{\Delta y^2} = \frac{\psi_{i,N} + \psi_{i,N+2}}{\Delta y^2} = 2\frac{\psi_{i,N}}{\Delta y^2} - 2\frac{u_b(x_i)}{\Delta y}.$$

Show that the Thom's formula is only first order accurate for $\omega$. Namely, verify that the local truncation error of $\omega_{0,j} = 2\frac{\psi_{1,j}}{\Delta x^2}$ for the equation

$\omega = \Delta\psi$ is only first order on the left boundary, on which the exact solution satisfies $u = v = 0$.

*Step IV: numerical validation.* Let $\|\mathbf{u}\|_\infty = \max\{\|u^n\|_\infty, \|v^n\|_\infty\}$ denote the maximum velocity. Use the time step $\Delta t = \min\{0.3\,\mathrm{Re}\,h^2, 0.3h/\|\mathbf{u}\|_\infty\}$ $(h = \Delta x = \Delta y)$. Adjust the last time step to reach exactly any given final time.

(1) *Test accuracy of the scheme on a smooth solution.* Consider a different exact solution $\omega(x,y,t) = -2\pi^2 \exp\left(-\frac{2\pi^2 t}{Re}\right)\sin(\pi x)\sin(\pi y)$ and $\psi(x,y,t) = \exp\left(-\frac{2\pi^2 t}{Re}\right)\sin(\pi x)\sin(\pi y)$. The normal velocity and $\psi$ are all zero along the boundary but the tangential velocity along the boundary is no longer constant zero. Implement the scheme above without using the Thom's formula to check the second order accuracy of the base scheme: use the exact solution along the boundary for the boundary values of $\omega$, and everything else stays the same. Set $\mathrm{Re} = 1000$ and list an error and order table at $T = 0.2$ for $1/\Delta x = 8, 16, 32, 64, 128$.

(2) Consider the same exact solution above. Use the Thom's formula for this problem. Set $\mathrm{Re} = 1000$ and list an error and order table at $T = 0.2$ for $1/\Delta x = 8, 16, 32, 64, 128$.

(3) Use the Thom's formula for the Lid Driven Cavity flow. Set $\mathrm{Re} = 10000$ and compute the solution at a given time like $T = 2$ or generate an animation of the solution changing in time. See Figure 4.3 for a numerical solution. Show your result on a $200 \times 200$ mesh or a finer mesh by giving the contour plot of stream line function using the exponentially distributed contour lines and 90 contour lines for the vorticity:

```
a=min(psi(:));
b=max(psi(:));
level1=-exp(linspace(log(-a),log(0.00001),10));
level2=exp(linspace(log(0.00001),log(b)-log(10), 10));
level3=linspace(0.1*b, b, 10);
level=[level1 level2 level3];
figure;contour(x,y,psi,level);axis equal;colorbar;
% exponentially distributed contour lines for psi
figure;contour(x,y,omega,90);axis equal;colorbar
```

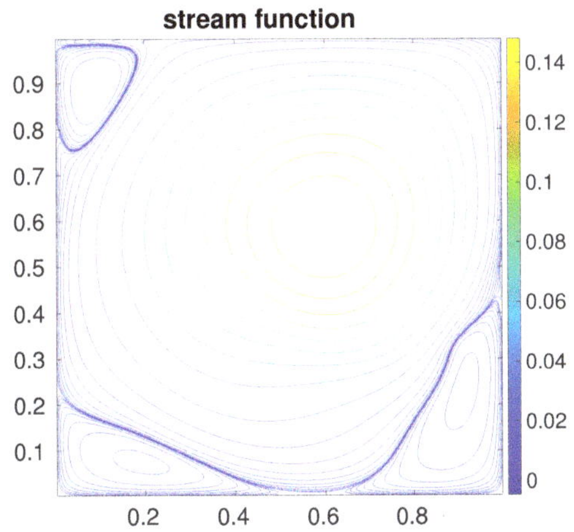


Fig. 4.3 The Lid Driven Cavity flow on a $300 \times 300$ grid for Re $= 10000$ at $T = 2$.

### 4.1.5 *Project: 3D Cahn-Hilliard in phase field models*

The Cahn-Hilliard equation is a fourth-order PDE used in the phase field modeling. A fourth order PDE is in general harder than solving a second order Poisson equation. However, when the boundary conditions are special, it can be solved by Poisson equations. For instance, the boundary value problem on the interval $[0, 1]$:

$$\frac{d^4}{dx^4}u(x) = f(x), \quad u(0) = u(1) = 0, \quad u''(0) = u''(1) = 0.$$

is equivalent to solving the Poisson equation twice:

$$\frac{d^2}{dx^2}v(x) = f(x), \quad v(0) = v(1) = 0,$$

$$\frac{d^2}{dx^2}u(x) = v(x), \quad u(0) = u(1) = 0.$$

We consider the Cahn-Hilliard equation on $\Omega = [-1, 1]^3$ with simple boundary conditions:

$$\begin{cases} \phi_t = m\Delta\left(-\epsilon\Delta\phi + \frac{1}{\epsilon}F'(\phi)\right) & \text{in } \Omega, \\ \partial_{\boldsymbol{n}}\phi = 0, \quad \partial_{\boldsymbol{n}}\Delta\phi = 0 & \text{on } \partial\Omega, \end{cases} \tag{4.7}$$

where $\phi(x, y, z, t)$ is a phase function with a thin, smooth transitional layer, whose thickness is proportional to a prescribed parameter $\epsilon > 0$, $m > 0$

is a prescribed mobility constant, and $F(\phi) = \frac{1}{4}(\phi^2 - 1)^2$ is a double-well potential function.

Due to the simplicity of the boundary conditions, we can avoid solving a fourth-order equation by reformulating (4.7) as a system of second-order equations after introducing the chemical potential $\mu := -\epsilon\Delta\phi + \frac{1}{\epsilon}F'(\phi)$, which is the variational derivative of the energy functional (see Section 5.5):

$$E(\phi) = \int_\Omega \frac{\epsilon}{2}|\nabla\phi|^2 + \frac{1}{\epsilon}F(\phi)\,\mathrm{d}\boldsymbol{x}. \tag{4.8}$$

Then the system (4.7) is equivalent to

$$\begin{cases} \phi_t - m\Delta\mu = 0 \quad \text{in } \Omega, \\ \mu = -\epsilon\Delta\phi + \frac{1}{\epsilon}F'(\phi) \quad \text{in } \Omega, \\ \partial_{\boldsymbol{n}}\phi = 0, \quad \partial_{\boldsymbol{n}}\mu = 0 \quad \text{on } \partial\Omega. \end{cases} \tag{4.9}$$

For time discretization, we use the second order backward differentiation formula (BDF-2) to the system (4.9):

$$\begin{cases} \frac{\alpha\phi_{n+1} - \hat{\phi}_n}{\tau} - m\Delta\mu_{n+1} = 0, \\ \mu_{n+1} = -\epsilon\Delta\phi_{n+1} + \frac{1}{\epsilon}F'(\bar{\phi}_n), \end{cases} \tag{4.10}$$

where $\tau$ is the time step size, $\alpha = \frac{3}{2}$, $\hat{\phi}_n = 2\phi_n - \frac{1}{2}\phi_{n-1}$, and $\bar{\phi}_n = 2\phi_n - \phi_{n-1}$. To solve this linear system, we can write it as

$$\begin{pmatrix} \alpha I & -m\tau\Delta \\ \epsilon\Delta & I \end{pmatrix} \begin{pmatrix} \phi_{n+1} \\ \mu_{n+1} \end{pmatrix} = \begin{pmatrix} \hat{\phi}_n \\ \frac{1}{\epsilon}F'(\bar{\phi}_n) \end{pmatrix},$$

and its solution is given by

$$\begin{pmatrix} \phi_{n+1} \\ \mu_{n+1} \end{pmatrix} = \begin{pmatrix} \mathcal{D}I & \mathcal{D}(m\tau\Delta) \\ \mathcal{D}(-\epsilon\Delta) & \alpha\mathcal{D}I \end{pmatrix} \begin{pmatrix} \hat{\phi}_n \\ \frac{1}{\epsilon}F'(\bar{\phi}_n) \end{pmatrix} = \begin{pmatrix} \mathcal{D}(\hat{\phi}_n + \frac{m\tau}{\epsilon}\Delta F'(\bar{\phi}_n)) \\ \mathcal{D}(-\epsilon\Delta\hat{\phi}_n + \frac{\alpha}{\epsilon}F'(\bar{\phi}_n)) \end{pmatrix}, \tag{4.11}$$

where $\mathcal{D} = (\alpha I + m\tau\epsilon\Delta^2)^{-1}$.

*Step I: efficient implementation with discrete Laplacian.* Let $\Delta_h$ be the discrete Laplacian from second order finite difference in Chapter 2. Then the second order finite difference with BDF2 for solving (4.9) can be implemented as

$$\begin{pmatrix} \phi_{n+1} \\ \mu_{n+1} \end{pmatrix} = \begin{pmatrix} \mathcal{D}_h(\hat{\phi}_n + \frac{m\tau}{\epsilon}\Delta_h F'(\bar{\phi}_n)) \\ \mathcal{D}_h(-\epsilon\Delta_h\hat{\phi}_n + \frac{\alpha}{\epsilon}F'(\bar{\phi}_n)) \end{pmatrix}, \tag{4.12}$$

where $\mathcal{D}_h = (\alpha I + m\tau\epsilon\Delta_h^2)^{-1}$. The eigenvalue decomposition $-\Delta_h = S\Lambda S^{-1}$ gives $\mathcal{D}_h = S(\alpha I + m\tau\epsilon\Lambda^2)^{-1}S^{-1}$. Thus (4.12) can be implemented in the same way as described in Section 2.9.

(1) Plug in the eigenvalue decomposition of both $\Delta_h$ and $\mathcal{D}_h$ to further simplify (4.12).
(2) The time step can be taken as $\Delta t = \mathcal{O}(\Delta x)$. Validate the second order accuracy of this scheme on a manufactured solution

$$\phi^* = \cos(\pi x)\cos(\pi y)\cos(\pi z)\exp(t).$$

This smooth function $\phi^*$ no longer satisfies the original equation but a modified equation with an extra source function $g(x, y, z, t)$:

$$\phi_t = m\Delta\left(-\epsilon\Delta\phi + \frac{1}{\epsilon}F'(\phi)\right) + g,$$

where $g := \phi_t^* - m\Delta\left(-\epsilon\Delta\phi^* + \frac{1}{\epsilon}F'(\phi^*)\right)$. In other words, implement second order finite difference with BDF2 for this modified equation to validate your code by checking second order accuracy.

*Step II: coalescence of two drops.* Once the code is validated for its accuracy, we can use it to study the coalescence of two droplets, described by the Cahn-Hilliard equation, within the same computational domain $\Omega = [-1, 1]^3$. We select the parameter settings as $\epsilon = 0.1$, the mobility constant $m = 0.02$, and the time step size $\tau = 0.002$ with a terminal time $T = 0.2$. At time $t = 0$, the domain is occupied by two neighboring spherical regions of the first material, while the second material fills the remaining space, as shown in Figure 4.4. Such an initial condition is given by

$$\phi_0(\boldsymbol{x}) = 1 - \tanh\frac{|\boldsymbol{x} - \boldsymbol{x}_1| - R}{\sqrt{2}\epsilon} - \tanh\frac{|\boldsymbol{x} - \boldsymbol{x}_2| - R}{\sqrt{2}\epsilon},$$

where $\boldsymbol{x}_1 = (x_1, y_1, z_1) = (0, 0, 0.37)$ and $\boldsymbol{x}_2 = (x_2, y_2, z_2) = (0, 0, -0.37)$ are the centers of the initial spherical regions of the first material, and $R = 0.35$ is the radius of these spheres. As time progresses under the

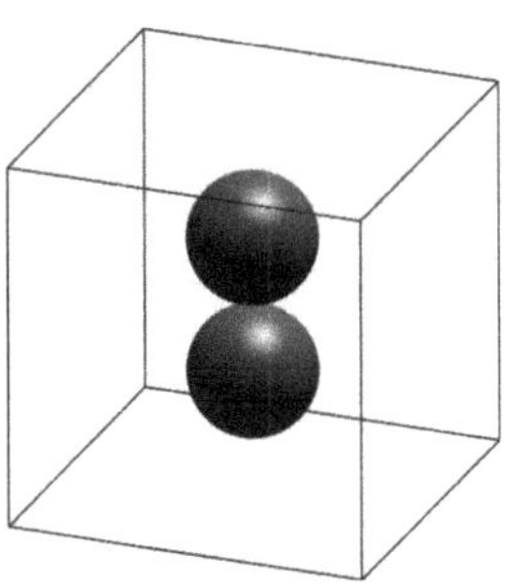

Fig. 4.4 The zero-isocontour of the phase function $\phi$ at time $t = 0$.

Cahn-Hilliard dynamics, these two spherical regions coalesce to form a single droplet. You can also replace the Neumann boundary conditions by periodic boundary conditions to use FFT in the eigenvector method. Compare your results to the ones in [Liu *et al.* (2024b)] and references therein.

## Matlab code for phase function initialization

```
% Plot the initial condition for two drops
eps = 2e-2; Lx = 1;  % the domain is [-L, L]^3
R1 = 0.35*Lx; R2 = 0.35*Lx;
z1 = -0.37*Lx; z2 = 0.37*Lx;
x1 = 0; x2 = 0; y1 = 0; y2 = 0;
x = linspace(-Lx,Lx,150)'; ex = ones(size(x));
phix1 = (x - x1).^2; phiy1 = (x - y1).^2; phiz1 = (x - z1).^2;
phix2 = (x - x2).^2; phiy2 = (x - y2).^2; phiz2 = (x - z2).^2;
phi0 = 1 - tanh((sqrt(squeeze(tensorprod(phix1, ex*ex'))...
       +squeeze(tensorprod(ex,phiy1*ex'))...
       +squeeze(tensorprod(ex,ex*phiz1')))-R1)/(sqrt(2)*eps))...
       - tanh((sqrt(squeeze(tensorprod(phix2, ex*ex'))...
       +squeeze(tensorprod(ex,phiy2*ex'))...
       +squeeze(tensorprod(ex,ex*phiz2')))-R2)/(sqrt(2)*eps));
[X, Y, Z] = meshgrid(x,x,x);
fig = figure;
isosurface(X,Y,Z,phi0,0),view(-68.718,22.6096); hold on;
plotEdges(Lx,2*Lx); hold off;
set(gca,'xtick',[],'ytick',[],'ztick',[]); axis off;
```

```
function plotEdges(L,HH)
A=[ L, HH-0.5*HH, L]; B=[-L, HH-0.5*HH, L];
C=[-L,0-0.5*HH, L]; D=[ L,0-0.5*HH, L];
E=[ L, HH-0.5*HH,-L]; F=[-L, HH-0.5*HH,-L];
G=[-L,0-0.5*HH,-L]; H=[ L,0-0.5*HH,-L];
color=[0 0 0];
ptx=[A(1),B(1),C(1),D(1),A(1)];
pty=[A(2),B(2),C(2),D(2),A(2)];
ptz=[A(3),B(3),C(3),D(3),A(3)];
plot3(ptx,pty,ptz,LineWidth=0.5,Color=color); hold on
ptx=[E(1),F(1),G(1),H(1),E(1)];
pty=[E(2),F(2),G(2),H(2),E(2)];
ptz=[E(3),F(3),G(3),H(3),E(3)];
plot3(ptx,pty,ptz,LineWidth=0.5,Color=color);
plot3([A(1) E(1)],[A(2) E(2)],[A(3) ...
    E(3)],LineWidth=0.5,Color=color)
plot3([B(1) F(1)],[B(2) F(2)],[B(3) ...
    F(3)],LineWidth=0.5,Color=color)
plot3([C(1) G(1)],[C(2) G(2)],[C(3) ...
    G(3)],LineWidth=0.5,Color=color)
```

```
plot3([D(1) H(1)],[D(2) H(2)],[D(3) ...
    H(3)],LineWidth=0.5,Color=color)
axis([-L*1.2, L*1.2, -HH*0.6, HH*0.6,-L*1.2, L*1.2]); axis on;
set(gca,'FontSize',12); pbaspect([1 1 1]); hold off;
end
```

## 4.2 Total Variation norm for image processing

The Total Variation (TV) norm minimization for image denoising has been proposed since early 1990s [Rudin *et al.* (1992); Sauer and Bouman (1992)]. We will first see how the model is defined, then discuss why Poisson equation is involved in popular optimization algorithms like ADMM (alternating direction method of multipliers) and Douglas-Rachford splitting for such a model. For solving TV norm minimization, another popular method is called Split Bregman [Goldstein and Osher (2009)], and its special version coincides with ADMM. Finally, an explicit formula will be given so that one can implement ADMM for TV norm minimization. See Appendix C for the definition of these splitting methods. Notice that it is not necessary to fully understand all concepts or algorithms in Appendix C for readers to be able to do the coding project in the section, which only requires the knowledge from Chapter 2.

### 4.2.1 *Continuum ROF image denoising model*

Consider a rectangular domain $\Omega = [0,1] \times [0,1]$, and a function $u(x,y) \in H^1(\Omega)$, which represents an image with infinite resolution. The discussion of a continuum setup here only serves as an intuitional guide for us to derive the discrete analogue later, because functions in $H^1(\Omega)$ represent a very small class of images. Then its total variation is defined as

$$\|u\|_{TV} = \iint_\Omega |\nabla u| \mathrm{d}x\mathrm{d}y,$$

where $\nabla u = (u_x, u_y)$ and $|\nabla u| = \sqrt{|u_x|^2 + |u_y|^2}$. With $L^2$-norm as $\|u\|_{L^2} = \sqrt{\iint_\Omega |u|^2 \mathrm{d}x\mathrm{d}y}$, for a given $a(x,y)$, the ROF (Rudin, Osher, and Fatemi, 1992) model [Rudin *et al.* (1992)] is to minimize

$$\|u\|_{TV} + \frac{1}{2}\lambda\|u-a\|^2_{L^2},$$

where $\lambda$ is a fixed parameter, and $u$ is in a proper function space. The function space that $u$ should belong to, is a subspace of $H^1(\Omega)$ with suitable boundary conditions. For instance, periodic or homogeneous Dirichlet boundary conditions make sense for MRI images, but not for a generic image. See Figure 4.5. For convenience, for a generic image, we just consider homogeneous Neumann boundary conditions, which will naturally emerge in the discrete setup as will be seen in the following subsections.

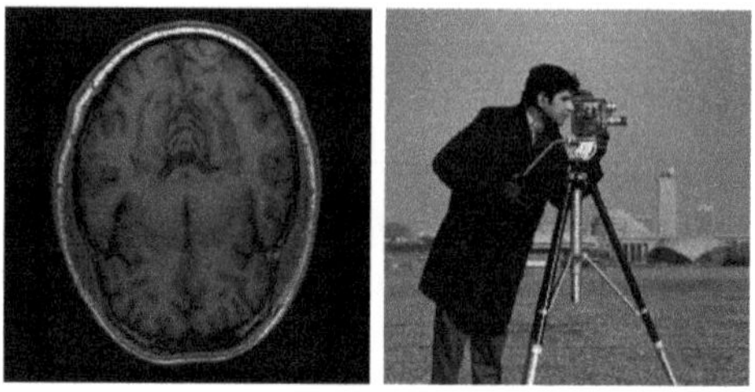

Fig. 4.5 Periodic or zero boundary conditions are suitable for MRI images, but not for a generic image.

To this end, we define a function space

$$\mathcal{H} = \{u \in H^1(\Omega) : \nabla u \cdot \mathbf{n}|_{\partial\Omega} = 0\},$$

where $\mathbf{n}$ is the unit normal vector of the boundary $\partial\Omega$. The gradient operator $\nabla$ is a linear mapping, and we use an abstract name for it $\mathcal{K} = \nabla$:

$$\begin{aligned} \mathcal{K} = \nabla : \mathcal{H} &\longrightarrow \mathcal{V} = (L^2(\Omega), L^2(\Omega)) \\ u &\longmapsto \nabla u = (u_x, u_y). \end{aligned}$$

To understand the adjoint operator of $\mathcal{K} = \nabla$, we need the $H(div)$-space:

$$H(div) = \{\mathbf{q} = (q^1, q^2) \in (L^2(\Omega), L^2(\Omega)) : \nabla \cdot (q^1, q^2) \in L^2(\Omega)\} \subset \mathcal{V}.$$

**Remark 4.1.** *Elements in $H(div)$ are not necessarily in $H^1(\Omega)$. For instance, consider the step function* $f(x) = \begin{cases} 1, & x \geq 0 \\ 0, & x < 0 \end{cases}$, *then*

$$\mathbf{q}(x, y) := (0, f(x)) \in H(div),$$

*but $f(x) \notin H^1(\Omega)$, because the weak derivative of $f(x)$ is $\delta(x)$, which is not a function but a functional, see Section 3.14.1.*

The divergence operator $\nabla\cdot$ is a linear mapping from $H(div)$ to $L^2(\Omega)$. If we assume suitable boundary conditions for smooth $\mathbf{q}$ so that boundary terms in the integration-by-parts vanish, then $\mathcal{K}^* = -\nabla\cdot$ is the adjoint operator of $\mathcal{K} = \nabla$ in the following sense: for the inner product in $L^2(\Omega)$ space

$$\langle u, v\rangle = \iint_\Omega uv \mathrm{d}x\mathrm{d}y,$$

the inner product in $\mathcal{V} = (L^2(\Omega), L^2(\Omega))$ satisfies

$$\forall \mathbf{q} \in (C_0^1(\Omega), C_0^1(\Omega)), \quad \langle \mathcal{K}u, \mathbf{q}\rangle := \iint_\Omega \nabla u \cdot \mathbf{q} \mathrm{d}x\mathrm{d}y = -\iint_\Omega u\nabla \cdot \mathbf{q}\mathrm{d}x\mathrm{d}y$$
$$= \langle u, -\nabla \cdot \mathbf{q}\rangle.$$

### 4.2.2 *Discrete ROF model*

Consider an image of size $n \times n$, corresponding to domain $[0,1] \times [0,1]$ and a uniform grid $x_i = (i-1)h, y_j = (j-1)h$, $j = 1, \cdots, n$ with $h = \frac{1}{n-1}$. In the literature, the grid spacing is usually taken as $h = 1$ for images, but here we assume the domain is $[0,1] \times [0,1]$ with the grid spacing $h = \frac{1}{n-1}$ so that readers can directly apply knowledge from Chapter 2. Notice that an image does not have any necessary association of a domain of size $[0,1] \times [0,1]$. Nonetheless, the assumption of $h = \frac{1}{n-1}$ does not affect the final implementation at all, if certain parameters are properly rescaled accordingly.

The following matrix, in which we add a zero row for convenience, can represent the forward difference approximation to the first order derivative:

$$D = \frac{1}{h}\bar{D} = \frac{1}{h}\begin{pmatrix} -1 & 1 & & & \\ & -1 & 1 & & \\ & & \ddots & \ddots & \\ & & & -1 & 1 \\ 0 & 0 & \cdots & 0 & 0 \end{pmatrix}_{n\times n}, \quad \bar{D} := \begin{pmatrix} -1 & 1 & & & \\ & -1 & 1 & & \\ & & \ddots & \ddots & \\ & & & -1 & 1 \\ 0 & 0 & \cdots & 0 & 0 \end{pmatrix}_{n\times n}.$$

Recall that $D$ is the finite difference matrix approximating first order derivative, then we have

$$D^\top = \frac{1}{h}\begin{pmatrix} -1 & & & & \\ 1 & -1 & & & \\ & \ddots & \ddots & & \\ & & 1 & -1 & 0 \\ & & & 1 & 0 \end{pmatrix}_{n\times n}, D^\top D = \frac{1}{h^2}\begin{pmatrix} 1 & -1 & & & \\ -1 & 2 & -1 & & \\ & \ddots & \ddots & \ddots & \\ & & -1 & 2 & -1 \\ & & & -1 & 1 \end{pmatrix}.$$

Use the same notation as in Chapter 2, for representing point values of a function $u(x, y)$, let $U$ be a 2D array with $U(j, i) = u(x_i, y_j)$, then

$$U \approx u \Rightarrow UD^\top \approx u_x, \quad DU \approx u_y.$$

For approximations of the norms or the integrals, we have

$$||u||_{TV} \approx \sum_i \sum_j h^2 \sqrt{|u_x(x_i, y_j)|^2 + |u_y(x_i, y_j)|^2},$$

$$||u - a||_{L^2}^2 \approx \sum_i \sum_j h^2 |u(x_i, y_j) - a(x_i, y_j)|^2.$$

Introduce $U_x := UD^\top$ and $U_y := DU$. Let $A$ be a 2D array with $A(j,i) = a(x_i, y_j)$, then the discrete ROF model is

$$\min_{U \in \mathbb{R}^{n\times n}} \sum_i \sum_j \left( h^2 \sqrt{U_x(j,i)^2 + U_y(j,i)^2} + h^2 \frac{1}{2} \lambda |U(j,i) - A(j,i)|^2 \right)$$

or equivalently

$$\min_{U \in \mathbb{R}^{n\times n}} \sum_i \sum_j \left( h \sqrt{[U\bar{D}^\top](j,i)^2 + [\bar{D}U](j,i)^2} + h^2 \frac{1}{2} \lambda |U(j,i) - A(j,i)|^2 \right), \tag{4.13}$$

where $D = \frac{1}{h}\bar{D}$ is the forward difference matrix. For a given image with fixed size $n \times n$, its minimizer does not depend on the choice of $h$ if $\lambda = \frac{C}{h}$ for some constant $C$. An image has no physical grid spacing anyway, so $h$ should be arbitrary, for which we should take $\lambda = \frac{C}{h}$. In practice, $C = 10$ usually produces a reasonable result. See Figure 4.6 and Figure 4.7.

There are some intuitional explanations why the minimizer to the discrete ROF model (4.13) could achieve certain denoising effects, e.g., minimizing TV norm favors a piecewise constant image, which can blur the noise out. On the other hand, if the noisy image is generated by adding random noise to a good image, then the recovery of the original image based only on the noisy image is a mathematically ill posed problem, e.g., we cannot expect the solution to be unique.

In Figure 4.7, we can see that smaller $\lambda$ gives blurry images with less noise and larger $\lambda$ gives better images but with more noise, and neither one is mathematically close to the original. Nonetheless, the minimizer of ROF model is useful to some extent, e.g., one might say that Figure 4.7 (d) is an improved image compared to the noisy image. The denoising problem is a good example of the famous quote *all models are wrong, but some are useful.*

### 4.2.3 *Primal, dual and primal-dual forms*

Using all notation above, the discrete ROF model can be written as a convex minimization problem

$$\min_{U \in \mathbb{R}^{n\times n}} f(\mathcal{K}U) + g(U), \tag{4.14a}$$

(a) Noisy Image.

(b) $C = 4$.

(c) $C = 8$.

(d) $C = 12$.

Fig. 4.6 The minimizer of ROF model (4.13) with different $\lambda = \frac{C}{h}$.

where $\mathcal{K} = \nabla_h : \mathbb{R}^{n\times n} \longrightarrow \mathbb{R}^{2(n\times n)}$ is a linear mapping

$$\mathcal{K}(U) = \nabla_h U = (UD^\top, DU), \tag{4.14b}$$

and

$$f(P,Q) = \sum_{i,j} h^2 \sqrt{P^2(i,j) + Q^2(i,j)}, \quad g(U) = \lambda \sum_{i,j} h^2 |U(i,j) - a(i,j)|^2. \tag{4.14c}$$

It is straightforward to verify that the adjoint operator of $\mathcal{K}$ is given by

$$\mathcal{K}^* = -\nabla_h \cdot : \mathbb{R}^{2(n\times n)} \longrightarrow \mathbb{R}^{n\times n}$$
$$(P,Q) \longmapsto (PD + D^\top Q).$$

The convex minimization (4.14a) is in *primal* form, and it has an equivalent *dual* form:

$$-\min_{\mathbf{P}\in\mathbb{R}^{2(n\times n)}} f^*(\mathbf{P}) + g^*(-\mathcal{K}^*\mathbf{P}), \tag{4.15}$$

where $f^*$ and $g^*$ are *convex conjugates* (a.k.a., Legendre transform) of the convex function $f$ and $g$ respectively. See Appendix C.3 for the definition

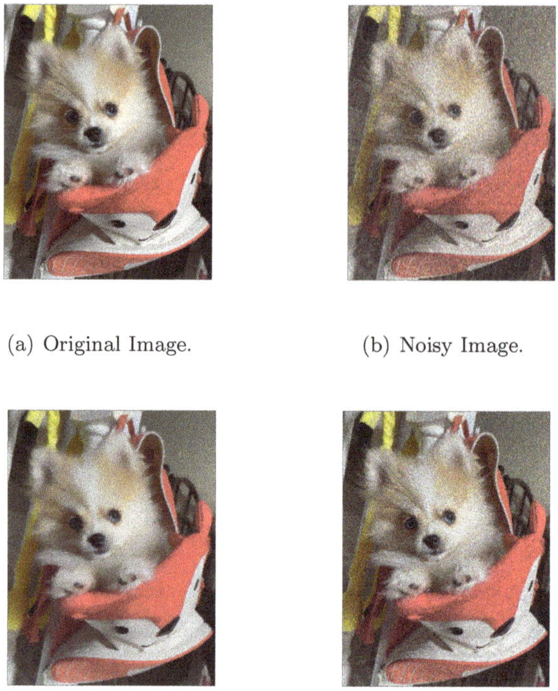

(a) Original Image. (b) Noisy Image.

(c) $C = 4$. (d) $C = 8$.

Fig. 4.7 The minimizer of ROF model (4.13) with different $\lambda = \frac{C}{h}$.

and properties of the convex conjugate functions. Both (4.14a) and (4.15) also have an equivalent *primal-dual* form:

$$\min_{U \in \mathbb{R}^{n\times n}} \max_{\mathbf{P} \in \mathbb{R}^{2(n\times n)}} \langle \mathcal{K}U, \mathbf{P}\rangle - f^*(\mathbf{P}) + g(U), \tag{4.16}$$

**Remark 4.2.** *The minimizer $U^*$ to* (4.14a) *and the minimizer $\mathbf{P}^*$ to* (4.15) *are related via the first order optimality condition:*

$$0 = \mathcal{K}^*\mathbf{P} + \nabla g(U),$$

*which gives* $0 = \mathcal{K}^*\mathbf{P} + \lambda(U - A) \Leftrightarrow U = A - \frac{1}{\lambda}\mathcal{K}^*\mathbf{P}$.

To be more explicit, for the TV-norm denoising problem of a 2D image $A \in \mathbb{R}^{n\times n}$, the primal problem can be written as

$$\min_{U \in \mathbb{R}^{n\times n}} \|\mathcal{K}U\|_1 + \frac{\lambda}{h}\|U - A\|_F^2,$$

where $\|\cdot\|_F$ is the Frobenius norm for a matrix $\|U - A\|_F = \sqrt{\sum_{i,j} |U(i,j) - a(i,j)|^2}$ and the 1-norm for a pair of matrices $\mathbf{V} = (P, Q)$ is

$$F(\mathbf{V}) = \|(P,Q)\|_1 = \sum_{i,j} \sqrt{P(i,j)^2 + Q(i,j)^2}.$$

Up to a constant shift, the dual problem can be written as

$$-\min F^*(\mathbf{V}) + \frac{h}{2\lambda}\|\mathcal{K}^*\mathbf{V} - \frac{\lambda}{h}A\|_F^2,$$

where the convex conjugate or Legendre transform of $F(\mathbf{V})$ is

$$F^*(\mathbf{V}) = F^*(P,Q) = \sum_{i,j} \iota_{\{x^2+y^2\leq 1\}}(P(i,j), Q(i,j)),$$

with a convex closed proper (extended) function $\iota_{\{x^2+y^2\leq 1\}}$ called *indicator function* (see Appendix C.1) of the unit ball $\{(x,y) : x^2 + y^2 \leq 1\}$:

$$\iota_{\{x^2+y^2\leq 1\}}(x,y) = \begin{cases} 0, & \text{if} \quad x^2 + y^2 \leq 1 \\ +\infty, & \text{if} \quad x^2 + y^2 > 1 \end{cases}.$$

### 4.2.4 *Poisson solver in ADMM for the primal problem*

The alternating direction method of multipliers (ADMM) introduced in [Glowinski and Marroco (1975); Gabay and Mercier (1976)] is a widely used popular algorithm for solving convex minimization problems like (4.14a). See Appendix C for some background knowledge about ADMM. For solving (4.14a), one particular version of ADMM with only one algorithm parameter $\sigma > 0$ is written as

$$\begin{aligned} &\mathbf{P}_{k+1} = \operatorname{argmin}_{\mathbf{P}} f(\mathbf{P}) + \langle \mathbf{Q}_k, \mathbf{P} - \mathcal{K}U_k \rangle + \frac{\sigma}{2}\|\mathbf{P} - \mathcal{K}U_k\|^2 \\ \text{(ADMM)}: \quad &U_{k+1} = \operatorname{argmin}_U g(U) + \langle \mathbf{Q}_k, \mathbf{P}_{k+1} - \mathcal{K}U \rangle + \frac{\sigma}{2}\|\mathbf{P}_{k+1} - \mathcal{K}U\|^2 \\ &\mathbf{Q}_{k+1} = \mathbf{Q}_k + \sigma(\mathbf{P}_{k+1} - \mathcal{K}U_{k+1}). \end{aligned}$$

This particular version of ADMM is equivalent to another popular and powerful algorithm, called *Douglas-Rachford splitting* (DRS) introduced in [Lions and Mercier (1979)]. ADMM and DRS are equivalent in the following sense (see Appendix C):

ADMM on primal $\Leftrightarrow$ Douglas-Rachford on dual,

ADMM on dual $\Leftrightarrow$ Douglas-Rachford on primal.

In particular, it was proven in [Lions and Mercier (1979)] that DRS with any positive step size converges to a minimizer for a convex problem, which implies the particular version of ADMM in this section converges with any positive parameter $\sigma > 0$.

We will see how to implement ADMM later. For now, we focus on why we need a Poisson solver, which is related to the second line in ADMM. By ignoring terms that are not dependent of $U$, the second line is written as

$$U_{k+1} = \text{argmin}_U g(U) - \langle \mathbf{Q}_k, \mathcal{K}U \rangle + \frac{\sigma}{2}\|\mathcal{K}U - \mathbf{P}_{k+1}\|^2.$$

Notice that $g(U)$ is a simple quadratic function, thus the minimizer is obtained by finding a critical point, for which we need to take derivative of $\|\mathcal{K}U - \mathbf{P}_{k+1}\|^2$ w.r.t. $U$:

$$\frac{\partial}{\partial U}\langle \mathcal{K}U - \mathbf{P}_{k+1}, \mathcal{K}U - \mathbf{P}_{k+1}\rangle = h^2(2\mathcal{K}^*\mathcal{K}U - 2\mathcal{K}^*\mathbf{P}_{k+1}).$$

So the second line can be equivalently written as

$$\lambda(U_{k+1} - A) - \mathcal{K}^*\mathbf{Q}_k + \sigma\mathcal{K}^*\mathcal{K}U_{k+1} - \sigma\mathcal{K}^*\mathbf{P}_{k+1} = 0$$

which is

$$(\lambda I + \sigma\mathcal{K}^*\mathcal{K})U_{k+1} = \lambda A + \mathcal{K}^*\mathbf{Q}_k + \sigma\mathcal{K}^*\mathbf{P}_{k+1}.$$

Notice that $\mathcal{K}^*\mathcal{K} = -\Delta_h$ is precisely the discrete Laplacian with purely Neumann boundary conditions, and $\lambda I - \sigma\Delta_h$ can be inverted efficiently in a simple way via the eigenvalue decomposition of the $K$ matrix as we have learned in Chapter 2. The multiplication of eigenvector matrix can be accelerated by using fast discrete sine, cosine or Fourier transforms (depending on the boundary condition of the discrete Laplacian operator) for large problems.

### 4.2.5 *Poisson solver in DRS on the dual problem*

We do not drive the explicit Douglas-Rachford splitting (DRS) on the dual problem here, since it can be transformed into exactly the ADMM for the primal problem, see Appendix C.4. Instead, we discuss how Poisson solver emerges.

DRS on the dual problem consists of the *proximal* operators of the functions $F^*$ and $G^*$. The proximal operator of a convex function $f(x)$ with a parameter $\eta > 0$ is defined as the following operator (see Appendix C):

$$\text{Prox}_f^\eta(x) = \text{argmin}_u f(u) + \frac{1}{2\eta}\|u - x\|^2.$$

We will see that the Poisson equation emerges in the *proximal* operator of the function $G^*(\mathbf{V}) = \frac{h}{2\lambda}\|\mathcal{K}^*\mathbf{V} - \frac{\lambda}{h}A\|_F^2$:

$$\mathrm{Prox}_{G^*}^{\eta}(\mathbf{W}) = \mathrm{argmin}_{\mathbf{V}} \frac{h}{2\lambda}\|\mathcal{K}^*\mathbf{V} - \frac{\lambda}{h}A\|_F^2 + \frac{1}{2\eta}\|\mathbf{V} - \mathbf{W}\|_F^2.$$

Let $\mathbf{V} = \mathrm{Prox}_{G^*}^{\eta}(\mathbf{W})$ and $\mathbb{I}$ be identity matrix or identity operator, then $\mathbf{V}$ is the minimizer and the critical point equation gives

$$\frac{h}{\lambda}\mathcal{K}(\mathcal{K}^*\mathbf{V} - \frac{\lambda}{h}A) + \frac{1}{\eta}(\mathbf{V} - \mathbf{W}) = 0$$

$$\Rightarrow (\frac{1}{\eta}\mathbb{I} + \frac{h}{\lambda}\mathcal{K}\mathcal{K}^*)\mathbf{V} = \mathcal{K}A + \frac{1}{\eta}\mathbf{W}.$$

We need to solve $\mathbf{V}$ in an equation in the form

$$\mathcal{K}\mathcal{K}^*\mathbf{V} + \beta\mathbf{V} = \mathbf{F}$$

where $\beta = \frac{\lambda}{\eta h}$ and $\mathbf{F} = \frac{\lambda}{h}\mathcal{K}A + \beta\mathbf{W}$ is some given vector field. At first glance, this corresponds to an equation

$$\nabla(-\nabla \cdot \vec{p}) + \beta\vec{p} = \vec{f},$$

which is a harder equation to solve due to the mixed second order derivatives, compared to $\mathcal{K}^*\mathcal{K} \approx -\Delta$ for the Poisson equation $-\Delta u = f$.

However, to solve this seemingly difficult equation, it turns out that we only need to compute $(-\Delta_h + \beta\mathbb{I})^{-1}$, which can be computed similarly for computing $(-\Delta_h)^{-1}$ as in Chapter 2, mainly due to a simple linear algebra fact:

**Lemma 4.1.** *For a linear operator $\mathcal{K} : \mathbb{R}^{n\times n} \longrightarrow \mathbb{R}^{2(n\times n)}$, assume $\mathcal{K}^*\mathcal{K}$ has an inverse or a right pseudo-inverse $(\mathcal{K}^*\mathcal{K})^{-1}$, then the solution to the equation $\mathcal{K}\mathcal{K}^*\mathbf{V} + \beta\mathbf{V} = \mathbf{F}$ can be written as*

$$\mathbf{V} = \frac{1}{\beta}[\mathbf{F} - \mathcal{K}(\beta\mathbb{I} + \mathcal{K}^*\mathcal{K})^{-1}\mathcal{K}^*\mathbf{F}].$$

*Proof.* The kernel of $\mathcal{K}^*$ is orthogonal to the range of $\mathcal{K}$ (column space of a matrix $K$ is orthogonal to the left null space of $K$), thus

$$\mathbb{R}^{2(n\times n)} = \mathrm{Kernel}(\mathcal{K}^*) \oplus \mathrm{Range}(\mathcal{K}),$$

which implies a very useful fact (corresponding to Helmholtz decomposition for suitable vector fields):

$$\mathbf{V} = \mathcal{K}W + \mathbf{G} = \nabla_h W + \mathbf{G}, \quad \text{where } \mathbf{G} \in \mathrm{Kernel}(\mathcal{K}^*), \text{ i.e., } \nabla_h \cdot \mathbf{G} = 0.$$

Apply $\mathcal{K}^*$ to both sides of the equation, we can first solve for $W$ as follows

$$\mathcal{K}^*\mathcal{K}\mathcal{K}^*\mathbf{V} + \beta\mathcal{K}^\top\mathbf{V} = K^*\mathbf{F} \Rightarrow \mathcal{K}^*\mathbf{V} = (\beta\mathbb{I} + \mathcal{K}^*\mathcal{K})^{-1}\mathcal{K}^*\mathbf{F},$$

$$\mathbf{V} = \mathcal{K}W + \mathbf{G} \Rightarrow \mathcal{K}^*\mathbf{V} = \mathcal{K}^*(\mathcal{K}W + \mathbf{G})$$

$$\Rightarrow W = (\mathcal{K}^*\mathcal{K})^{-1}\mathcal{K}^*\mathbf{V} = (\mathcal{K}^*\mathcal{K})^{-1}(\beta\mathbb{I} + \mathcal{K}^*\mathcal{K})^{-1}\mathcal{K}^*\mathbf{F}.$$

Then we can solve $\mathbf{G}$ by

$$\mathcal{K}\mathcal{K}^*(\mathcal{K}W + \mathbf{G}) + \beta(\mathcal{K}W + \mathbf{G}) = \mathbf{F} \Rightarrow \mathbf{G} = \frac{1}{\beta}[\mathbf{F} - \mathcal{K}\mathcal{K}^*\mathcal{K}W] - \mathcal{K}W.$$

Finally we get

$$\mathbf{V} = \mathcal{K}W + \mathbf{G} = \frac{1}{\beta}[\mathbf{F} - \mathcal{K}\mathcal{K}^*\mathcal{K}W] = \frac{1}{\beta}[\mathbf{F} - \mathcal{K}(\beta\mathbb{I} + \mathcal{K}^*\mathcal{K})^{-1}\mathcal{K}^*\mathbf{F}].$$

□

**Remark 4.3.** *It is not a surprise that the seemingly more difficult equation (with very special boundary conditions)* $\nabla(-\nabla\cdot\vec{p}) + \beta\vec{p} = \vec{f}$ *can actually be solved by computing* $(-\Delta_h + \beta\mathbb{I})^{-1}$*, since the DRS on the dual problem is equivalent to ADMM on the primal problem, which involves solving* $(-\Delta_h + \beta\mathbb{I})^{-1}$*, see [Gil Torres* et al. *(2025)] and references therein.*

### 4.2.6 *Project: ADMM for TV norm minimization*

To implement ADMM in Section 4.2.4 for ROF model (4.13), follow the steps below.

*Step I: load the Shepp Logan Phantom imagc in MATLAB and generate a noisy image to process.*

```
A = phantom('Modified Shepp—Logan',500); % image of size 500^2
variance = 0.03; B = imnoise(A,'Gaussian',0,variance);
imshow(B); axis image; % show noisy image
```

*Step II: derive the explicit formula of the first line in ADMM.* By adding and subtracting terms that are independent of $\mathbf{P}$, the first line in ADMM can be rewritten as

$$\begin{aligned}\mathbf{P}_{k+1} &= \text{argmin}_{\mathbf{P}} f(\mathbf{P}) + \langle \mathbf{Q}_k, \mathbf{P} - \mathcal{K}U_k\rangle + \frac{\sigma}{2}\|\mathbf{P} - \mathcal{K}U_k\|^2 \\ &= \text{argmin}_{\mathbf{P}} f(\mathbf{P}) + \frac{\sigma}{2}\|\mathbf{P} - \mathcal{K}U_k + \frac{1}{\sigma}\mathbf{Q}_k\|^2 \\ &= \text{argmin}_{\mathbf{P}} f(\mathbf{P}) + \frac{1}{2\eta}\|\mathbf{P} - \mathbf{V}\|^2,\end{aligned}$$

where $\mathbf{V} := \mathcal{K}U_k - \frac{1}{\sigma}\mathbf{Q}_k$ and $\eta = \frac{1}{\sigma}$. Let $\mathbf{V} = (V, W)$, then it can be further explicitly written as

$$\text{argmin}_{P,Q} \sum_{i,j} h^2 \sqrt{P(i,j)^2 + Q^2(i,j)} + \frac{1}{2\eta} h^2 (|P(i,j) - V(i,j)|^2 + |Q(i,j) - W(i,j)|^2)$$

$$= \text{argmin}_{P,Q} \sum_{i,j} h^2 [\sqrt{P(i,j)^2 + Q^2(i,j)} + \frac{1}{2\eta} (|P(i,j) - V(i,j)|^2 + |Q(i,j) - W(i,j)|^2)]$$

$$\Rightarrow \mathbf{P}_{k+1}(i,j) = \text{argmin}_{(x,y)} [\sqrt{x^2 + y^2} + \frac{1}{2\eta} (|x - V(i,j)|^2 + |y - W(i,j)|^2)],$$

which is precisely the proximal operator of the bivariate function

$$F(x,y) = \sqrt{x^2 + y^2}.$$

Its convex conjugate function $F^*$ is the indicator function of the unit ball

$$F^*(x,y) = \iota_{\{x^2+y^2 \leq 1\}}(x,y).$$

By the definition of the proximal operator, $\text{Prox}_{F^*}^{\eta}(u,v)$ is the Euclidean projection of $(u,v)$ onto the unit ball $u^2 + v^2 \leq 1$ in $\mathbb{R}^2$. Drive the formula for $\text{Prox}_F^{\eta}(\mathbf{x})$ using *Moreau's formula*

$$\text{Prox}_F^{\eta}(\mathbf{x}) + \eta \, \text{Prox}_{F^*}^{1/\eta}(\mathbf{x}/\eta) = \mathbf{x}.$$

So we obtain the implementation of first line in ADMM as

$$\mathbf{P}_{k+1}(i,j) = \text{Prox}_F^{\eta}[V(i,j), W(i,j)].$$

*Step III: implement the second line of ADMM using a fast Poisson solver.* This was already explained in Section 4.2.4. Follow Section 2.7 to implement the fast Poisson solver.

*Step IV: choose suitable parameters.* The initial guess of ADMM iteration can be the given noisy image and its gradient. For a fixed $\lambda = \frac{C}{h}$, numerically try different ADMM parameters $\sigma = 0.001, 0.01, 0.1, 1, 10, 1000$, and one of them should give you faster convergence of ADMM than the others. The fastest choice parameter should give you very satisfying results within 100 ADMM iterations, i.e., no more than inverting Laplacian 100 times.

*Step V: color images.* Each color image is represented by three matrices for red, green and blue colors. For each color matrix, apply the ROF denoising by ADMM.

*Bonus: TV compressive sensing problems.* Read [Gil Torres *et al.* (2025)] to extend your code to solve the TV norm minimization for compressive sensing problems.

## 4.3 Sobolev gradient flow for Gross-Pitaevskii eigenvalue problem

A Bose–Einstein condensate (BEC) is a state of matter formed by a gas of bosons in low density and low temperature in condensed matter physics. One popular BEC model equation is the Gross–Pitaevskii equation, which is a cubic nonlinear Schrödinger equation:

$$\mathrm{i}\partial_t\psi = -\frac{1}{2}\Delta\psi + V(x)\psi + \beta\psi|\psi|^2.$$

Here $\psi(x,t)$ is the macroscopic wave function of the condensate with a normalization constraint $\int |\psi|^2\,\mathrm{d}x = 1$, $V(x)$ is a trapping potential, which usually is harmonic, e.g., $V(x) = |x|^2$, and $\beta$ is a parameter with $\beta > 0$ for the defocusing case and $\beta < 0$ for the focusing case. For finding a stationary solution $\psi(x,t) = e^{-\mathrm{i}\lambda t}\phi(x)$, inserting it into the equation, we get a nonlinear eigenvalue problem:

$$-\frac{1}{2}\Delta\phi + V(x)\phi + \beta\phi|\phi|^2 = \lambda\phi,$$

in which $V(x)$ and $\beta$ are given and fixed, and we need to find an unknown scalar $\lambda$ and an unknown complex-valued function $\phi(x)$. We will see how a Poisson solver is useful and critical in algorithms for computing its smallest eigenvalue and the corresponding eigenfunction, which is also called *ground state.*

### 4.3.1 *The ground state eigenvalue problem*

We will only focus on the *ground state* of the case $\beta > 0$, for which there are many numerical algorithms using a Poisson solver. Numerical methods for *excited states* might be very different and the focusing case $\beta < 0$ is usually much harder to solve.

We first describe an alternative way to deriving the same ground state eigenvalue problem. For modeling the equilibrium states in Bose–Einstein condensation for $N$ identical bosons with a scattering length $a$ and an external potential $V(x)$, and finding the ground state energy per unit volume of a dilute, thermodynamically infinite, homogeneous gas, it is a standard model to minimize the Gross-Pitaevskii (GP) energy functional defined as

$$\mathcal{E}^{\mathrm{GP}}(\phi) = \int_{\mathbb{R}^3} \left(|\nabla\phi(x)|^2 + V(x)|\phi(x)|^2 + 4\pi a|\phi(x)|^4\right)\mathrm{d}x,$$

and the GP energy, denoted by $E^{\mathrm{GP}}(N,a)$, is defined as the infimum of $\mathcal{E}^{\mathrm{GP}}$ under normalization $\int_{\mathbb{R}^3} |\phi(x)|^2\,\mathrm{d}x = N$. Typical values in experiments of $a$ are around $10^{-3}$, while $N$ varies from $10^3$ to $10^7$.

It is usually an acceptable assumption that we can approximate this energy on a finite domain due to the fast decay rate of the wave function at infinity. Let $\mathcal{E}_L^{\mathrm{GP}}$ denote the energy functional defined on a box domain $\Omega = [-L, L]^d$ and $E_L^{GP}(N, a)$ denote the corresponding GP energy, then

$$\lim_{L\to\infty} E_L^{GP}(N,a) = E^{GP}(N,a),$$

see [Lieb *et al.* (2001)].

The GP energy satisfies a scaling relation $E^{GP}(N,a) = NE^{GP}(1, Na)$, thus we can just consider the minimization of a simplified rescaled model

$$E(\phi) = \frac{1}{2}\int_\Omega \left(|\nabla\phi(x)|^2 + V(x)|\phi(x)|^2\right) \mathrm{d}x + \frac{\beta}{4}\int_\Omega |\phi(x)|^4 \,\mathrm{d}x, \quad \Omega = [-L, L]^d,$$

over the constraint set $\left\{\phi \in H_0^1(\Omega) : \int_\Omega |\phi(x)|^2 \,\mathrm{d}x = 1\right\}$, where $d = 1, 2, 3$, $V(x) \geq 0$ and $\beta > 0$.

The existence, uniqueness, and regularity of the GP ground states have been proven in [Lieb *et al.* (2001)]. For $\beta > 0$, $E(\phi)$ has the unique positive minimizer $\phi(x) > 0$, which is also the ground state eigenfunction to the nonlinear eigenvalue problem

$$-\Delta u(x) + V(x)u(x) + \beta|u(x)|^2 u(x) = \lambda u(x), \quad \int_\Omega |u(x)|^2 \,\mathrm{d}x = 1, u(x)|_{\partial\Omega} = 0. \tag{4.17}$$

In other words, even though the original wave function is complex-valued, one only needs to consider a real positive valued wave function for finding the ground state eigenfunction for the defocusing case $\beta > 0$.

Notice that (4.17) should be understood in the sense of weak derivative as we briefly introduced in Chapter 3, i.e., the variational form of (4.17) is to seek $\lambda \in \mathbb{R}$ and $u \in H_0^1(\Omega)$ satisfying

$$(\nabla u, \nabla v) + (Vu, v) + \beta(|u|^2 u, v) = \lambda(u, v), \quad \forall\, v \in H_0^1(\Omega), \tag{4.18}$$

where $(u, v) = \int_\Omega u(x)v(x)\,\mathrm{d}x$. Let $u^*$ be the ground state to $E(\cdot)$, then by setting $u = v = u^*$ in (4.18), the corresponding eigenvalue should satisfy

$$\lambda^* = 2E(u^*) + \frac{\beta}{2}\bar{\rho}, \quad \bar{\rho} = \int_\Omega |u^*(x)|^4 \,\mathrm{d}x.$$

### 4.3.2 *The finite element method with quadrature*

We consider the finite element method with quadrature using $Q^k$ basis on a rectangular mesh in Chapter 3. Beginners can just focus on the $Q^1$

case, which equivalently gives the second order finite difference scheme in Chapter 2.

Consider a uniform rectangular mesh for the rectangular domain $\Omega$. Let $V_0^h \subset H_0^1(\Omega)$ be continuous piecewise $Q^k$ polynomial space with zero boundary values.

The Galerkin finite element method for the eigenvalue problem (4.17) is to seek $\lambda_h \in \mathbb{R}$ and $u_h \in V_0^h$ satisfying

$$(\nabla u_h, \nabla v_h)+(Vu_h, v_h)+\beta(|u_h|^2 u_h, v_h) = \lambda_h (u_h, v_h), \quad \forall\, v_h \in V_0^h. \tag{4.19}$$

The corresponding discrete energy can be given as

$$E(u_h) = \frac{1}{2}(\nabla u_h, \nabla u_h) + \frac{1}{2}(Vu_h, u_h) + \frac{\beta}{4}(u_h^2, u_h^2).$$

The convergence of the finite element method for the nonlinear eigenvalue problem (4.17) was discussed in [Cancès *et al.* (2010)]. The standard *a priori* error estimates for a linear eigenvalue problem, e.g., $\beta = 0$ in (4.19) is $2k$-order for eigenvalues and $k$-order for eigenvector in $H^1$-norm, under suitable regularity assumptions.

Next, for the $Q^k$ basis, we replace all integrals in (4.19) by $(k+1)$-point Gauss-Lobatto quadrature in each dimension, exactly like what we did in Chapter 3, and such a method is sometimes called the $Q^k$ spectral element method. Standard *a priori* finite element method error estimates still hold, see [Ciarlet (2002)] and references therein. Let $\langle \cdot, \cdot \rangle$ denote that integrals are replaced by quadrature, then the method is to find $u_h \in V_0^h$ satisfying

$$\langle \nabla u_h, \nabla v_h\rangle+\langle Vu_h, v_h\rangle+\beta\langle |u_h|^2 u_h, v_h\rangle = \lambda_h \langle u_h, v_h\rangle, \quad \forall\, v_h \in V_0^h. \tag{4.20}$$

The corresponding discrete energy is given as

$$E_h(u_h) = \frac{1}{2}\langle \nabla u_h, \nabla u_h\rangle + \frac{1}{2}\langle Vu_h, u_h\rangle + \frac{\beta}{4}\langle u_h^2, u_h^2\rangle. \tag{4.21}$$

### 4.3.3 *The matrix-vector form*

Assume a uniform mesh of $N_c^d$ cubic cells for the cubic domain $\Omega = [-L, L]^d$. Then there are in total $(N_c k + 1)^d$ Gauss-Lobatto points. Any $Q^k$ polynomial on a cubic element $e$ can be represented as a Lagrangian interpolation polynomial at $(k+1)^d$ Gauss-Lobatto points, thus the $Q^k$ spectral element method (4.20) (i.e., $Q^k$ basis using $(k+1)^d$ Gauss-Lobatto quadrature) also becomes a finite difference scheme on all Gauss-Lobatto nodes. For $Q^1$ and $Q^2$ bases, all the Gauss-Lobatto points form a uniform grid, as we have seen in Chapter 3. For $k \geq 3$, the Gauss-Lobatto points are not uniform in each element. For the homogeneous Dirichlet boundary condition, the boundary points are not unknown. Thus the total number of unknowns is the interior grid points with the number $N = n^d$ where $n = N_c k - 1$. To derive an equivalent matrix form of the scheme (4.20), let $\phi_i(x) \in V_0^h$ ($i = 1, \cdots, N$) be the continuous piecewise $Q^k$ Lagrangian basis at all Gauss-Lobatto points $x_i$ ($i = 1, \cdots, N$) in the interior of the domain, i.e., $\phi_i(x_j) = \delta_{ij}$. For any piecewise polynomial $u_h(x) \in V_0^h$, let $u_i = u_h(x_i)$. Then $u_h(x) = \sum_{i=1}^N u_i \phi_i(x)$. Let $\mathbf{u} = \begin{pmatrix} u_1 & \cdots & u_N \end{pmatrix}^\top$ and $w_i$ be the quadrature weight at $x_i$.

With the notation above, we have

$$\langle V u_h, v_h \rangle = \sum_{i=1}^{N} w_i V_i u_i v_i = \mathbf{v}^\top \mathbb{M} \mathbb{V} \mathbf{u}, \tag{4.22}$$

where $\mathbb{M} = \operatorname{diag}\{w_1, \cdots, w_N\}$ and $\mathbb{V} = \operatorname{diag}\{V_1, \cdots, V_N\}$ are diagonal matrices and $V_i = V(x_i)$. We also have

$$\langle \nabla u_h, \nabla v_h \rangle = \mathbf{v}^\top \mathbb{S} \mathbf{u}, \tag{4.23}$$

where $\mathbb{S}$ is the stiffness matrix given by $\mathbb{S}_{ij} = \langle \nabla \phi_j, \nabla \phi_i \rangle = \langle \nabla \phi_i, \nabla \phi_j \rangle$. Using (4.22) and (4.23), the matrix form of (4.20) is to find $\mathbf{u} \in \mathbb{R}^N$ satisfying

$$\mathbf{v}^\top \mathbb{S} \mathbf{u} + \mathbf{v}^\top \mathbb{M} \mathbb{V} \mathbf{u} + \beta \mathbf{v}^\top \mathbb{M} \mathbf{u}^3 = \lambda_h \mathbf{v}^\top \mathbb{M} \mathbf{u}, \quad \forall\, \mathbf{v} \in \mathbb{R}^N,$$

or equivalently

$$\mathbb{S} \mathbf{u} + \mathbb{M} \mathbb{V} \mathbf{u} + \beta \mathbb{M} \mathbf{u}^3 = \lambda_h \mathbb{M} \mathbf{u}, \tag{4.24}$$

where $\mathbf{u}^3 = \begin{pmatrix} u_1^3 & \cdots & u_N^3 \end{pmatrix}^\top$. Let $\Delta_h = -\mathbb{M}^{-1} \mathbb{S}$, then (4.24) can also be written as

$$-\Delta_h \mathbf{u} + \mathbb{V} \mathbf{u} + \beta \mathbf{u}^3 = \lambda_h \mathbf{u}. \tag{4.25}$$

In the formulation (4.24), all the matrices are symmetric positive definite, but the discrete Laplacian $\Delta_h = -\mathbb{M}^{-1}\mathbb{S}$ is in general not symmetric in (4.25) except the special case of second order finite difference, which however does not affect numerical implementations since the symmetric form (4.24) should be implemented instead of the form (4.25).

For the nonlinear eigenvalue problem like (4.25) with second order finite difference or $P^1$ finite element on simplicial mesh, it can be proven (4.25) has a unique positive ground state, see [Chen *et al.* (2025)].

### 4.3.4 *Explicit finite difference formulation*

We now give explicit formulation for the $Q^k$ finite element method with quadrature.

We start with the $Q^1$ case, which is equivalent to finite difference in Chapter 2. For a one-dimensional uniform grid $-L = x_0 < x_1 < \cdots < x_n < x_{n+1} = L$ with grid spacing $h = \frac{2L}{n+1}$, for any vector $u_h \in V_0^h$ with $P^1$ polynomial basis, it can be represented by $\mathbf{u} = \begin{pmatrix} u_1 & u_2 & \cdots & u_n \end{pmatrix}^\top$ with $u_0 = u_{n+1} = 0$. The discrete inner product is given by

$$\langle u_h, v_h \rangle = \frac{1}{2} h u_0 v_0 + \sum_{i=1}^{n} u_i v_i h + \frac{1}{2} h u_n v_n = h\mathbf{u}^\top \mathbf{v}.$$

Define $M = h\mathbb{I}_{n\times n}$ and $K = M^{-1}S$ where $S = \frac{1}{h}\begin{pmatrix} 2 & -1 & & & \\ -1 & 2 & -1 & & \\ & \ddots & \ddots & \ddots & \\ & & -1 & 2 & -1 \\ & & & -1 & 2 \end{pmatrix}_{n\times n}$.

Then the matrices in (4.24) and (4.25) for 1D are $\mathbb{M} = M$, $\mathbb{S} = S$, and $-\Delta_h = K$.

For a two-dimensional problem on a uniform grid for the domain $[-L, L]^2$, assume there are $n \times n$ interior grid points. Let $U$, $V$ and $F$ denote 2D arrays of size $n \times n$ consisting of point values of $u_h(x_1, x_2), V(x_1, x_2), f(x_1, x_2)$ at grid points. Let $\mathrm{vec}(U)$ be the vector generated by arranging $U$ column by column. The scheme (4.24) becomes

$$[S\otimes M + M\otimes S + \mathrm{diag}(\mathrm{vec}(V))]\mathrm{vec}(U) + \beta M\otimes M\mathrm{vec}(U^3) = \lambda_h M\otimes M\mathrm{vec}(U),$$

where $U^3$ denotes the entrywise cubic power. In 2D, (4.25) can be written as

$$[K\otimes I + I\otimes K + \mathrm{diag}(\mathrm{vec}(V))]\mathrm{vec}(U) + \beta\mathrm{vec}(U^3) = \lambda_h \mathrm{vec}(U).$$

With the property $(B^\top \otimes A)\mathrm{vec}(X) = \mathrm{vec}(AXB)$, it can be equivalently expressed as

$$KU + UK^\top + V\circ U + \beta U^3 = \lambda_h U,$$

where $\circ$ represents Hadamard product, i.e., entrywise product. With similar notation as in Section 2.9, the three-dimensional case of (4.25) can be expressed as

$$[K\otimes I\otimes I + I\otimes K\otimes I + I\otimes I\otimes K + \mathrm{diag}(\mathrm{vec}(V))]\mathrm{vec}(U) + \beta\mathrm{vec}(U^3) = \lambda_h\mathrm{vec}(U).$$

Next, for the $Q^k$ case. Let $S$ be the stiffness matrix and $M$ be the mass matrix $Q^k$ spectral element method in one dimension. In two dimensions (4.24) can be written as

$$[S \otimes M + M \otimes S + (M \otimes M)\text{diag}(\text{vec}(V))]\text{vec}(U) = \lambda_h (M \otimes M)\text{vec}(U).$$

Define $H = M^{-1}S$, the scheme (4.25) in two dimensions can be written as

$$[H \otimes I + I \otimes H + \text{diag}(\text{vec}(V))]\text{vec}(U) + \beta\text{vec}(U^3) = \lambda_h \text{vec}(U),$$

or equivalently $HU + UH^\top + V \circ U + \beta U^3 = \lambda_h U$, and in three dimensions it is

$$[H \otimes I \otimes I + I \otimes H \otimes I + I \otimes I \otimes H + \text{diag}(\text{vec}(V))]\text{vec}(U) + \beta\text{vec}(U^3) = \lambda_h \text{vec}(U).$$

See expressions of matrices $S, M, H$ for $Q^2$ case in Chapter 3.

### 4.3.5 *Gradient flow methods*

Now we focus on how to numerically solve (4.25), and we can focus on the second order finite difference, for which $-\Delta_h$ is given explicitly as follows:

$$-\Delta_h = \begin{cases} K, & d = 1, \\ K \otimes I + I \otimes K, & d = 2, \\ K \otimes I \otimes I + I \otimes K \otimes I + I \otimes I \otimes K, & d = 3. \end{cases} \tag{4.26}$$

Before we discuss a nonlinear eigenvalue problem, we need to know how to solve a linear eigenvalue problem. For instance, given a real symmetric positive definite matrix $A \in \mathbb{R}^{n \times n}$ with eigenvalues $\lambda_1 > \lambda_2 > \cdots > \lambda_n$ and associated orthonormal eigenvectors $v_i$, the *Power Iteration* is a simple iterative method for finding $v_1$:

$$u_{k+1} = \frac{Au_k}{\|Au_k\|}.$$

For a nonlinear eigenvalue problem like (4.25), the naive *Self Consistent Field* (SCF) method is an iteration of linearized eigenvalue problems: given $\mathbf{u}_k$, find $\lambda_{k+1}, \mathbf{u}_{k+1}$ as the smallest eigenvalue and eigenvector, for the matrix $-\Delta_h + \mathbb{V} + \beta\text{diag}\{\mathbf{u}_k^2\}$. In other words, the SCF iteration satisfies

$$-\Delta_h \mathbf{u}_{k+1} + \mathbb{V}\mathbf{u}_{k+1} + \beta \mathbf{u}_k^2 \circ \mathbf{u}_{k+1} = \lambda_{k+1}\mathbf{u}_{k+1}. \tag{4.27}$$

However, the naive SCF does not converge, and its convergence can be improved by various relaxations, see [Henning and Jarlebring (2025)] and references therein. On the other hand, the linearized eigenvalue problem (4.27) is still difficult to solve directly on computers since $\Delta_h$ is a huge matrix in 3D. If solving each iteration (4.27) only approximately and not too accurately, which is some kind of relaxation, then it can greatly improve the convergence of SCF. In particular, we can use just one Power Iteration for the matrix $[-\Delta_h+\mathbb{V}+\beta\text{diag}\{\mathbf{u}_k^2\}]^{-1}$ for approximating (4.27), then we get the following convergent algorithm,

$$\tilde{\mathbf{u}}_{n+1}=(-\Delta_h+\mathbb{V}+\beta|\mathbf{u}_n|^2)^{-1}\mathbf{u}_n,\quad \mathbf{u}_{n+1}=\frac{\tilde{\mathbf{u}}_{n+1}}{\|\tilde{\mathbf{u}}_{n+1}\|_2},\tag{4.28}$$

with the discrete 2-norm defined as an approximation to $L^2$-norm, i.e., $\|\mathbf{u}\|_2:=\sqrt{h^d\mathbf{u}\cdot\mathbf{u}}$ in dimension $d$. The method (4.28) is a relaxed version of SCF but also the $A_u$ gradient flow method with step size 1. Recall that (4.25) has a unique positive ground state. With a positive initial guess $u_0$, it can be rigorously proven that (4.28) will converge to this unique positive ground state. See [Henning and Jarlebring (2025)] for an exposition of these gradient flow methods and the convergence of (4.28).

Notice that implementation of (4.28) requires inversion of the matrix $-\Delta+V+\beta|\mathbf{u}_n|^2$, which is in general more difficult than inverting only $-\Delta+a\mathbb{I}$ for some number $a$. A computationally easier algorithm is the $H^1$-norm gradient flow method, which can also be regarded as Riemannian gradient descent method under $H^1$-norm induced metric for the unit sphere [Chen *et al.* (2025)]. Its form is given as follows. First, define the gradient from the $H^1$-metric with a fixed parameter $\alpha>0$ as

$$\nabla E_h(\mathbf{u}):=(-\Delta_h+\alpha\mathbb{I})^{-1}(-\Delta_h+\mathbb{V}+\beta\text{diag}(\mathbf{u})^2)\mathbf{u}.\tag{4.29a}$$

Second, the corresponding Riemannian gradient on unit sphere is given as

$$\nabla^R E_h(\mathbf{u}):=\nabla E_h(\mathbf{u})-\frac{\langle\mathbf{u},\nabla E_h(\mathbf{u})\rangle}{\langle\mathbf{u},(-\Delta_h+\alpha\mathbb{I})^{-1}\mathbf{u}\rangle}(-\Delta_h+\alpha\mathbb{I})^{-1}\mathbf{u}.\tag{4.29b}$$

Third, the Riemannian gradient descent method or the Sobolev gradient flow under the $H^1$-metric is hence given by

$$\mathbf{u}^{n+1}=\frac{\mathbf{u}^n-\tau\nabla^R E_h(\mathbf{u}^n)}{\|\mathbf{u}^n-\tau\nabla^R E_h(\mathbf{u}^n)\|_2},\tag{4.29c}$$

where $\tau>0$ is a step size. In each iteration of (4.29), one needs to compute $(-\Delta_h+\alpha\mathbb{I})^{-1}$ at least twice.

### 4.3.6 *Project: gradient flows for the ground state*

Now we are ready to implement the two gradient flow methods. The $H^1$-norm gradient flow (4.29) has two parameters $\alpha>0$ and a step size $\tau>0$. For a fixed $\alpha\leq 1$, (4.29) converges with a small enough step size $\tau$, see

[Chen *et al.* (2025)]. The $A_u$ gradient flow (4.28) can be more efficient if adding a variable step size, which will not be considered here only for simplicity.

We consider an exact solution to the nonlinear eigenvalue problem (4.17) on $\Omega = [-1,1]^d$ with a potential $V(\mathbf{x}) = \beta(1 - |u^*(\mathbf{x})|^2)$ where the ground state is

$$u^*(\mathbf{x}) = \begin{cases} \beta \sin\left(\pi\frac{x+1}{2}\right), & d=1 \\ \beta \sin\left(\pi\frac{x+1}{2}\right)\sin\left(\pi\frac{y+1}{2}\right), & d=2 \\ \beta \sin\left(\pi\frac{x+1}{2}\right)\sin\left(\pi\frac{y+1}{2}\right)\sin\left(\pi\frac{z+1}{2}\right), & d=3 \end{cases} \tag{4.30}$$

and the eigenvalue and energy are $\lambda^* = d\frac{\pi^2}{4} + \beta, \quad E(u^*) = \frac{1}{2}\lambda^* - \frac{\beta}{4}\left(\frac{3}{4}\right)^d.$

*Step I: 1D test.* Use second order finite difference to implement both $A_u$ gradient flow (4.28) and $H^1$-norm gradient flow (4.29) for the problem (4.30) in 1D. When inverting matrices, you can simply use backslash in MATLAB, as in Section 2.2.3. Test the accuracy and convergence order of your scheme and implementation as in Section 2.2.3.

*Step II: 2D test.* Use second order finite difference to implement both $A_u$ gradient flow (4.28) and $H^1$-norm gradient flow (4.29) for the problem (4.30) in 2D. For inverting $-\Delta_h + \alpha\mathbb{I}$, use the eigenvector method in Section 2.7. For inverting $-\Delta + V + \beta|\mathbf{u}_n|^2$, use the Precondition Conjugate Gradient method in Appendix B.4, with $(-\Delta_h + \mathbb{I})^{-1}$ as the preconditioner. Test the accuracy and convergence order of your scheme and implementation.

*Step III: 3D problems.* Use second order finite difference to implement $H^1$-norm gradient flow (4.29) for the problem (4.30) in 3D. For inverting $-\Delta_h + \alpha\mathbb{I}$, use the eigenvector method in Section 2.9. Test the accuracy and convergence order of your scheme and implementation.

Then use your code to solve the following problem on $\Omega = [-16,16]^3$ with a potential:

$$V(\mathbf{x}) = \sin\left(\frac{\pi}{4}x\right)^2 \sin\left(\frac{\pi}{4}y\right)^2 \sin\left(\frac{\pi}{4}z\right)^2. \tag{4.31}$$

For (4.31) and $\beta = 10$, implement the modified $H^1$ scheme (4.29) with $\alpha = 0.15$ and $\tau = 1$.

*Bonus: extensions to high order finite element methods and GPU.* Use the code in [Liu *et al.* (2024b), Appendix] to implement the high order $Q^k$ element method with $(k+1)$-point Gauss-Lobatto quadrature. Compare your results to those in [Chen *et al.* (2025)]. If you have access to suitable GPU cards, follow [Liu *et al.* (2024b)] to run your code on GPU.

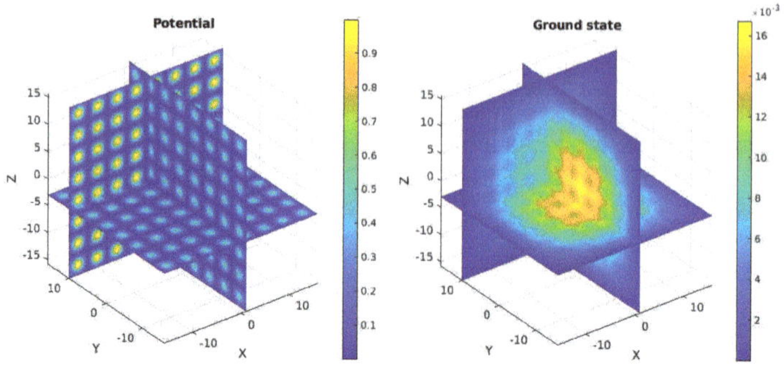


(a) The potential function. (b) $\beta = 10$

Fig. 4.8 The potential function (4.31) and its ground state.

## 4.4 Optimal Transport problems

The *Monge* mass transfer problem can be described as finding a transport plan for moving a source distribution of mass to a target distribution, with a minimal transport cost, which is usually proportional to the distance of moving mass, see Figure 4.9. Such a problem was first considered by Monge in 1781 and later formalized by Kantorovich in 1942, and it is often called the Monge–Kantorovich optimal transport problem. In this section, we will see how the Poisson equation naturally arises in its dynamical formulation.

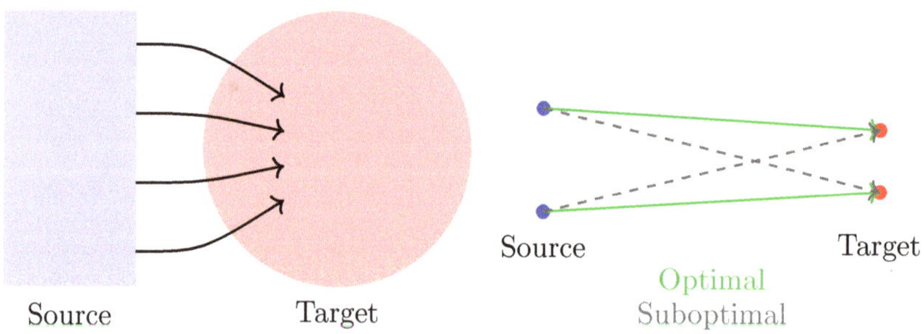


(a) From a rectangle to a disk. (b) From two points to two points.

Fig. 4.9 An illustration of *Monge* mass transfer problem, in which transport cost is proportional to distance of moving mass. For the example on the right: the green arrows show the optimal transport plan which pairs each source point with its closest target point thus minimizes the total distance.

### 4.4.1 *Dynamical formulation*

The Monge-Kantorovich problem in $L^2$ norm can be stated as follows. Given two density functions $\rho_0(\mathbf{x}) \geq 0$ for the source distribution and $\rho_1(\mathbf{x}) \geq 0$ for the target distribution with total mass one $\int_{\mathbb{R}^d} \rho_0(\mathbf{x})\,\mathrm{d}\mathbf{x} = \int_{\mathbb{R}^d} \rho_1(\mathbf{x})\,\mathrm{d}\mathbf{x} = 1$, we seek a mapping $T : \mathbb{R}^d \to \mathbb{R}^d$ to minimize the cost

$$\inf \int_{\mathbb{R}^d} \|T(\mathbf{x}) - \mathbf{x}\|^2 \rho_0(\mathbf{x})\,\mathrm{d}\mathbf{x}.$$

under the constraint that it is a transport plan: for any bounded set $S$, $T$ satisfies

$$\int_S \rho_1(\mathbf{x})\,\mathrm{d}\mathbf{x} = \int_{T(\mathbf{x})\in S} \rho_0(\mathbf{x})\,\mathrm{d}\mathbf{x}.$$

In [Benamou and Brenier (2000)], Benamou and Brenier proposed an equivalent dynamical formulation, in which the variables are lifted from space to space-time. In the Benamou-Brenier dynamical formulation, one seeks a time-dependent density function $\rho(\mathbf{x}, t)$ and velocity field $\mathbf{v}(\mathbf{x}, t)$ to minimize the following cost,

$$\min \int_0^1 \int_{\mathbb{R}^d} \rho(\mathbf{x}, t) \frac{\|\mathbf{v}(\mathbf{x}, t)\|^2}{2}\,\mathrm{d}\mathbf{x}\,\mathrm{d}t \tag{4.32}$$

under the constraints that $\rho(\mathbf{x}, t) \geq 0$ and $\mathbf{v}$ should satisfy an equation (continuity equation for fluids, see (4.1)):

$$\rho_t + \nabla \cdot (\rho \mathbf{v}) = 0, \quad \forall t \in (0, 1)$$

with both initial and terminal conditions

$$\rho(\mathbf{x}, 0) = \rho_0(\mathbf{x}), \quad \rho(\mathbf{x}, 1) = \rho_1(\mathbf{x}).$$

The minimizer to this formulation will naturally give the optimal transport plan as illustrated in the following examples.

**Example 4.1.** *Consider an example of the translation of a Gaussian distribution. This problem will be formulated by choosing $\rho_0 \sim \mathcal{N}(\frac{1}{3}, 0.01)$ to be a Gaussian with mean value $\frac{1}{3}$ and variance $\sigma^2 = 0.01$, while $\rho_1 \sim \mathcal{N}(\frac{2}{3}, 0.01)$. Notice that $\int_0^1 \rho_0(x)\,\mathrm{d}x = \int_0^1 \rho_1(x)\,\mathrm{d}x$, so no normalization is needed. The optimal transport plan will translate $\rho_0$ to $\rho_1$ with no deformation. See Figure 4.10 for the minimizer $\rho$ to the dynamical formulation.*

**Example 4.2.** *The second example is the merging of two Gaussians into one. The distributions are $\rho_0 \sim \mathcal{N}(0.1, 0.01) + \mathcal{N}(0.9, 0.01)$ and $\rho_1 \sim \mathcal{N}(0.5, 0.0025)$, with the normalization on the interval $[0, 1]$ so that $\int_0^1 \rho_0(x)\,\mathrm{d}x = \int_0^1 \rho_1(x)\,\mathrm{d}x$. See Figure 4.11 for the minimizer $\rho$ to the dynamical formulation, which gives the optimal transport plan.*

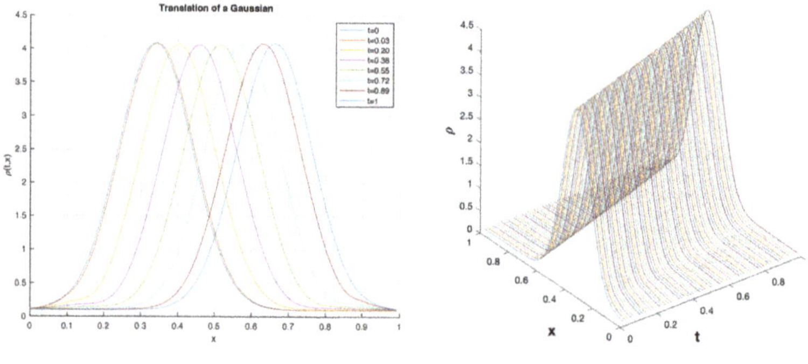


(a) $\rho(x,t)$ at different time. (b) $\rho(x,t)$ at different time.

Fig. 4.10 Translating a Gaussian to another: an illustration of the minimizer $\rho(x,t)$ to the dynamical formulation of the optimal transport problem.

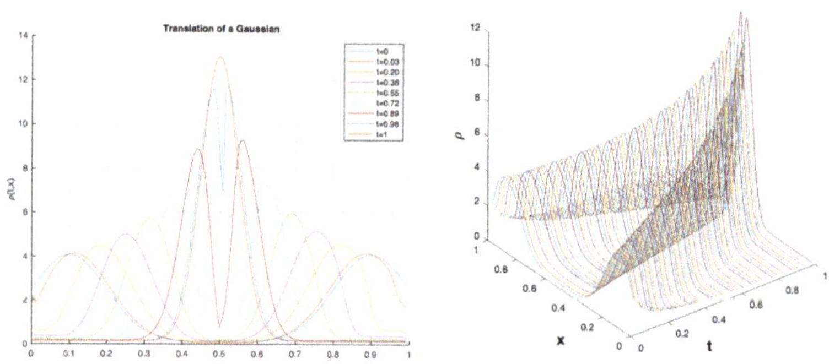


(a) $\rho(x,t)$ at different time. (b) $\rho(x,t)$ at different time.

Fig. 4.11 Merging two Gaussians into one: an illustration of the minimizer $\rho(x,t)$ to the dynamical formulation of the optimal transport problem.

**Example 4.3.** *The third example shown in Figure 4.12 is transforming one Pikachu picture denoted by $\rho(x,y,0)$ to another one denoted by $\rho(x,y,1)$ in the sense of optimal transport. See Figure 4.13 for the minimizer $\rho(x,y,t)$ at different time to the dynamical formulation, which gives the optimal transport plan.*

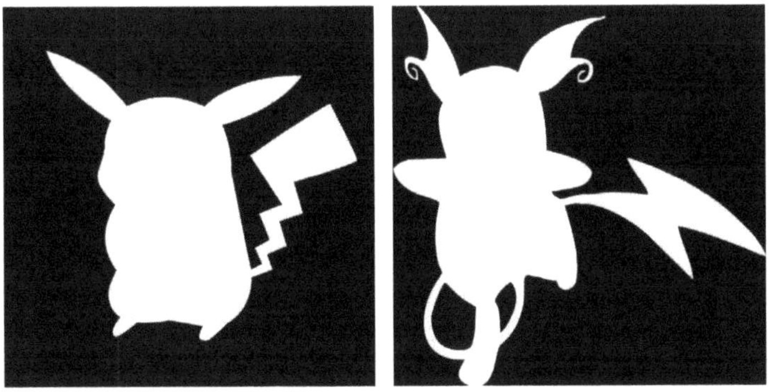

(a) The distribution $\rho_0(x, y)$. (b) The distribution $\rho_1(x, y)$.

Fig. 4.12 A two dimensional optimal transport problem for transporting a distribution $\rho_0(x, y)$ to another $\rho_1(x, y)$.

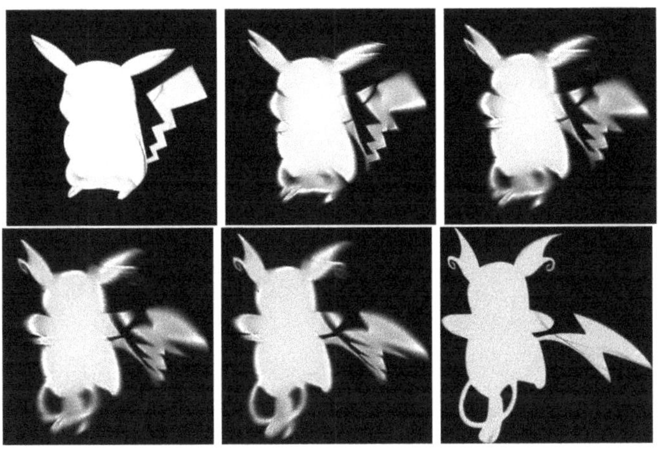

Fig. 4.13 The minimizer $\rho(x, y, t)$ to the dynamical formulation for the two dimensional optimal transport problem in Figure 4.12: $\rho(x, y, t)$ at $t = 0.02, 0.2, 0.39, 0.59, 0.78, 0.98$.

### 4.4.2 *The convex formulation*

By introducing the momentum $\mathbf{m} = \rho\mathbf{v}$, the minimization above is equivalent to the following convex minimization problem,

$$\min_{\rho,\mathbf{m}} \int_0^1 \int_{\mathbb{R}^d} f(\rho, \mathbf{m})\, \mathrm{d}\mathbf{x}\, \mathrm{d}t, \tag{4.33a}$$

$$\rho_t + \nabla \cdot \mathbf{m} = 0, \quad \rho(\mathbf{x}, 0) = \rho_0(\mathbf{x}), \quad \rho(\mathbf{x}, 1) = \rho_1(\mathbf{x}), \tag{4.33b}$$

where

$$f(\rho,\mathbf{m}) = \begin{cases} \frac{\|\mathbf{m}\|^2}{2\rho} & \text{if } \rho > 0 \\ 0 & \text{if } \rho = 0, \mathbf{m} = 0 \\ +\infty & \text{else} \end{cases}.$$

To see how a Poisson solver is hidden in such a problem, we consider the one dimensional case as an example. For $d = 1$, (4.33b) becomes $\rho_t + m_x = 0$, which is the divergence free condition $(\partial_t, \partial_x) \cdot (\rho, m) = 0$. We have seen that a Poisson equation is naturally connected to the divergence free condition in the example of Navier-Stokes equations.

Next we consider a discretization to approximate all the derivative operators so that (4.33) can be easily solved by the ADMM method. To this end, we first consider a toy problem of approximating the divergence free condition $u'(t) = 0$ on the domain $[0, 1]$ with the boundary conditions $u(0) = u(1) = a$. On a uniform grid $t_i$ with $t_i = \frac{i}{n+1} = i\Delta t$, the forward difference gives $Du = b$, where

$$D = \frac{1}{\Delta t}\begin{pmatrix} 1 & & & \\ -1 & 1 & & \\ & \ddots & \ddots & \\ & & -1 & 1 \\ & & & -1 \end{pmatrix}_{(n+3)\times(n+2)}, \quad u = \begin{pmatrix} u_0 \\ u_1 \\ \vdots \\ u_n \\ u_{n+1} \end{pmatrix}, \quad b = \begin{pmatrix} a \\ 0 \\ \vdots \\ 0 \\ -a \end{pmatrix}.$$

Notice $Du = b$ approximates both b.c. $u(0) = u(1) = a$, and the equation $u'(t) = 0$ at $n + 1$ grid points $t_i$ $(i = 0, 1, \cdots, n)$.

**Remark 4.4.** *Let* $\mathbf{1}$ *be the constant one column vector. Then* $D^\top \mathbf{1} = 0$ *implies that the column space of* $D$ *is orthogonal to* $\mathbf{1}$. *Since* $b$ *is also orthogonal to* $\mathbf{1}$, $b$ *is in the column space of* $D$ *thus* $Du = b$ *is a consistent linear system. This is an important feature of this simple approximation.*

We now consider approximating (4.33b) with $d = 1$, $x \in [0, 1]$, and $t \in [0, 1]$. For convenience, we can further add homogeneous boundary conditions for the momentum $m(0) = m(1) = 0$. We use the same scheme to approximate both the equation $\rho_t + m_x = 0$ and the initial-terminal condition, which leads to slightly different grids for representing unknown variable locations for $\rho$ and $m$. With a uniform grid $(x_i, t_j)$ for time and space, the unknowns of $\rho$ are defined at $(x_i, t_j)$ for $i = 0, 1, \ldots, n_x; j = 1, \ldots, n_t$, while the unknowns of $m$ are defined at $(x_i, t_j)$ for $i = 1, \ldots, n_x; j = 0, 1, \ldots, n_t$, as shown in Figure 4.14.

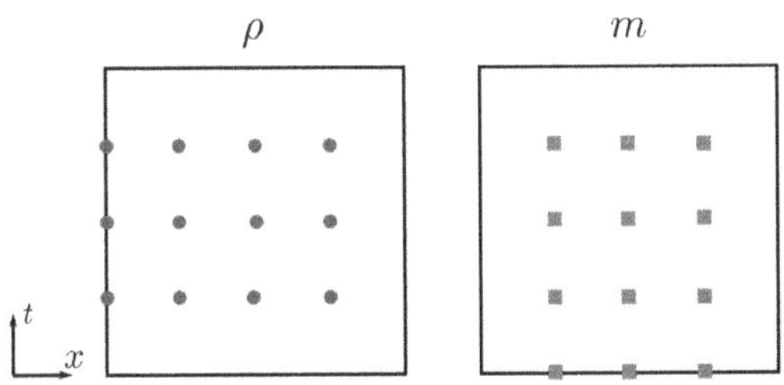


Fig. 4.14 Grid points for the density and momentum variables when $n_t = n_x = 3$.

Following the notation in Chapter 2, let $\rho_h$ be an array of size $(n_t + 2) \times (n_x + 1)$ with entries $\rho_h(j, i) = \rho(x_i, t_j)$ and $m_h$ be an array of size $(n_t + 1) \times (n_x + 2)$ with entries $m_h(j, i) = m(x_i, t_j)$. Notice that not all the entries of $\rho_h, m_h$ are unknowns shown in Figure 4.14, since we have included some boundary entries in these two arrays for convenience. For the initial-terminal condition, define an array

$$b = \begin{pmatrix} \rho_0(x_0) & \cdots & \rho_0(x_{n_x}) \\ 0 & \cdots & 0 \\ \vdots & & \vdots \\ 0 & \cdots & 0 \\ -\rho_1(x_0) & \cdots & -\rho_1(x_{n_x}) \end{pmatrix},$$

then the same method in the toy problem for approximating $\rho_t + m_x = 0$ with the initial condition $\rho_0$ and terminal condition $\rho_1$ can be written as:

$$I_x \otimes D_t \text{vec}(\rho_h) + D_x \otimes I_t \text{vec}(m_h) = \text{vec}(b).$$

This can be formulated as a linear constraint $Au = d$ by letting $A = \left(I_x \otimes D_t \; D_x \otimes I_t\right)$, $u = \begin{pmatrix} \text{vec}(\rho_h) \\ \text{vec}(m_h) \end{pmatrix}$, and $d = \text{vec}(b)$.

Such a discretized version of dynamical formulation of optimal transport problem (4.33) can now be written as

$$\min_{\rho_h, m_h} F(\rho_h, m_h) + \iota_{\{u: Au=d\}}(\rho_h, m_h) \tag{4.34}$$

where

$$F(\rho_h, m_h) = \Delta x \Delta t \sum_{i=0}^{n_t} \sum_{j=0}^{n_x} f\left[\rho_h(i, j), m_h(i, j)\right],$$

$$f(\rho, m) = \begin{cases} \frac{m^2}{2\rho} & \text{if } \rho > 0 \\ 0 & \text{if } \rho = 0, m = 0 \\ +\infty & \text{else} \end{cases},$$

and the indicator function (see Appendix C.1) is

$$\iota_{\{u:Au=d\}}(\rho_h, m_h) = \begin{cases} 0 & \text{if} \quad I_x \otimes D_t \mathrm{vec}(\rho_h) + D_x \otimes I_t \mathrm{vec}(m_h) = \mathrm{vec}(b) \\ +\infty & \text{otherwise} \end{cases}.$$

### 4.4.3 *ADMM and Poisson solver*

The proximal operator (see Appendix C) for the indicator function is just a projection to the set $\{u : Au = d\}$.

$$\mathrm{Prox}^{\gamma}_{\iota_{\{u:Au=d\}}}(u_h) = \mathrm{Proj}_{\{u:Au=d\}} = u_h - A^\top (AA^\top)^+ (Au_h - d),$$

where $(AA^\top)^+$ denotes the pseudo-inverse of $AA^\top$. In particular, we have

$$A = \begin{pmatrix} I_x \otimes D_t & D_x \otimes I_t \end{pmatrix} \Rightarrow A^\top A = I_x \otimes D_t^\top D_t + D_x^\top D_x \otimes I_t,$$

thus $A^\top A$ is a discrete Laplacian type operator here, which can be inverted following the method in Section 2.7.

For convenience, let $G(\rho_h, m_h) := \iota_{\{u:Au=d\}}(\rho_h, m_h)$, then the discretized optimal transport dynamical formulation is simply

$$\min F(\rho_h, m_h) + G(\rho_h, m_h).$$

For this problem, the ADMM method in the form of (C.9) in the Appendix with one step size $\tau > 0$ becomes

$$\begin{aligned} (\rho, m)^{k+1} &= \mathrm{Prox}^{\frac{1}{\tau}}_{F}\left( (\widetilde{\rho}, \widetilde{m}^k) - \frac{\delta^k}{\tau} \right) \\ (\widetilde{\rho}, \widetilde{m})^{k+1} &= \mathrm{Proj}_{\{u:Au=d\}}\left( (\rho, m)^{k+1} + \frac{\delta^k}{\tau} \right) \\ \delta^{k+1} &= \delta^k + \tau \left( (\widetilde{\rho}, \widetilde{m})^{k+1} - (\rho, m)^{k+1} \right) \end{aligned}$$

**Remark 4.5.** *We leave it as a linear algebra exercise to readers for verifying that $\{u : Au = d\}$ is not empty if the discretized initial and terminal conditions satisfy $\sum_i \rho_0(x_i) = \sum_i \rho_1(x_i)$, by extending Remark 4.4 to the 2D case.*

**Remark 4.6.** *One important property of $\rho$ is that it is a probability distribution, thus it is non-negative and its total mass should be one. For the ADMM method and our chosen spatial discretization, such a property is preserved in the following sense. As will be shown in the next subsection, $\rho^{k+1} \geq 0$ is ensured by the definition of $F$ and its proximal operator. Assume $\sum_i \Delta x \rho_0(x_i) = \sum_i \Delta x \rho_1(x_i) = 1$, which can be achieved by shifting the numerical initial and terminal conditions by constants. In the second step of ADMM, the projection onto the set $\{u : Au = d\}$ can render*

$\sum_i \tilde{\rho}^{k+1}(x_i, t_j) = 1$ *because the chosen numerical scheme for the continuity equation reads*

$$\frac{\rho(x_i, t_{j+1}) - \rho(x_i, t_j)}{\Delta t} + \frac{m(x_{i+1}, t_j) - m(x_i, t_j)}{\Delta x} = 0,$$

*which with zero boundary condition for* $m$ *implies* $\sum_i \rho(x_i, t_{j+1}) = \sum_i \rho(x_i, t_j)$. *Thus during the ADMM iteration, the first step enforces non-negativity and the second step enforces total mass, and if ADMM iteration stops at k-th iteration with error* $\epsilon$, *then we get a numerical* $\rho^k \geq 0$ *with total mass error proportional to* $\epsilon$.

### 4.4.4 *Project: implement ADMM for optimal transport*

*Step I: derive the formula for the first line in ADMM.* We list the main idea here and leave the details to readers to finish. For $F$, we should plug in initial condition $\rho_h(j, 0) = \rho_0(x_j)$ and boundary condition $m_h(0, i) = 0$, with which the function $F$ turns into

$$F(\rho_h, m_h) = \Delta t \Delta x \sum_{j=1}^{n_x} \left( \sum_{i=1}^{n_t} f\left[\rho_h(j, i), m_h(j, i)\right] + f\left[\rho_0(x_j), m_h(j, 0)\right] \right),$$

and its proximal operator can be computed in an element-wise fashion. Thus we only need to consider the proximal operator for $f(\rho, m)$. Since $f(\rho, m) \geq 0$ and $f(\rho, m)$ is strongly convex and $C^1$ when $\rho > 0$, its proximal operator has a unique solution,

$$(\rho_*, m_*) = \mathrm{Prox}_f^{\gamma}(\widetilde{\rho}, \widetilde{m}) = \mathrm{argmin}_{(\rho,m)}\, f(\rho, m) + \frac{1}{2\gamma}\|(\rho, m) - (\widetilde{\rho}, \widetilde{m})\|^2.$$

If $\rho_* > 0$, then the critical point equation implies

$$\begin{cases} \gamma \frac{m_*}{\rho_*} + m_* - \widetilde{m} = 0 \\ -\frac{\gamma}{2} \frac{\|m_*\|^2}{\rho_*^2} + \rho_* - \widetilde{\rho} = 0, \end{cases}$$

which gives us a cubic equation to solve for $\rho_*$. Let $P(x) := (x - \widetilde{\rho})(x + \gamma)^2 - \frac{\gamma}{2}\|\widetilde{m}\|^2 = 0$. Then the optimality conditions above turn into

$$P(\rho_*) = 0 \quad \text{and} \quad m_* = \frac{\rho_* \widetilde{m}}{\rho_* + \gamma}.$$

Therefore, if $P(x) = 0$ has at least one strictly positive real root $\rho_*$, it is necessarily unique. Otherwise, $\rho_* = 0$ and $m_* = 0$ by the definition of $f$. This formula works for all the interior points $i = 1, \ldots, n_t; j = 1, \ldots, n_x$. For the boundary terms, e.g. at time $t_0 = 0$, similarly we have

$$m_* = \mathrm{argmin}_m\, f(\rho_0, m) + \frac{1}{2\gamma}(m - \widetilde{m}) = \begin{cases} \frac{\rho_0 \widetilde{m}}{\rho_0 + \gamma} & \text{if } \rho_0 > 0 \\ 0 & \text{otherwise} \end{cases}.$$

Notice that our definition of $F$ does not include the variable $\rho_h(x,t)$ for $t=0$, and the update for those locations is simply $\rho_* = \text{Prox}(\widetilde{\rho}) = \widetilde{\rho}$.

*Step II: efficient projection formula for the second line in ADMM.* For the projection onto the set $\{u : Au = d\}$, recall that you need to shift the discretized initial and terminal conditions by a constant to satisfy the condition $\sum_i \Delta x \rho_0(x_i) = \sum_i \Delta x \rho_1(x_i) = 1$ so that $\{u : Au = d\}$ is not an empty set, see Remark 4.5. Then use the same approach in Section 2.7 for inverting the big matrix

$$A^\top A = I_x \otimes D_t^\top + D_t D_x^\top D_x \otimes I_t.$$

You will need eigenvalue decomposition of $D^\top D$, which can be computed numerically by the *eig* function in MATLAB if you cannot derive the exact ones.

*Step III: test your code for the two examples in this section.* If your code works well for 1D optimal transport, extend it to 2D.

# Part II
# Unstructured Meshes and Point Clouds

## Chapter 5

# A Brief Introduction to Differential Geometry

In Part II, we will switch gears to discuss numerical methods for Laplacian defined on surfaces and point clouds. This chapter serves as an introduction to the essential elements in differential geometry, which are needed for formulating and solving PDEs on surfaces. Differential geometry provides a mathematical framework for describing and analyzing curved spaces. While calculus in Euclidean space concerns functions and differential operators defined on flat domains, many phenomena in science and engineering naturally occur on curved surfaces or manifolds. Examples include diffusion on biological membranes, heat transfer on curved interfaces, fluid motion on the Earth's surface, and shape analysis in computer graphics and medical imaging. To model and solve partial differential equations (PDEs) in such settings, one must first understand how to generalize fundamental notions to manifolds, such as differentiation, integration, and vector fields.

In this chapter, we begin by defining a *smooth manifold*, which generalizes the concept of a surface, i.e., surfaces are two-dimensional smooth manifolds. Manifolds locally resemble Euclidean spaces but can possess rich global geometry and topology. We then introduce functions on manifolds and how to measure, average, and differentiate quantities defined on manifolds. Building upon these foundations, we can define differential operators such as the gradient, divergence, and Laplace–Beltrami operator, which is the Laplacian operator on a manifold.

From a computational point of view, these geometric constructions enable one to translate PDEs on curved surfaces into concrete coordinate expressions or intrinsic formulations. It provides crucial insight for designing numerical algorithms by understanding how geometry influences these operators through the metric, curvature, and coordinate transformations.

By the end of this chapter, the reader will have a taste of the fundamental geometric tools to interpret and solve PDEs on manifolds. The concepts introduced here will serve as a bridge between geometry and analysis, laying the foundation for subsequent chapters on numerical methods for PDEs on surfaces and its applications.

**Further reading:** This section is by no means a complete introduction to differential geometry. Our goal here is only to establish the minimal geometric background necessary for the forthcoming chapters, where we formulate and approximate PDEs on curved surfaces and manifolds. Readers interested in a more comprehensive treatment of differential geometry are encouraged to read standard references such as [do Carmo (1976); Millman and Parker (1977); Lee (2013, 2018)].

## 5.1 Defining surfaces and manifolds

### 5.1.1 *Definition by level-set representation*

We first start with a special case. Some manifolds, but not all manifolds, can often be described as the set of points in Euclidean space satisfying one or more equations. This representation, known as the *implicit* or *level-set* method, may sometimes be preferred in computational mathematics since it allows an easy representation of complex geometries without explicit parameterizations.

Formally, such a $d$ dimensional manifold $\mathcal{M} \subset \mathbb{R}^D$ can be defined as the common zero set of $D-d$ smooth functions

$$f_i : \mathbb{R}^D \to \mathbb{R}, \qquad i = 1, \ldots, D-d,$$

that is,

$$\mathcal{M} = \{x \in \mathbb{R}^D : f_i(x) = 0 \quad \forall i = 1, \ldots, D-d\}.$$

Equivalently, we may write $\mathcal{M} = F^{-1}(0)$ where

$$F = (f_1, \ldots, f_{D-d}) : \mathbb{R}^D \to \mathbb{R}^{D-d}.$$

For convenience, we assume that $f_i$ are differentiable functions, and their gradients $\nabla f_1(x), \ldots, \nabla f_{D-d}(x)$ are linearly independent at each $x \in \mathcal{M}$. Such an assumption is equivalent to that 0 is a *regular value* of $F$, i.e., $\text{rank}[\nabla F(x)] = D - d$ for any $x \in F^{-1}(0)$. Under this assumption, the *Implicit Function Theorem* guarantees that $\mathcal{M}$ defined above is a smooth embedded $d$-dimensional manifold.

The level-set formulation provides an intrinsic description of geometry: the manifold $\mathcal{M}$ is *embedded* in $\mathbb{R}^D$ and inherits the geometric structure from the ambient space. For instance, if $D-d=1$, then $\mathcal{M}$ is a hypersurface given by $f(x)=0$, and the normalized gradient

$$\hat{n}(x) = \frac{\nabla f(x)}{\|\nabla f(x)\|}$$

defines a smooth *unit normal field* along $\mathcal{M}$. This simple geometric quantity plays a key role in defining surface differential operators and curvature.

The implicit method is especially useful for *numerical simulation of evolving surfaces.* When a surface moves with a normal velocity $V_n$, one can evolve the level-set function $f(x,t)$ via the Hamilton–Jacobi equation

$$f_t + V_n \|\nabla f\| = 0,$$

so that the zero level set of $f$ naturally tracks the moving interface. This approach avoids explicit mesh tracking and automatically handles topological changes such as merging or splitting—making it widely used in computational geometry, image processing, and fluid–structure interface problems [Osher and Sethian (1988); Osher and Fedkiw (2003); Sethian (1999)].

Despite its elegance, the level-set approach has some limitations:

(1) When the ambient dimension $n$ is much larger than the intrinsic dimension $d$, the representation involves too many functions $f_i$, which can increase computational complexity.
(2) Operations such as integration or differential operators on $\mathcal{M}$ must be carefully reformulated in the ambient space, often involving projection operators or regularization.
(3) The level-set representation is not convenient for formulating or analyzing *variational PDEs on manifolds*, where one often needs intrinsic quantities such as the metric tensor, gradient, and Laplace–Beltrami operator in local coordinates. Variational problems involving surface energies or differential operators typically require explicit expressions of these geometric quantities, which are more naturally obtained through coordinate charts or parametrizations.

For these reasons, while the level-set framework is powerful for interface evolution, this book will primarily focus on the *explicit* approach, i.e. using local coordinate charts and parameterizations. This provides direct access to intrinsic geometric quantities essential for the analysis and numerical solution of PDEs on manifolds.

### 5.1.2 *Definition by coordinate charts*

The *explicit method* defines a manifold by locally parameterizing its points using Euclidean coordinates. Unlike the implicit level-set representation, which defines a manifold as the zero set of a function, the explicit approach describes the manifold directly in terms of coordinate charts. This not only clarifies its geometric structure but also provides a convenient framework for computation, especially when formulating or discretizing PDEs on curved surfaces. To formalize this idea, we begin by recalling the notion of open sets, which form the basic building blocks of local coordinate domains.

**Definition 5.1** (Open set). *A subset $\Omega \subset \mathbb{R}^D$ is* open *if for every $x \in \Omega$ there exists $\delta > 0$ such that*

$$B(x;\delta) := \{\, y \in \mathbb{R}^D : \|x - y\|_2 < \delta \,\} \subset \Omega. \tag{5.1}$$

*Intuitively, this means that $\Omega$ contains a small neighborhood around each of its points. Open sets serve as the coordinate domains where local parametrizations of a manifold are defined.*

With this topological concept, we can now define what it needs to locally describe a manifold inside a higher-dimensional Euclidean space.

**Definition 5.2** (Coordinate patch or chart). *Let $\Omega \subset \mathbb{R}^d$ be an open set. A $C^k$* coordinate patch *(or* chart*) is an injective $C^k$ map*

$$X : \Omega \to \mathbb{R}^D$$

*whose differential has full rank d at every point, i.e.*

$$\operatorname{rank}\big(\nabla X(u)\big) = \operatorname{rank}\begin{pmatrix} \frac{\partial X_1}{\partial u_1} & \cdots & \frac{\partial X_1}{\partial u_d} \\ \vdots & \ddots & \vdots \\ \frac{\partial X_D}{\partial u_1} & \cdots & \frac{\partial X_D}{\partial u_d} \end{pmatrix} = d \quad \text{for all } u \in \Omega. \tag{5.2}$$

If the image $X(\Omega)$ belongs to some manifold embedded in $\mathbb{R}^D$, then $X(\Omega)$ represents a local *patch* of the manifold. In practice, $u = (u_1, \ldots, u_d)$ provides local coordinates for points $x = X(u)$ on the manifold. Coordinate patches allow us to represent smooth surfaces locally in terms of parameters. The following examples illustrate how common surfaces can be described by simple coordinate maps.

**Example 5.1** (Monge patch). *Let $\Omega \subset \mathbb{R}^2$ and let $f : \Omega \to \mathbb{R}$ be a $C^k$ function. The map*

$$X(u_1, u_2) = \big(u_1,\, u_2,\, f(u_1, u_2)\big)$$

*defines a local surface in* $\mathbb{R}^3$*. Its Jacobian matrix is*

$$\nabla X = h \begin{pmatrix} 1 & 0 \\ 0 & 1 \\ f_{u_1} & f_{u_2} \end{pmatrix}$$

*whose columns span the tangent plane at each point of the surface. A specific example is the upper half of the unit sphere*

$$\mathbb{S}^2_+ = \{\, (x_1, x_2, x_3) \in \mathbb{R}^3 : x_1^2 + x_2^2 + x_3^2 = 1,\ x_3 > 0 \,\}$$

*which can be represented over the open unit disk* $\Omega = B(0,1)$ *by*

$$X(u_1, u_2) = \big(u_1,\ u_2,\ \sqrt{1-(u_1^2+u_2^2)}\big).$$

*This parameterization covers only the upper hemisphere but not the whole unit sphere, showing that a single chart may not be enough to describe an entire manifold.*

Not all surfaces can be represented globally by a single function as in the Monge patch. For this reason, more general coordinate maps, such as stereographic projections, are often used to construct atlases that cover the entire manifold.

**Example 5.2** (Stereographic projection of $\mathbb{S}^2$). *Consider the unit sphere*

$$\mathbb{S}^2 = \{\, (x_1, x_2, x_3) \in \mathbb{R}^3 : x_1^2 + x_2^2 + x_3^2 = 1 \,\}.$$

*Let* $x^S = (0, 0, -1)$ *be the south pole. For each* $x \in \mathbb{S}^2 \setminus \{x^S\}$*, the line through* $x$ *and* $x^S$ *intersects the plane* $x_3 = 0$ *at a unique point* $(u_1, u_2)$*. This defines a smooth map:*

$$X(u_1, u_2) = \frac{1}{\|u\|_2^2 + 1}\big(2u_1,\ 2u_2,\ 1 - \|u\|_2^2\big), \qquad \|u\|_2^2 = u_1^2 + u_2^2. \tag{5.3}$$

*Near* $(u_1, u_2) = (0, 0)$*, this coincides locally with the Monge patch. Two stereographic projections—one from the south pole and one from the north pole—suffice to cover* $\mathbb{S}^2$ *smoothly.*

Another common class of parameterizations arises when a curve is revolved around an axis to form a surface.

**Example 5.3** (Surface of revolution). *Let*

$$\Omega = \{\, (u_1, u_2) \in \mathbb{R}^2 : a < u_1 < b,\ 0 < u_2 < 2\pi \,\}.$$

*Given smooth functions* $f : (a, b) \to \mathbb{R}_{\geq 0}$ *and* $g : (a, b) \to \mathbb{R}$*, define*

$$X(u_1, u_2) = \big(f(u_1)\cos u_2,\ f(u_1)\sin u_2,\ g(u_1)\big).$$

*This parameterization generates a surface of revolution about the* $z$*-axis. Classical examples include the sphere (*$f(u_1) = \sin u_1$, $g(u_1) = \cos u_1$*) and the torus (*$f(u_1) = R + r\cos u_1$, $g(u_1) = r\sin u_1$*).*

When multiple charts are needed to cover a manifold, we must ensure they fit together smoothly. This is achieved through transition maps between overlapping coordinate domains.

**Definition 5.3** (Coordinate transformation). *Suppose $\varphi_1 : U_1 \to \mathcal{M}$ and $\varphi_2 : U_2 \to \mathcal{M}$ are two coordinate charts. The* coordinate transformation *(or* transition map*) between them is*

$$h = \varphi_2^{-1} \circ \varphi_1 : \varphi_1^{-1}(\varphi_1(U_1) \cap \varphi_2(U_2)) \to \varphi_2^{-1}(\varphi_1(U_1) \cap \varphi_2(U_2)).$$

*We require that $h$ is a $C^k$ diffeomorphism, i.e., both $h$ and $h^{-1}$ are smooth and $\nabla h \cdot \nabla(h^{-1}) = I$. This compatibility condition ensures that the manifold can be traversed smoothly from one coordinate patch to another.*

By piecing together such compatible coordinate patches, we obtain the general concept of a smooth manifold.

**Definition 5.4** (Smooth manifold). *A subset $\mathcal{M} \subset \mathbb{R}^D$ is called a $C^k$ $d$-dimensional* smooth manifold *(or a* surface *when $d = 2$) if for every point $p \in \mathcal{M}$, there exists an open set $U \subset \mathbb{R}^d$ and a $C^k$ coordinate chart $X : U \to \mathcal{M}$ such that $p \in X(U)$. Moreover, if $X : U \to \mathcal{M}$ and $Y : V \to \mathcal{M}$ are two charts with overlapping images $X(U) \cap Y(V) \neq \varnothing$, then the transition map*

$$Y^{-1} \circ X : X^{-1}(X(U) \cap Y(V)) \to Y^{-1}(X(U) \cap Y(V))$$

*is a $C^k$ diffeomorphism. The collection of all such compatible charts forms an* atlas*, which provides a complete coordinate description of $\mathcal{M}$.*

Beyond classical surfaces, many important spaces in applied mathematics are manifolds of matrices.

**Example 5.4** (Matrix manifolds). *Common examples include:*

*(1)* $\mathrm{Mat}(n, m) = \{A \in \mathbb{R}^{n\times m}\}$, *the Euclidean space of matrices;*
*(2)* $\mathrm{Sym}(n) = \{A \in \mathbb{R}^{n\times n} : A = A^\top\}$, *the space of symmetric matrices;*
*(3)* $\mathrm{GL}(n) = \{A \in \mathbb{R}^{n\times n} : \det A \neq 0\}$, *the general linear group;*
*(4)* $\mathrm{O}(n) = \{A \in \mathbb{R}^{n\times n} : A^\top A = I\}$, *the orthogonal group.*

*Some matrix manifolds, such as $\mathrm{GL}(n)$ and $\mathrm{O}(n)$, carry a natural group structure, forming* Lie groups*. These structures play an important role in geometric optimization and algorithm design on manifolds.*

Finally, not every subset of $\mathbb{R}^D$ is a manifold. The following examples illustrate situations where local Euclidean structure fails.

**Example 5.5** (Sets that are not manifolds).

$$S_1 = \{(x, y, z) \in \mathbb{R}^3 : x = 0, y = 0\} \cup \{(x, y, z) \in \mathbb{R}^3 : z = 0\}, \tag{5.4}$$

$$S_2 = \{(x, y) \in \mathbb{R}^2 : x \leq 0, y = -1\} \cup \{(x, y) \in \mathbb{R}^2 : x^2 + y^2 = 1\}. \tag{5.5}$$

*Both sets fail to be locally Euclidean at their intersection points, thus they are not manifolds.*

**Remark 5.1** (Coordinate invariance). *Although computations are often performed in local coordinates, geometric quantities on a manifold should be* coordinate-invariant, *which means that a property defined in one valid chart remains valid under all smooth coordinate transformations. Coordinate descriptions are a tool for analysis and computation, not an intrinsic part of the geometry itself.*

Once a manifold has been defined through local coordinate charts, the next step is to describe functions defined on it. Just as in Euclidean space, we are often interested in how quantities vary smoothly across the surface, such as temperature on a curved interface, probability density on a sphere, and potential energy on a configuration manifold. To formalize this idea, we must generalize the notion of smoothness from $\mathbb{R}^D$ to manifolds.

**Definition 5.5** (Smooth functions on a manifold). *Let $\mathcal{M}$ be a $C^k$ smooth manifold. A function $f : \mathcal{M} \to \mathbb{R}$ is said to be $C^k$* smooth at a point $p \in \mathcal{M}$ *if there exists an open set $U \subset \mathbb{R}^d$ and a coordinate chart $\varphi : U \to \mathcal{M}$ with $p \in \varphi(U)$ such that the composition*

$$f \circ \varphi : U \to \mathbb{R}$$

*is $C^k$ smooth in the usual Euclidean sense. If $f$ is smooth at every point of $\mathcal{M}$, we simply call $f$ a $C^k$* smooth function on $\mathcal{M}$.

This definition reflects the local nature of manifolds: smoothness is checked in each coordinate chart, just as differentiability in $\mathbb{R}^D$ is verified in small neighborhoods.

It is important to emphasize that smoothness, like other geometric notions, is independent of the chosen chart. Namely, smoothness is *coordinate-invariant.* If $\varphi$ and $\psi$ are two overlapping charts on $\mathcal{M}$, then for any function $f : \mathcal{M} \to \mathbb{R}$ we have

$$f \circ \psi = (f \circ \varphi) \circ (\varphi^{-1} \circ \psi).$$

Since the transition map $\varphi^{-1} \circ \psi$ is $C^k$, the smoothness of $f \circ \varphi$ implies the smoothness of $f \circ \psi$. Therefore, the notion of a smooth function on a manifold is globally consistent and does not depend on any particular coordinate system.

**Example 5.6** (Smooth functions on simple manifolds). *The Euclidean space $\mathbb{R}^D$ itself is a smooth manifold with a single global chart, and the definition of smoothness on this simple manifold coincides with the standard notion of a $C^k$ function. On the unit sphere $\mathbb{S}^2$, a scalar field $f : \mathbb{S}^2 \to \mathbb{R}$ is smooth if its coordinate representation $f \circ \varphi$ is smooth for each stereographic projection chart $\varphi$. For instance, the function $f(x) = x_3$ (the height above the equatorial plane) is globally smooth, whereas $f(x) = |x_3|$ is not smooth along the equator.*

Beyond scalar-valued functions, many geometric and physical problems involve mappings between manifolds, such as embeddings, coordinate transformations, or constraint manifolds in optimization problems. We therefore extend the notion of smoothness to a map between manifolds.

**Definition 5.6** (Smooth map between manifolds). *Let $\mathcal{M}$ and $\mathcal{N}$ be $C^k$ smooth manifolds. A map $F : \mathcal{M} \to \mathcal{N}$ is $C^k$* smooth at a point $p \in \mathcal{M}$ *if there exist coordinate charts $\varphi : U \to \mathcal{M}$ and $\psi : V \to \mathcal{N}$ with $p \in \varphi(U)$ and $F(p) \in \psi(V)$ such that the local expression*

$$\psi^{-1} \circ F \circ \varphi : U \to V$$

*is $C^k$ smooth in the Euclidean sense. If $F$ is smooth at all points of $\mathcal{M}$, we call it a $C^k$* map between manifolds.

This definition ensures that the smoothness of a mapping is determined entirely by its local coordinate representation. The smoothness of the manifold itself is encoded in its coordinate transition maps since a manifold with $C^k$-smooth transition functions is called a $C^k$ manifold. Hence, the smoothness of functions and maps cannot exceed that of the underlying coordinate structure. Geometrically, it allows us to treat maps between curved spaces using standard multivariable calculus in each chart.

**Example 5.7** (Smooth and nonsmooth embeddings). *Consider the standard embedding of the unit circle in the plane,*

$$X : S^1 \hookrightarrow \mathbb{R}^2, \qquad \theta \mapsto (\cos\theta, \sin\theta),$$

*which is $C^\infty$ smooth. Now define $F : S^1 \to \mathbb{R}$ by $F(\cos\theta, \sin\theta) = |\theta|$ for $\theta \in (-\pi, \pi]$. Although $F$ is continuous, its coordinate representation $|\theta|$ is not differentiable at $\theta = 0$, so $F$ fails to be smooth. This example highlights how local coordinate expressions determine smoothness globally.*

## 5.2 Metrics of a manifold

Having introduced smooth manifolds and smooth functions, we now develop the geometric framework needed to measure directions, lengths, and angles on a manifold. These notions are fundamental for defining differential operators such as the gradient, divergence, and Laplace–Beltrami operator, and they also describe how curvature and intrinsic geometry influence PDEs on manifolds.

**Tangent space.** At each point of a smooth manifold, we can approximate the surface locally by a linear space that "touches" it at that point—this is the *tangent space.* It serves as the local Euclidean model of the manifold, providing the setting where differentiation and directional motion are defined.

**Definition 5.7** (Tangent space). *Let $\mathcal{M}$ be a d-dimensional $C^1$ manifold embedded in $\mathbb{R}^D$, and let $p \in \mathcal{M}$ have a local chart $X : U \subset \mathbb{R}^d \to \mathcal{M} \subset \mathbb{R}^D$. The* tangent space *of $\mathcal{M}$ at $p$ is the d-dimensional vector space*

$$T_p\mathcal{M} := \mathrm{Span}\left\{\frac{\partial X}{\partial u_1}, \frac{\partial X}{\partial u_2}, \ldots, \frac{\partial X}{\partial u_d}\right\}. \tag{5.6}$$

*The vectors $\frac{\partial X}{\partial u_i}$ are called* coordinate tangent vectors.

**Remark 5.2.** *For the tangent space to be well-defined, $\mathcal{M}$ must be at least $C^1$ near $p$ so that these derivatives exist. The set $\{\frac{\partial X}{\partial u_i}\}_{i=1}^d$ forms a basis of $T_p\mathcal{M}$, called the* coordinate basis.

Although the tangent space is defined using a particular chart, it turns out to be intrinsic to the manifold itself. The *chart independence* of the tangent space can be stated as follows. If $X$ and $Y$ are two charts at $p \in \mathcal{M}$, then

$$\mathrm{Span}\left\{\frac{\partial X}{\partial u_1}, \frac{\partial X}{\partial u_2}, \ldots, \frac{\partial X}{\partial u_d}\right\} = \mathrm{Span}\left\{\frac{\partial Y}{\partial u_1}, \frac{\partial Y}{\partial u_2}, \ldots, \frac{\partial Y}{\partial u_d}\right\}.$$

Indeed, if $X = Y \circ \phi$ for a coordinate transformation $\phi$, then $\nabla X = (\nabla\phi)(\nabla Y)$ by the chain rule. Since $\nabla\phi$ is invertible, the two spans coincide [Lee (2013)].

**Tangent bundle.** Once the tangent space is defined at each point, it is natural to gather all these spaces into a single geometric object, called the *tangent bundle.* This bundle collects all tangent vectors across the manifold and serves as the domain for vector fields and differential operators.

**Definition 5.8** (Tangent bundle). *The* tangent bundle *of a smooth manifold* $\mathcal{M}$ *is the disjoint union*

$$T\mathcal{M} := \cup_{p\in\mathcal{M}} T_p\mathcal{M} = \{(p, v) : p \in \mathcal{M},\, v \in T_p\mathcal{M}\}.$$

The tangent bundle $T\mathcal{M}$ is also $2d$-dimensional smooth manifold, and the projection map $\pi : T\mathcal{M} \to \mathcal{M}$, given by $\pi(p, v) = p$, is smooth. Intuitively, $T\mathcal{M}$ represents all possible directions of motion on $\mathcal{M}$. A *vector field* is a mapping from a manifold $\mathcal{M}$ to its tangent bundle $T\mathcal{M}$.

**Riemannian metric.** The tangent space provides directions, but not yet a way to measure them. To define geometric quantities such as lengths, angles, and areas, we equip each tangent space with an inner product that varies smoothly from point to point.

**Definition 5.9** (Riemannian metric and Riemannian manifold). *A* Riemannian metric *(or* metric tensor*) on* $\mathcal{M}$ *assigns to each point* $p \in \mathcal{M}$ *an inner product*

$$g_p : T_p\mathcal{M} \times T_p\mathcal{M} \to \mathbb{R}, \qquad (u, v) \mapsto \langle u, v\rangle_{g_p},$$

*that depends smoothly on* $p$. *In local coordinates, the metric is represented by a symmetric positive-definite matrix* $g(p) = (g_{ij}(p))_{i,j=1}^d$, *where*

$$g_{ij}(p) = \langle \partial_{u_i} X(p), \partial_{u_j} X(p)\rangle_{g_p}.$$

*With a Riemannian metric* $g$, *the manifold* $\mathcal{M}$ *is then called a* Riemannian manifold, *represented by a pair* $(\mathcal{M}, g)$.

Since inner products are not unique, for the same manifold we can have many different *Riemannian metrics*, one of which is the *induced metric.*

**Induced metric.** When a manifold is embedded in Euclidean space, it naturally inherits the metric structure of its ambient space. This provides a convenient and geometrically meaningful way to measure distances along the surface.

**Definition 5.10** (Induced metric). *If a manifold* $\mathcal{M}$ *is a subset of* $\mathbb{R}^D$, *the* induced metric *is simply the inner product in* $\mathbb{R}^D$*:*

$$g_p(u, v) := \langle u, v\rangle_{\mathbb{R}^D}, \qquad u, v \in T_p\mathcal{M}.$$

*In local coordinates* $u^1, \ldots, u^d$, *if* $X : U \to \mathcal{M}$ *is a local chart, then*

$$g_{ij}(p) = \left\langle \frac{\partial X}{\partial u^i}(p), \frac{\partial X}{\partial u^j}(p) \right\rangle_{\mathbb{R}^D}.$$

The induced metric measures distances *along* the manifold rather than through the ambient space. It captures the intrinsic geometry that governs geodesics, curvature, and PDE behavior on $\mathcal{M}$.

**Metric matrix.** For surfaces in $\mathbb{R}^3$, the components of the metric can be assembled into a $2 \times 2$ symmetric matrix that encodes local geometric information.

**Definition 5.11** (Matrix representation of the metric). *For a parametrized surface $X(u^1, u^2)$ in $\mathbb{R}^3$, the metric matrix at $p = X(u^1, u^2)$ is*

$$g_p = \begin{pmatrix} \langle X_{u^1}, X_{u^1} \rangle_{\mathbb{R}^3} & \langle X_{u^1}, X_{u^2} \rangle_{\mathbb{R}^3} \\ \langle X_{u^2}, X_{u^1} \rangle_{\mathbb{R}^3} & \langle X_{u^2}, X_{u^2} \rangle_{\mathbb{R}^3} \end{pmatrix}.$$

*If $u = \alpha_1 X_{u^1} + \alpha_2 X_{u^2}$ and $v = \beta_1 X_{u^1} + \beta_2 X_{u^2}$ are tangent vectors at $p$, then*

$$\langle u, v \rangle_{g_p} = (\alpha_1, \alpha_2)\, g_p \begin{pmatrix} \beta_1 \\ \beta_2 \end{pmatrix}.$$

**Example 5.8.** *Consider the Monge patch of the upper hemisphere*

$$X(u^1, u^2) = \left(u^1,\, u^2,\, \sqrt{1 - (u^1)^2 - (u^2)^2}\right).$$

*Then*

$$X_{u^1} = \left(1,\, 0,\, \frac{-u^1}{\sqrt{1 - (u^1)^2 - (u^2)^2}}\right),$$
$$X_{u^2} = \left(0,\, 1,\, \frac{-u^2}{\sqrt{1 - (u^1)^2 - (u^2)^2}}\right),$$

*and the components of the induced metric are*

$$g_{11} = 1 + \frac{(u^1)^2}{1 - (u^1)^2 - (u^2)^2},$$
$$g_{12} = \frac{u^1 u^2}{1 - (u^1)^2 - (u^2)^2},$$
$$g_{22} = 1 + \frac{(u^2)^2}{1 - (u^1)^2 - (u^2)^2}.$$

*The matrix $g \in \mathbb{R}^{2\times 2}$ encodes the intrinsic geometry of the hemisphere inherited from $\mathbb{R}^3$, capturing how lengths and angles are distorted relative to the flat plane.*

Like all geometric quantities, the metric tensor is intrinsic to the manifold and independent of the coordinate system used to express it, i.e., the metric is *coordinate invariant.* For example, under a coordinate change $(u^1, u^2) \mapsto (\tilde{u}^1, \tilde{u}^2)$ with Jacobian matrix $J$, the metric transforms as

$$g_p = J^\top \, g_p|_{\tilde{x}} \, J,$$

which is the standard $(0, 2)$-tensor transformation law. This confirms that the Riemannian metric is coordinate-invariant: its geometric meaning does not depend on the chosen parameterization.

## 5.3 Integration on manifolds

With the notion of a Riemannian metric, we can now introduce integration on manifolds. In Euclidean space, integrals measure length, area, and volume. On manifolds, the same concepts are generalized using the metric tensor, which provides a local way to measure infinitesimal distances and areas. These constructions form the foundation for defining variational energies and weak formulations of PDEs on curved domains.

**Curves on manifolds.** We begin with one-dimensional integration along curves. Let $\mathcal{M}$ be a smooth surface parametrized by $X : U \subset \mathbb{R}^2 \to \mathcal{M} \subset \mathbb{R}^3$ with local coordinates $(u^1, u^2)$.

**Definition 5.12** (Smooth curve on $\mathcal{M}$). *A smooth curve $\alpha$ on $\mathcal{M}$ is a smooth map*

$$\alpha : [0, 1] \longrightarrow U \longrightarrow \mathcal{M}, \tag{5.7}$$

$$t \longmapsto \left(u^1(t), u^2(t)\right) \longmapsto X\left(u^1(t), u^2(t)\right), \tag{5.8}$$

*where $u^1(t)$ and $u^2(t)$ are smooth functions of $t$. This expresses a curve on the surface in terms of its parameter-space coordinates.*

The Riemannian metric induces a natural notion of length, generalizing the Euclidean arc-length formula. Given a Riemannian metric $g$, the *length* of a smooth curve $\alpha$ is

$$L(\alpha) = \int_\alpha \mathrm{d}s := \int_0^1 \left\| \frac{\mathrm{d}\alpha}{\mathrm{d}t} \right\|_g \mathrm{d}t = \int_0^1 \sqrt{\left\langle \frac{\mathrm{d}\alpha}{\mathrm{d}t}, \frac{\mathrm{d}\alpha}{\mathrm{d}t} \right\rangle_g} \, \mathrm{d}t \tag{5.9}$$

$$= \int_0^1 \sqrt{\left(\dot{\alpha}^1(t) \; \dot{\alpha}^2(t)\right) \begin{pmatrix} g_{11} & g_{12} \\ g_{21} & g_{22} \end{pmatrix} \begin{pmatrix} \dot{\alpha}^1(t) \\ \dot{\alpha}^2(t) \end{pmatrix}} \, \mathrm{d}t, \tag{5.10}$$

where $s$ denotes the arc-length parameter, satisfying $\mathrm{d}s = \|\dot{\alpha}\|_g \mathrm{d}t$. This definition coincides with the Euclidean notion when $g_{ij} = \delta_{ij}$ and extends it to curved surfaces [do Carmo (1976); Lee (2013)].

**Area and volume forms.** Next we consider integration over surfaces. The metric again provides the necessary local scaling factor that converts coordinate differentials into true geometric area.

**Definition 5.13** (Area or volume form). *For a surface $\mathcal{M}$ embedded in $\mathbb{R}^3$, the infinitesimal area element is given by*

$$\|X_{u^1} \times X_{u^2}\| \, \mathrm{d}u^1 \, \mathrm{d}u^2 = \sqrt{\det(g_{ij})} \, \mathrm{d}u^1 \, \mathrm{d}u^2. \tag{5.11}$$

*If $f : \mathcal{M} \to \mathbb{R}$ is continuous, the integral of $f$ over the parametrized patch $X(U)$ is*

$$\int_{X(U)} f(p) \, \mathrm{dvol}_g = \int_U f \circ X(u^1, u^2) \sqrt{\det(g_{ij})} \, \mathrm{d}u^1 \, \mathrm{d}u^2. \tag{5.12}$$

*The factor $\sqrt{\det(g_{ij})}$ serves as the* Jacobian determinant *accounting for how the parameter domain is stretched or compressed under the mapping $X$. For a d-dimensional manifold, this expression generalizes naturally, with $\sqrt{\det(g_{ij})}$ representing the local volume-scaling factor.*

**Global integration via partition of unity.** Integrals defined on a single coordinate patch can be extended globally across the entire manifold using a partition of unity, defined via an open cover on a manifold which is simply a collection of open sets on the manifold. This construction ensures that integration is independent of the particular atlas chosen.

**Definition 5.14** (Partition of unity). *A* partition of unity *subordinate to an open cover $\{U_\alpha\}$ of $\mathcal{M}$ is a collection of smooth functions $\{\rho_\alpha\}_\alpha$ such that:*

- $\rho_\alpha \in C^\infty(\mathcal{M})$ *for all $\alpha$,*
- $\mathrm{supp}(\rho_\alpha) \subset U_\alpha$, *for each open set $U_\alpha$,*
- *For each $p \in \mathcal{M}$, only finitely many $\rho_\alpha(p)$ are nonzero,*
- $\sum_\alpha \rho_\alpha(p) = 1$ *for all $p \in \mathcal{M}$.*

Partitions of unity provide the technical machinery that allows globally defined integrals and differential forms on $\mathcal{M}$. Let $f \in C(\mathcal{M})$ be a continuous function. By multiplying $f$ with a partition of unity, the integral of $f$

can be decomposed into a finite sum of patchwise integrals:

$$\begin{aligned}\int_{\mathcal{M}} f(p)\,\mathrm{dvol}_g &= \int_{\mathcal{M}} f(p)\left(\sum_{\alpha}\rho_\alpha(p)\right)\mathrm{dvol}_g\\ &= \sum_{\alpha}\int_{\mathcal{M}} f(p)\rho_\alpha(p)\,\mathrm{dvol}_g\\ &= \sum_{\alpha}\int_{U_\alpha} f(p)\rho_\alpha(p)\,\mathrm{dvol}_g.\end{aligned}$$

Since each $\rho_\alpha$ has support in $U_\alpha$, each term can be computed locally in coordinates and summed to yield a global, coordinate-invariant definition of the integral over $\mathcal{M}$.

## 5.4 Differential operators on manifolds

In Section 4.2, we have seen that the gradient operator $\nabla$ and negative divergence operator $-\mathrm{div}$ are *adjoint* to each other in $\mathbb{R}^D$ with the standard inner product. This adjoint relation, together with the associated integration by parts identity, will be the guiding principles for defining their manifold analogues.

Let $\Omega \subset \mathbb{R}^D$ be a smooth bounded domain, $f \in C^\infty(\overline{\Omega})$, and $v \in C^\infty(\overline{\Omega};\mathbb{R}^D)$. Then integration by parts gives

$$\int_{\Omega}(\mathrm{div}\, v)\, f\,\mathrm{d}x = -\int_{\Omega}\nabla f\cdot v\,\mathrm{d}x + \int_{\partial\Omega} f\, v\cdot n\,\mathrm{d}S, \tag{5.13}$$

where $n$ is the outward unit normal on $\partial\Omega$. Thus, with respect to the Euclidean $L^2$ inner product, $-\nabla$ is the formal adjoint operator of div. When $\partial\Omega = \varnothing$ or suitable boundary conditions hold, the boundary term vanishes.

On a Riemannian manifold $(\mathcal{M}, g)$, the gradient operator $\nabla_{\mathcal{M}}$ and the divergence operator $\mathrm{div}_{\mathcal{M}}$ are defined so that an analogue of (5.13) still holds with the metric inner product and the Riemannian volume form. This guarantees the right variational structure for PDEs and weak formulations on $\mathcal{M}$.

**Differentials on manifolds.** A smooth function $f$ on $\mathcal{M}$ admits a differential *df* that acts linearly on tangent vectors. In a local chart $(u^1, \ldots, u^d)$ with parametrization $X : U \to \mathcal{M}$,

$$\mathrm{d}f = \sum_{i=1}^{d}\frac{\partial f}{\partial u^i}\,\mathrm{d}u^i,$$

and this expression is invariant under smooth coordinate changes by the chain rule (cf. [Lee (2013)]). In local coordinates $(u^i)$, we write $\partial_{u^i} = X_{u^i}, i = 1, \cdots, d$. The $\mathrm{d}u^i$ are examples of 1-forms, and they are linear functionals on $T_p\mathcal{M}$ satisfying $\mathrm{d}u^i(\partial_{u^j}) = \delta_{ij} = \begin{cases} 1 & i = j, \\ 0 & i \neq j, \end{cases}$.

**Theorem 5.1** (Coordinate invariance of $df$). *Let $f : \mathcal{M} \to \mathbb{R}$ be smooth, and let $X(u^1, \cdots, u^d)$ be a local chart. Then*

$$\mathrm{d}f = \sum_{i=1}^{d} \frac{\partial f}{\partial u^i} \, \mathrm{d}u^i$$

*is invariant under smooth coordinate transformations.*

*Proof.* Let $\phi : (u^1, \cdots, u^d) \mapsto (v^1, \cdots, v^d)$ be a change of coordinates with Jacobian $J\phi = \left(\frac{\partial v^i}{\partial u^j}\right)$. Applying the chain rule,

$$\frac{\partial f}{\partial u^k} = \sum_{i=1}^{d} \frac{\partial f}{\partial v^i} \frac{\partial v^i}{\partial u^k}.$$

Similarly, $\mathrm{d}v^i = \sum_{j=1}^{d} \frac{\partial v^i}{\partial u^j} \mathrm{d}u^j$. Combining these,

$$\begin{aligned} \mathrm{d}f &= \sum_{i=1}^{d} \frac{\partial f}{\partial v^i} \, dv^i = \sum_{i=1}^{d} \frac{\partial f}{\partial v^i} \left( \sum_{j=1}^{d} \frac{\partial v^i}{\partial u^j} \mathrm{d}u^j \right) \\ &= \sum_{j=1}^{d} \left( \sum_{i=1}^{d} \frac{\partial f}{\partial v^i} \frac{\partial v^i}{\partial u^j} \right) \mathrm{d}u^j = \sum_{j=1}^{d} \frac{\partial f}{\partial u^j} \, \mathrm{d}u^j. \end{aligned}$$

Thus $df$ has the same expression in both coordinate systems. □

**Gradient (coordinate-free definition).** The Riemannian metric $g$ provides an inner product $\langle \cdot, \cdot \rangle_g$ on each $T_p\mathcal{M}$, allowing us to identify differentials with tangent vectors. Recall that a vector field is a mapping from $\mathcal{M}$ to its tangent bundle $T\mathcal{M}$.

**Definition 5.15** (Gradient on a manifold). *For $f \in C^\infty(\mathcal{M})$, the* gradient *$\nabla_{\mathcal{M}} f$ is the unique vector field on $\mathcal{M}$ such that*

$$\mathrm{d}f(v) = \langle \nabla_{\mathcal{M}} f, \, v \rangle_g \qquad \textit{for any smooth vector field } v.$$

In local coordinates $(u^i)$, we write $\partial_{u^i} = X_{u^i}, i = 1, \cdots, d$, the metric matrix $g_{ij} = \langle \partial_{u^i}, \partial_{u^j} \rangle_g$ and its inverse matrix entries $g^{ij}$. With some unknown coefficients $a_i$, we can write $\nabla_{\mathcal{M}} f$ as

$$\nabla_{\mathcal{M}} f = \sum_{i=1}^{d} a^i \partial_{u^i}.$$

By considering $\mathrm{d}f$ on $\mathcal{M}$, one can derive $\nabla_{\mathcal{M}} f$. For $i = 1, \cdots, d$, we consider

$$\frac{\partial f}{\partial u^i} = \mathrm{d}f(\partial_{u^i}) = \left\langle \sum_{j=1}^{d} a^j \partial_{u^j}, \partial_{u^i} \right\rangle_g.$$

This implies:

$$\begin{pmatrix} \frac{\partial f}{\partial u^1} \\ \vdots \\ \frac{\partial f}{\partial u^d} \end{pmatrix} = \begin{pmatrix} g_{11} & \cdots & g_{1d} \\ \vdots & \ddots & \vdots \\ g_{d1} & \cdots & g_{dd} \end{pmatrix} \begin{pmatrix} a^1 \\ \vdots \\ a^d \end{pmatrix} \Longrightarrow \begin{pmatrix} a^1 \\ \vdots \\ a^d \end{pmatrix} = \begin{pmatrix} g^{11} & \cdots & g^{1d} \\ \vdots & \ddots & \vdots \\ g^{d1} & \cdots & g^{dd} \end{pmatrix} \begin{pmatrix} \frac{\partial f}{\partial u^1} \\ \vdots \\ \frac{\partial f}{\partial u^d} \end{pmatrix},$$

where $g^{ij}$ denote the entries of the inverse matrix to metric matrix $g_{ij}$.

Therefore, we get

$$\nabla_{\mathcal{M}} f = \sum_{i,j=1}^{d} g^{ij} \frac{\partial f}{\partial u^j} \frac{\partial}{\partial u^i}. \tag{5.14}$$

**Divergence (adjoint of $-\nabla_{\mathcal{M}}$).** In order to derive the correct divergence on a Riemannian manifold $(\mathcal{M}, g)$, which is the adjoint operator of the negative gradient operator, we can integrate $\nabla_{\mathcal{M}} f$ over the whole manifold $\mathcal{M}$. Recall that we can reduce such an integral into each integration over an open set $U \subset \mathcal{M}$, if using a partition of unity. Thus without loss of generality, we can just focus on the integration on $U$ with the assumption that the integrand vanishes on the boundary of $U$, since the partition of unity function has compact support on $U$.

To retain the Euclidean adjoint relation, we define divergence so that an integration by parts identity holds with the Riemannian volume form $\mathrm{dvol}_g$. In local coordinates $(u^i)$ on $U$, denoting $G := \det(g_{ij})$, then the Riemannian volume form can be written as $\mathrm{dvol}_g = \sqrt{G}\, \mathrm{d}u^1 \wedge \cdots \wedge \mathrm{d}u^d$, where $\wedge$ denotes the wedge product, see [Spivak (1965)]. Then given a smooth vector field $v = \sum_i v^i \partial_{u^i}$, we have

$$\begin{aligned}
&\int_U \langle \nabla_{\mathcal{M}} f, v \rangle_g \sqrt{G}\, \mathrm{d}u^1 \wedge \cdots \wedge \mathrm{d}u^d \\
&= \int_U \left\langle \sum_{i,j} g^{ij} \frac{\partial f}{\partial u^j} \frac{\partial}{\partial u^i}, \sum_k v^k \frac{\partial}{\partial u^k} \right\rangle_g \sqrt{G}\, \mathrm{d}u^1 \wedge \cdots \wedge \mathrm{d}u^d \\
&= \int_U \sum_{i,j,k} g^{ij} \frac{\partial f}{\partial u^j} v^k g_{ik} \sqrt{G}\, \mathrm{d}u^1 \wedge \cdots \wedge \mathrm{d}u^d \\
&= \int_U \sum_{j,k} \delta_{jk} \frac{\partial f}{\partial u^j} v^k \sqrt{G}\, \mathrm{d}u^1 \wedge \cdots \wedge \mathrm{d}u^d \\
&= -\int_U f \sum_k \frac{\partial \sqrt{G} v^k}{\partial u^k}\, \mathrm{d}u^1 \wedge \cdots \wedge \mathrm{d}u^d \\
&= -\int_U f \sum_k \frac{1}{\sqrt{G}} \frac{\partial \sqrt{G} v^k}{\partial u^k} \sqrt{G}\, \mathrm{d}u^1 \wedge \cdots \wedge \mathrm{d}u^d.
\end{aligned}$$

Thus we obtain the divergence operator for a vector field $v = \sum_i v^i \partial_{u^i}$ on a manifold as

$$\operatorname{div}_{\mathcal{M}} v = \frac{1}{\sqrt{G}} \sum_{i=1}^d \frac{\partial}{\partial u^i} (\sqrt{G}\, v^i). \tag{5.15}$$

The following adjoint relation is also true on manifolds. For a smooth vector field $v$ on $\mathcal{M}$, the *divergence* $\operatorname{div}_{\mathcal{M}} v$ is the unique function satisfying

$$\int_{\mathcal{M}} (\operatorname{div}_{\mathcal{M}} v)\, \phi \,\mathrm{dvol}_g = -\int_{\mathcal{M}} \langle \nabla_{\mathcal{M}} \phi, v \rangle_g \,\mathrm{dvol}_g + \int_{\partial \mathcal{M}} \phi\, v \cdot \eta \,\mathrm{d}S_g$$

for all $\phi \in C^\infty(\mathcal{M})$, where $\eta$ is the outward co-normal to the boundary $\partial\mathcal{M}$ and $dS_g$ the induced boundary measure. This is exactly the expression that we can derive by integrating by parts in coordinates, and it ensures the adjoint relation with respect to $\mathrm{dvol}_g = \sqrt{G}\, \mathrm{d}u^1 \cdots \mathrm{d}u^d$.

**Laplace–Beltrami operator.** With gradient and divergence in hand, the canonical second-order scalar operator is their composition.

**Definition 5.16** (Laplace–Beltrami operator)**.** *The Laplace–Beltrami operator on a Riemannian manifold* $(\mathcal{M}, g)$ *is defined as*

$$\Delta_{\mathcal{M}} f := \operatorname{div}_{\mathcal{M}} (\nabla_{\mathcal{M}} f), \quad \forall f \in C^\infty(\mathcal{M}).$$

In local coordinates, using (5.14) and (5.15), we may express it as

$$\Delta_{\mathcal{M}} f \;=\; \frac{1}{\sqrt{G}} \sum_{i,j=1}^{d} \frac{\partial}{\partial u^i}\big(\sqrt{G}\, g^{ij}\, \frac{\partial f}{\partial u^j}\big). \tag{5.16}$$

This is the coordinate form that is most useful for computations on parameterized patches.

**Example 5.9.** *On* $\mathbb{R}^D$ *with* $g_{ij} = \delta_{ij}$*, we have* $G = 1$ *and also* $g^{ij} = \delta_{ij}$*, so*

$$\nabla_{\mathbb{R}^D} f = \sum_i \frac{\partial f}{\partial u^i}\frac{\partial}{\partial u^i}, \qquad \mathrm{div}_{\mathbb{R}^D} X = \sum_i \frac{\partial X^i}{\partial u^i}, \qquad \Delta_{\mathbb{R}^D} f = \sum_i \frac{\partial^2 f}{\partial (u^i)^2},$$

*which is consistent with the classical Laplacian operator.*

For $f, h \in C^\infty(\mathcal{M})$, the adjoint relation implies the following manifold version of the Green's identities:

$$\int_{\mathcal{M}} \langle \nabla_{\mathcal{M}} f, \nabla_{\mathcal{M}} h\rangle_g \,\mathrm{dvol}_g = -\int_{\mathcal{M}} h\,\Delta_{\mathcal{M}} f \,\mathrm{dvol}_g \;+\; \int_{\partial\mathcal{M}} h\,\partial_\eta f \,\mathrm{d}S_g, \tag{5.17}$$

$$\int_{\mathcal{M}} (f\,\Delta_{\mathcal{M}} h - h\,\Delta_{\mathcal{M}} f)\,\mathrm{dvol}_g = \int_{\partial\mathcal{M}} (f\,\partial_\eta h - h\,\partial_\eta f)\,\mathrm{d}S_g, \tag{5.18}$$

where $\partial_\eta f := \langle \nabla_{\mathcal{M}} f, \eta\rangle_g$ is the co-normal derivative.

## 5.5 Energy on manifolds

Having established notions of differentiation and integration on manifolds, we now turn to *variational formulations*, which lie at the heart of many physical and geometric problems. In Euclidean space, equilibrium configurations often minimize an energy functional, and the corresponding Euler–Lagrange equation provides the governing PDE. On a manifold, the same principle applies, with the Riemannian metric playing a crucial role in defining energy, variation, and gradient flow. This section also provides an explanation for how the Laplacian operator emerges in the minimization of many physical energy functionals, e.g., Section 4.3.

**Dirichlet energy.** Among the simplest and most important functionals is the Dirichlet energy, which measures the total squared magnitude of the gradient of a function. Given a Riemannian manifold $(\mathcal{M}, g)$ and a smooth function $u : \mathcal{M} \to \mathbb{R}$, the *Dirichlet energy* of $u$ is defined as

$$E(u) = \frac{1}{2}\int_{\mathcal{M}} \|\nabla_{\mathcal{M}} u(p)\|_g^2 \,\mathrm{dvol}_g. \tag{5.19}$$

Geometrically, $E(u)$ quantifies the smoothness of $u$ on $\mathcal{M}$: functions with small Dirichlet energy vary slowly, while minimizing $E$ corresponds to finding the most "harmonic" configuration consistent with boundary data.

**First variation.** To obtain the Euler–Lagrange equation associated with an energy, we first define the notion of variation and compute how the energy changes under infinitesimal perturbations.

Let $E : C^\infty(\mathcal{M}) \to \mathbb{R}$ be a functional, and let $\eta \in C_0^\infty(\mathcal{M})$ be a smooth test function with compact support. The *first variation* of $E$ at $u$ in the direction $\eta$ is

$$\delta_\eta E(u) = \lim_{\varepsilon \to 0} \frac{E(u+\varepsilon\eta) - E(u)}{\varepsilon} = \frac{\mathrm{d}}{\mathrm{d}\varepsilon} E(u+\varepsilon\eta)|_{\varepsilon=0}. \tag{5.20}$$

This directional derivative measures the instantaneous rate of change of the energy under perturbation by $\eta$. It is also called the *Gâteaux derivative* of $E$ in the direction $\eta$, which should not be confused with the *functional derivative* below.

To express variations in a coordinate-free and integral form, it is convenient to represent the first variation as an $L^2$ pairing between a function and the perturbation. More precisely, given $E : C^\infty(\mathcal{M}) \to \mathbb{R}$, its *functional derivative* $\frac{\delta E}{\delta u} : \mathcal{M} \to \mathbb{R}$ is defined by the relation

$$\delta_\eta E(u) = \left\langle \frac{\delta E}{\delta u}, \eta \right\rangle_{L^2(\mathcal{M})} = \int_{\mathcal{M}} \frac{\delta E}{\delta u}\, \eta \,\mathrm{dvol}_g, \quad \forall \eta \in C_0^\infty(\mathcal{M}). \tag{5.21}$$

The function $\frac{\delta E}{\delta u}$ thus plays the role of a *gradient of the functional* in the $L^2$ sense, and vanishes at critical points of $E$.

We now compute the first variation of $E(u)$ defined above. Using the bilinearity of the inner product and differentiating under the integral sign, we have

$$\begin{aligned}\delta_\eta E(u) &= \frac{1}{2}\frac{\mathrm{d}}{\mathrm{d}\varepsilon} \int_{\mathcal{M}} \langle \nabla_{\mathcal{M}}(u+\varepsilon\eta), \nabla_{\mathcal{M}}(u+\varepsilon\eta)\rangle_g \,\mathrm{dvol}_g|_{\varepsilon=0} \\ &= \int_{\mathcal{M}} \langle \nabla_{\mathcal{M}} u, \nabla_{\mathcal{M}} \eta\rangle_g \,\mathrm{dvol}_g,\end{aligned}$$

in which the second-order term in $\varepsilon$ vanishes at $\varepsilon = 0$, leaving only the linear term in $\eta$.

To express this variation in a form involving $\eta$ alone, we apply the Green's identity (5.17):

$$\int_{\mathcal{M}} \langle \nabla_{\mathcal{M}} u, \nabla_{\mathcal{M}} \eta\rangle_g \,\mathrm{dvol}_g = -\int_{\mathcal{M}} (\Delta_{\mathcal{M}} u)\, \eta \,\mathrm{dvol}_g + \int_{\partial\mathcal{M}} \eta\, \partial_\eta u \,\mathrm{d}S_g. \tag{5.22}$$

If $\mathcal{M}$ has no boundary or if $\eta$ vanishes on $\partial\mathcal{M}$, the boundary term disappears, yielding

$$\delta_\eta E(u) = -\int_{\mathcal{M}} (\Delta_{\mathcal{M}} u)\, \eta \,\mathrm{dvol}. \tag{5.23}$$

**Euler–Lagrange equation.** Comparing this expression with the definition of the functional derivative, we obtain

$$\frac{\delta E}{\delta u} = -\,\Delta_{\mathcal{M}} u. \tag{5.24}$$

Therefore, the Euler–Lagrange equation corresponding to the Dirichlet energy is

$$\Delta_{\mathcal{M}} u = 0.$$

This generalizes the familiar Laplace equation from Euclidean space to Riemannian manifolds.

Functions satisfying $\Delta_{\mathcal{M}} u = 0$ are called *harmonic functions* on $\mathcal{M}$. They represent the stationary points (and in fact local minimizers) of the Dirichlet energy. Harmonic functions play a central role across mathematics and physics: they describe steady-state heat distributions, electrostatic potentials, minimal surfaces, and even serve as coordinate functions in geometric analysis [Chavel (2006)]. In the next chapters, we will see how similar energy principles lead to the formulation of diffusion equations, mean curvature flows, and other geometric PDEs on manifolds.

## 5.6 PDEs of interest

Having introduced differential operators and variational principles on manifolds, we now turn to several fundamental *partial differential equations (PDEs)* that naturally arise in this setting. These equations generalize their Euclidean counterparts by replacing the standard Laplacian $\Delta$ with the *Laplace–Beltrami operator* $\Delta_{\mathcal{M}}$, which encodes the intrinsic geometry through the Riemannian metric $g$. PDEs on manifolds are ubiquitous in geometry, physics, and data science, appearing in models of diffusion on curved surfaces, deformation of elastic membranes, and spectral representations of shape and data [Chavel (2006)].

**Poisson equation.** One of the most fundamental elliptic PDEs on manifolds is the Poisson equation, which represents equilibrium or steady-state phenomena.

**Definition 5.17** (Poisson equation). *Given a compact Riemannian manifold $(\mathcal{M}, g)$ with a smooth* forcing function $f : \mathcal{M} \to \mathbb{R}$ *and an unknown* solution function $u : \mathcal{M} \to \mathbb{R}$, *the* Poisson equation *with Neumann boundary condition is*

$$-\Delta_{\mathcal{M}} u = f,$$
$$\langle \nabla_{\mathcal{M}} u, \hat{n} \rangle|_{\partial\mathcal{M}} = 0,$$

*where $\hat{n}$ is the outward unit normal vector to the boundary $\partial\mathcal{M}$. The Neumann condition prescribes the normal derivative of $u$ along the boundary, enforcing zero flux across $\partial\mathcal{M}$. The forcing function must satisfy the compatibility condition $\int_{\mathcal{M}} f \, \mathrm{dvol}_g = 0$ to ensure the existence of solutions.*

The Poisson equation models a broad range of steady-state processes: electrostatic potential with charge density $f$, equilibrium displacement of a membrane, or stationary temperature distribution in heat conduction. The geometry of $\mathcal{M}$ influences the solution through the metric tensor in $\Delta_{\mathcal{M}}$, modifying both diffusion rate and propagation direction compared with the flat Euclidean case.

**Laplace–Beltrami eigenvalue problem.** The spectral structure of $\Delta_{\mathcal{M}}$ plays a central role in analysis and geometry. When the forcing term is proportional to $u$, the Poisson equation reduces to an eigenvalue problem.

If $f = \lambda u$ for some $\lambda \in \mathbb{R}$, then

$$-\Delta_{\mathcal{M}} u = \lambda u, \tag{5.25}$$

which defines the *Laplace–Beltrami eigenvalue problem.* In an eigenvalue problem, both the scalar $\lambda$ and the function $u$ are unknown. The eigenpairs $(\lambda_k, u_k)$ satisfy the following properties (for compact $\mathcal{M}$ without boundary):

- The eigenvalues form a discrete sequence

$$0 = \lambda_0 < \lambda_1 \leq \lambda_2 \leq \cdots \to \infty.$$

- The eigenfunctions $\{u_k\}$ form an orthonormal basis of $L^2(\mathcal{M})$.
- Each eigenvalue admits the *Rayleigh quotient* characterization:

$$\lambda_k = \min_{\substack{V \subset H^1(\mathcal{M}) \\ \dim V = k+1}} \max_{u \in V \setminus \{0\}} \frac{\int_{\mathcal{M}} \|\nabla_{\mathcal{M}} u\|_g^2 \, \mathrm{dvol}_g}{\int_{\mathcal{M}} u^2 \, \mathrm{dvol}_g}.$$

These eigenpairs encode intrinsic geometric information about the manifold. For instance, the *Weyl asymptotic law* relates the growth of eigenvalues to the total volume of $\mathcal{M}$, while the nodal structure of eigenfunctions reflects curvature and topology. Spectral decompositions of $\Delta_{\mathcal{M}}$ form the foundation for manifold signal processing, shape analysis, and diffusion geometry.

**Heat equation.** The time-dependent analogue of the Poisson equation is the heat equation, which models diffusion of scalar quantities over a curved domain.

**Definition 5.18** (Heat equation). *Given* $(\mathcal{M}, g)$, *a forcing term* $f : \mathcal{M} \to \mathbb{R}$, *and an initial condition* $g_0 : \mathcal{M} \to \mathbb{R}$, *the* heat equation *with Dirichlet boundary condition for* $u(\cdot, t) : \mathcal{M} \to \mathbb{R}$ *is*

$$\begin{aligned} \frac{\partial u}{\partial t} &= \Delta_{\mathcal{M}} u + f, \\ u(\cdot, 0) &= g_0, \\ u(\cdot, t)|_{\partial\mathcal{M}} &= 0, \quad \forall t \geq 0. \end{aligned}$$

The Dirichlet condition prescribes the value of $u$ along the boundary for all times $t$. When $f \equiv 0$, this PDE describes diffusion of heat (or any scalar field) along the manifold, where the local diffusion rate and anisotropy are determined by the metric tensor. Variationally, the heat equation is the *gradient flow* of the Dirichlet energy:

$$\frac{\partial u}{\partial t} = -\frac{\delta E}{\delta u} = \Delta_{\mathcal{M}} u.$$

Thus, the evolution of $u$ monotonically decreases the Dirichlet energy until a harmonic steady state is reached.

**Spectral expansion.** Both the Poisson and heat equations are intimately connected to the Laplace–Beltrami eigenproblem:

- The Poisson equation can be solved by expanding $f$ in the eigenbasis $\{u_k\}$ of $\Delta_{\mathcal{M}}$:

$$f = \sum_{k=0}^{\infty} \langle f, u_k \rangle \, u_k, \qquad u = \sum_{k=1}^{\infty} \frac{\langle f, u_k \rangle}{\lambda_k} \, u_k.$$

- The heat equation admits the spectral solution

$$u(p, t) = \sum_{k=0}^{\infty} e^{-\lambda_k t} \, \langle g_0, u_k \rangle \, u_k(p), \quad \forall p \in \mathcal{M},$$

  which expresses diffusion as exponential damping of eigenmodes.

- The associated *heat kernel*

$$K_t(p, q) = \sum_{k=0}^{\infty} e^{-\lambda_k t}\, u_k(p)\, u_k(q)$$

satisfies $\partial_t K_t = \Delta_{\mathcal{M}} K_t$ and encodes rich geometric information: its short-time asymptotics capture curvature invariants, while its long-time behavior reveals global topology and connectivity.

Chapter 6

# Laplacian on Triangular Mesh Discretization of Surfaces

Partial differential equations (PDEs) on surfaces arise naturally in a wide range of scientific and engineering disciplines. Many physical processes are inherently constrained to evolve along a surface rather than in a Euclidean space. Examples include the diffusion of heat or chemicals across a biological membrane, wave propagation along a shell or interface, fluid flow on the surface of a droplet, and geometric evolution of shapes in computer graphics. These problems are mathematically formulated as PDEs defined not in Euclidean domains, but on two-dimensional manifolds embedded in three-dimensional space.

In practice, such manifolds are often approximately represented by *triangular meshes*. A triangular mesh provides a piecewise linear approximation of a smooth surface using vertices, edges, and triangular faces. This representation is widely adopted due to its simplicity, flexibility, and compatibility with modern computational pipelines. From computer graphics and geometric modeling to finite element analysis and machine learning, triangular meshes serve as the de facto standard for digital geometry.

The study of PDEs on triangular meshes blends ideas from differential geometry, numerical analysis, and scientific computing. On the one hand, we should take into account the intrinsic geometry of the underlying surface, such as its metric, curvature, and topological structure. On the other hand, it is necessary to design efficient and robust discrete operators that act directly on the mesh, which is inherently irregular and may suffer from issues such as varying triangle size, non-uniform sampling, and the presence of boundaries.

A key challenge lies in discretizing differential operators such as the surface gradient, divergence, and the Laplace–Beltrami operator. In the continuum setting, these operators are defined using the Riemannian metric of the surface. In the discrete setting, careful construction is required to ensure properties like symmetry, positive semi-definiteness, and convergence to their smooth counterparts. The development of such operators forms the foundation for solving PDEs on triangular meshes.

Over the past two decades, significant progress has been made in this area. Approaches such as the finite element method (FEM), finite volume method (FVM), and spectral methods based on the Laplace–Beltrami eigenfunctions have enabled the efficient simulation of PDEs on meshes. These tools not only provide accurate approximations of classical PDEs, such as the Poisson, heat, and wave equations, but also open the door to novel applications in geometry processing, shape analysis, and physics-based simulation.

The goal of this chapter is to introduce the mathematical and computational framework for solving PDEs on triangular-mesh–represented surfaces. We will begin with the basics of surface representation and discrete operators, then move on to finite element discretization, numerical solvers, and practical applications. By the end of this chapter, the reader will be equipped with both the theoretical understanding and practical tools needed to model and compute PDEs on discrete surfaces.

### *Further reading*

The literature for PDEs on surfaces is vast. Early systematic treatments can be found in the works of Dziuk and Elliott [Dziuk and Elliott (2013)], who developed the finite element method for surface PDEs. In computer graphics, discrete differential geometry approaches such as the cotangent Laplacian [Pinkall and Polthier (1993); Desbrun *et al.* (1999)] have been widely adopted. See [Bonito *et al.* (2020)] and references therein for more on the finite element method. Comprehensive introductions to geometry processing and mesh-based computation are given in [Botsch *et al.* (2010); Wardetzky *et al.* (2007)]. These references provide both theoretical foundations and practical algorithms, and serve as excellent starting points for further study.

## 6.1 Surfaces and triangular mesh representation

In order to solve PDEs on surfaces, first we need a precise description of the underlying geometry. In the smooth setting, a surface is a two-dimensional manifold embedded in $\mathbb{R}^3$, equipped with a Riemannian metric that allows us to measure lengths, angles, areas, and to define differential operators. In many applications, however, surfaces are often approximated by *triangular meshes*, which provide a piecewise linear representation that is amenable to easy and efficient computation. This section describes the basic building blocks: triangles, barycentric coordinates, simplices, and their assembly into triangular meshes.

### 6.1.1 *Barycentric coordinates in a triangle*

We first define a triangle in $\mathbb{R}^n$.

**Definition 6.1** (Triangle). *Given three points $p_1, p_2, p_3 \in \mathbb{R}^n$ with $n \geq 2$ such that two vectors $\overrightarrow{p_2p_1}$ and $\overrightarrow{p_3p_1}$ are linearly independent, the (closed) triangle $T$ spanned by these vertices is defined as the convex hull:*

$$T = \triangle(p_1, p_2, p_3) := \left\{ x \in \mathbb{R}^n : x = \sum_{i=1}^{3} \lambda_i p_i, \ \sum_{i=1}^{3} \lambda_i = 1, \ \lambda_i \geq 0 \right\}.$$

Any point $p \in T \subset \mathbb{R}^n$ can therefore be expressed as a convex combination of the three vertices $p_i \in \mathbb{R}^n$, and the coefficients $\{\lambda_i\}$ are called the *barycentric coordinates* of the point $p \in T$. The notion of a triangle generalizes to a *simplex* as follows:

**Definition 6.2** (Simplex). *Given $d + 1$ vertices $p_1, \ldots, p_{d+1} \in \mathbb{R}^n$ with $n \geq d$ such that $d$ vectors $\overrightarrow{p_2p_1}, \cdots, \overrightarrow{p_{d+1}p_1}$ are linearly independent, the $d$-simplex is the convex hull*

$$T := \Big\{ x = \sum_{i=1}^{d+1} \lambda_i p_i : \sum_{i=1}^{d+1} \lambda_i = 1, \ \lambda_i \geq 0 \Big\}.$$

*Local coordinates $(x^1, \ldots, x^d)$ are defined on the reference simplex*

$$U = \{(x^1, \ldots, x^d) \in \mathbb{R}^d : x^i \geq 0, \ \sum_{i=1}^{d} x^i \leq 1\}.$$

Triangles are simplices for the case $d = 2$. In numerical PDEs, simplices are the natural finite element building blocks, since piecewise linear functions are easily defined via *barycentric coordinates* $\lambda_i$. Though many discussions in this chapter can also be extended to triangles or simplices in $\mathbb{R}^n$, for simplicity, we only consider triangles in $\mathbb{R}^3$ in the rest of this chapter.

For a triangle $T = \triangle(p_1, p_2, p_3)$ in $\mathbb{R}^3$, we may use a *local parameterization* $(x^1, x^2)$ defined on a reference triangle

$$U = \{(x^1, x^2) \in \mathbb{R}^2 : x^1 \geq 0,\ x^2 \geq 0,\ x^1 + x^2 \leq 1\},$$

which can also be regarded as a local chart if regarding $T$ as a manifold. A point $p \in T \subset \mathbb{R}^3$ can be parameterized as

$$p(x^1, x^2) = p_1 + x^1(p_2 - p_1) + x^2(p_3 - p_1) \in \mathbb{R}^3.$$

The tangent vectors of this local chart are

$$\frac{\partial p}{\partial x^1} = p_2 - p_1 \in \mathbb{R}^3, \qquad \frac{\partial p}{\partial x^2} = p_3 - p_1 \in \mathbb{R}^3.$$

The induced metric tensor on $T$ is given by

$$g = \left( \langle \frac{\partial p}{\partial x^i}, \frac{\partial p}{\partial x^j} \rangle \right)_{i,j=1,2} = \begin{pmatrix} \|p_2 - p_1\|^2 & \langle p_2 - p_1, p_3 - p_1 \rangle \\ \langle p_2 - p_1, p_3 - p_1 \rangle & \|p_3 - p_1\|^2 \end{pmatrix} \in \mathbb{R}^{2\times 2}.$$

The metric $g$ encodes intrinsic geometric quantities such as edge lengths, triangle area, and angles. For instance, the area of the triangle is

$$A_T = \frac{1}{2}\sqrt{\det g}.$$

For a triangle $T = \triangle(p_i, p_j, p_k) \subset \mathbb{R}^3$ with edge vectors $e_{ij} = p_j - p_i \in \mathbb{R}^3$, the area is also equal to $A_T = \frac{1}{2}\|e_{ij} \times e_{ik}\|$. If we further specify an orientation for $T$ via the order $i, j, k$, then the unit vector $n_T \in \mathbb{R}^3$ normal to this oriented surface $T$ is defined by

$$n_T = \frac{e_{ij} \times e_{ik}}{\|e_{ij} \times e_{ik}\|}.$$

Equivalently, if $a = \|e_{jk}\|$, $b = \|e_{ik}\|$, $c = \|e_{ij}\|$ and $s = \frac{1}{2}(a + b + c)$, then Heron's formula gives $A_T = \sqrt{s(s-a)(s-b)(s-c)}$. The edge co-normal vector in $T$ opposite to vertex $i$ (or $p_i$) is

$$N_i^T = n_T \times e_{jk}. \tag{6.1}$$

### 6.1.2 *Functions, vector fields, and operators on a triangle*

In practice, usually only values at mesh vertices are available. To evaluate or differentiate such data within a triangle, interpolation schemes are needed. The simplest and most common one is the linear interpolation.

Given some scalar values $u_i$ defined only at the vertices $p_i \in \mathbb{R}^3$ of a triangle $T \subset \mathbb{R}^3$. The *linear interpolation* at any point $p \in T$ with barycentric coordinates $(1 - x_1 - x_2, x_1, x_2)$ is given by

$$u(p) = (1 - x_1 - x_2)u_1 + x_1 u_2 + x_2 u_3.$$

For the linear interpolation function $u(p)$ defined on $T$, its gradient is a constant vector over the interior of the triangle, which can be written as follows using local coordinates,

$$\nabla_T u = \begin{pmatrix} u_2 - u_1 \\ u_3 - u_1 \end{pmatrix}^\top g^{-1} \begin{pmatrix} p_2 - p_1 \\ p_3 - p_1 \end{pmatrix} \in \mathbb{R}^3.$$

Similarly, given tangent vectors $\vec{v}_i \in \mathbb{R}^3$ defined only at vertices $p_i$, the linear interpolation is

$$\vec{v}(p) = (1 - x_1 - x_2)\vec{v}_1 + x_1\vec{v}_2 + x_2\vec{v}_3.$$

We compute

$$\begin{pmatrix} v_1 \\ v_2 \end{pmatrix} = g^{-1} \begin{pmatrix} \vec{v} \cdot (p_2 - p_1) \\ \vec{v} \cdot (p_3 - p_1) \end{pmatrix}$$

and its divergence is given in local coordinates by

$$\nabla_T \cdot \vec{v} = \frac{1}{\sqrt{\det g}} \sum_{i=1}^{2} \frac{\partial}{\partial x^i} \left( \sqrt{\det g}\, v_i \right).$$

Since linear functions have constant gradients, the Laplacian of the linear interpolation function $u$ on a single triangle vanishes:

$$\Delta_T u = \nabla_T \cdot (\nabla_T u) = 0.$$

For a continuous piecewise linear function defined on a triangular mesh, it can be uniquely represented by its values at the vertices in the mesh, and its gradient in the sense of *weak derivatives* as defined in Chapter 3 is a piecewise constant vector.

### 6.1.3 *Triangular meshes for surfaces*

A surface can be discretized as a collection of triangles that are consistently glued together. A *triangular mesh* $\mathcal{M}_h$ is a collection of triangles $\{\mathcal{T}_i\} \subset \mathbb{R}^3$ such that the intersection of any two triangles is either empty, a shared vertex, or a shared edge. See Figure 6.1 for two such examples. Throughout the rest of this book, the notation $\mathcal{M}_h$ is used to emphasize the mesh-based (discrete) surface, in contrast to its continuous counterpart $\mathcal{M}$.

For computation, we need a data structure. A triangular mesh in $\mathbb{R}^3$ can be described by a tuple $\mathcal{M}_h = (\mathcal{P}, \mathcal{T})$, where

- $\mathcal{P} = \{p_1, \dots, p_n\}$ is the set of $n$ vertices, represented by a matrix $P \in \mathbb{R}^{n\times 3}$;
- $\mathcal{T}$ is the set of $f$ triangular faces, each specified by three vertex indices, and $\mathcal{T}$ is represented by a matrix $\mathcal{T} \in \mathbb{N}^{f\times 3}$.

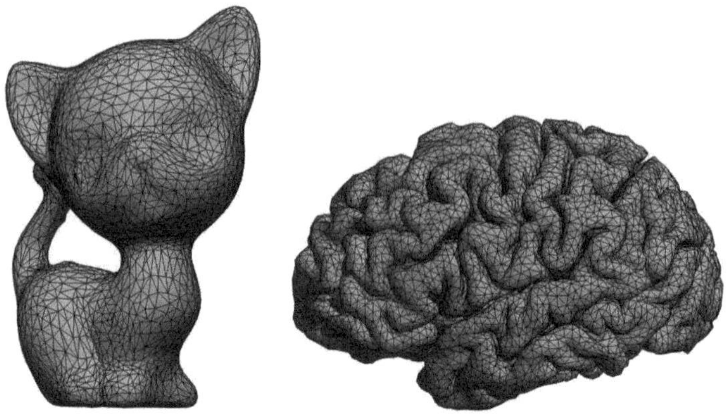

Fig. 6.1 Example of triangular meshes for surfaces.

Although triangular meshes are widely used, their quality significantly affects numerical PDE solvers, e.g.,

- Mesh quality: highly skewed or degenerate triangles reduce accuracy. The element condition number of a triangle, i.e., its angles, depends on the shape. Near-degenerate angles amplify numerical error and should be avoided. See Figure 6.2.
- Manifoldness: for approximating PDEs on a surface, nonmanifold edges or inconsistent orientation can cause breakdowns.
- Adaptivity: in regions of high curvature or solution variability, adaptive refinement and remeshing improve stability.

Despite these challenges, triangular meshes remain the standard tool for approximating smooth surfaces, serving as the foundation for constructing discrete differential operators and PDE solvers.

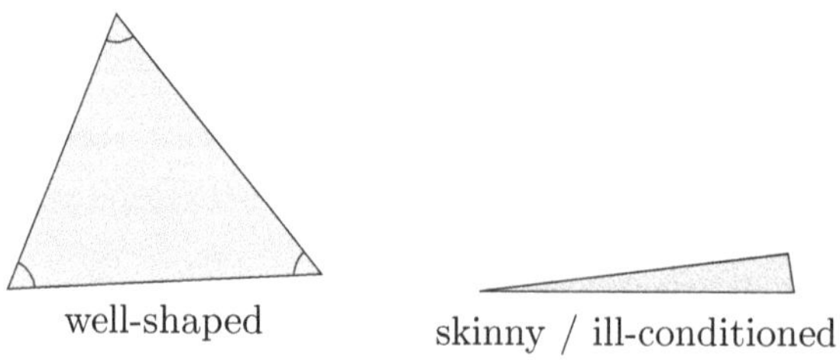


Fig. 6.2 Triangle shape quality matters. Near-degenerate angles deteriorate interpolation accuracy and conditioning of linear systems in FEM/FVM discretizations. A poor aspect ratio leads to large condition numbers in mass/stiffness matrices.

More background on polygon meshes, barycentric coordinates, and discrete differential geometry can be found in [Botsch *et al.* (2010); Meyer *et al.* (2003); Dziuk and Elliott (2013); Crane (2018); Wardetzky *et al.* (2007)]. We remark that the many concepts discussed in this section naturally extend to simplicial complex approximations of higher-dimensional manifolds. For clarity of exposition, however, the discussion here is restricted to the 2-dimensional case.

## 6.2 Discrete differential operators on triangular meshes

Let $\mathcal{M}_h = (\mathcal{P}, \mathcal{T})$ be a triangulated surface approximating a smooth surface $\mathcal{M} \subset \mathbb{R}^3$, with $P = \{p_1, \ldots, p_n\} \subset \mathbb{R}^3$ and oriented faces $\mathcal{T} = \{T = (i,j,k)\}$. For each face $T$, its area is denoted by $A_T$ and its unit normal is denoted by $n_T$. We store scalar data at vertices, $u\big|_{\mathcal{P}} : \mathcal{P} \to \mathbb{R}$ with samples $u_i = u(p_i)$. The standard piecewise linear *hat function* basis $\{\phi_i\}_{i=1}^n$ satisfy that $\phi_i$ is linear on each face and $\phi_i(p_j) = \begin{cases} 1, & i = j \\ 0, & i \neq j \end{cases}$.

For $p_i \in \mathcal{P}$, its *first ring* $\mathcal{R}_i$ is defined as the set of vertices that share a triangle with $p_i$, see Figure 6.3. Let $\mathcal{T}_i := \{\, T \in \mathcal{T} : \; T \ni p_i \,\}$ be the set of incident faces containing $p_i$.

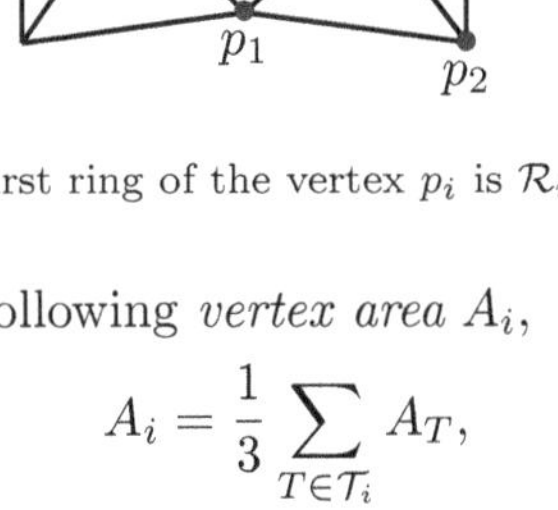


Fig. 6.3 The first ring of the vertex $p_i$ is $\mathcal{R}_i = \{1, 2, 3, 4, 5\}$.

Next, we define the following *vertex area* $A_i$,

$$A_i = \frac{1}{3} \sum_{T \in \mathcal{T}_i} A_T, \tag{6.2}$$

which will be used to normalize vertex-wise operators. We remark that a different definition of $A_i$ defined by mixed *Voronoi* area may also be considered for improving spectral accuracy [Meyer *et al.* (2003); Botsch *et al.* (2010); Wardetzky *et al.* (2007)]. For simplicity, we only consider the vertex area defined by (6.2) in this book.

### 6.2.1 *Gradient on a triangular mesh*

On each face or triangle $T = (i, j, k)$, the gradients of barycentric basis functions are constant vectors. By using (6.1), they can be written as,

$$\nabla\phi_i\big|_T = \tfrac{n_T\times(p_k-p_j)}{2A_T}, \quad \nabla\phi_j\big|_T = \tfrac{n_T\times(p_i-p_k)}{2A_T}, \quad \nabla\phi_k\big|_T = \tfrac{n_T\times(p_j-p_i)}{2A_T}. \tag{6.3}$$

Hence, the gradient of a linear interpolant $u = \sum_\ell u_\ell\phi_\ell$ can be written as

$$\nabla u\big|_T = u_i\,\nabla\phi_i\big|_T + u_j\,\nabla\phi_j\big|_T + u_k\,\nabla\phi_k\big|_T, \tag{6.4}$$

which is constant over $T$.

**Definition 6.3** (Vertex Gradient). *Given vertex values $u_i = u(p_i)$, define the vertex gradient at $p_i$ by an area-weighted average of incident face gradients:*

$$\nabla^{\mathrm{vtx}}_{\mathcal{M}_h} u(p_i) := \frac{\displaystyle\sum_{T\in\mathcal{T}_i} A_T\,\nabla u\big|_T}{\displaystyle\sum_{T\in\mathcal{T}_i} A_T}, \qquad \textit{with } \nabla u\big|_T \textit{ computed by (6.4).} \tag{6.5}$$

**Remark 6.1.** *The vertex gradient formula (6.5) is also the $L^2$ projection of per-face gradients to the vertex using lumped mass.*

**Example 6.1** (Centered difference on a planar cross). *Consider the set of vertices on a plane*

$$P = \{p_0 = (0,0),\ p_1 = (0,1),\ p_2 = (0,-1),\ p_3 = (1,0),\ p_4 = (-1,0)\},$$

*which gives a triangular mesh $T = \{(0,1,4),(0,4,2),(0,2,3),(0,3,1)\}$. Then (6.5) yields*

$$\nabla^{\mathrm{vtx}}_{\mathcal{M}_h} f(0,0) = \frac{1}{2}\,(f_3 - f_4,\ f_1 - f_2),$$

*which recovers the centered difference for approximating the gradient in Chapter 2.*

### 6.2.2 *Discrete divergence*

We define a vertex divergence as the adjoint of the per-face gradient, which mirrors discrete integration-by-parts and is compatible with perspectives from Discrete Exterior Calculus (DEC) and the finite volume method.

**Definition 6.4** (Discrete Divergence). *For any tangent vector field $X$ which is constant $X_T \in \mathbb{R}^3$ on each face $T$ and satisfies $X_T \cdot n_T = 0$, its discrete divergence at the vertex $p_i$ is defined as*

$$(\nabla^{\mathrm{vtx}}_{\mathcal{M}_{\mathrm{h}}} \cdot X)_i := -\frac{1}{A_i} \sum_{T \in \mathcal{T}_i} A_T \left\langle X_T,\ \nabla \phi_i \big|_T \right\rangle, \tag{6.6}$$

*where $A_i$ is the given vertex area* (6.2).

If a vector field is given only at vertices $p_i$ as $v_i \in \mathbb{R}^3$, a common choice is to consider a simple and crude interpolation $X_T = \frac{1}{3}(v_i + v_j + v_k)$ on a face $T = (i, j, k)$ then apply (6.6) to the simple interpolant $X_T$.

## 6.3 A finite element method for Laplacian on surface

We give a description of a finite element method (FEM) on a triangulated surface $\mathcal{M}_h = (\mathcal{P}, \mathcal{T})$, which involves the Galerkin weak form, piecewise linear hat function basis, element mass and stiffness matrices, etc. Our exposition follows classical FEM references [Dziuk and Elliott (2013)], together with discrete differential geometry sources for concrete P1 formulae on triangular meshes.

### 6.3.1 *Nodal basis, linear interpolation, and projection*

Let the mesh vertices be $\mathcal{P} = \{p_1, \ldots, p_n\} \subset \mathbb{R}^3$ and faces be $\mathcal{T} = \{T = (i, j, k)\}$. We use the standard P1 *nodal* basis $\{\phi_i\}_{i=1}^n$ defined by

$$\phi_i : \mathcal{M}_h \to \mathbb{R}, \qquad \phi_i(p_j) = \delta_{ij} \quad \text{and} \quad \phi_i \text{ is affine on each } T \in \mathcal{T}. \tag{6.7}$$

We have $\phi_i(x) + \phi_j(x) + \phi_k(x) \equiv 1$ for any $x$ on the face $T = (i, j, k)$, and $\phi_\ell \equiv 0$ for $\ell \notin \{i, j, k\}$. The piecewise linear finite element space is

$$V_h := \operatorname{Span}\{\phi_i\}_{i=1}^n.$$

Given a smooth function $u \in C^\infty(\mathcal{M})$ with its nodal values $u_i = u(p_i)$, its nodal interpolation is

$$\sum_{i=1}^n u(p_i)\, \phi_i \in V_h,$$

which we may also represent by the coefficient vector $(u_1, \ldots, u_n)^\top$.

The nodal basis $\{\phi_i\}$ is *universal* since it does not adapt to any particular PDE. Its accuracy comes from mesh refinement. On the other hand, there are other choices of bases for the sake of model reduction. For instance, problem-adapted linear trial spaces such as reduced bases and Proper Orthogonal Decomposition (POD) can yield much smaller systems while remaining linear in the coefficients. Data-driven nonlinear bases and operators such as operator inference and neural operators provide nonlinear model reduction beyond linear subspaces [Brunton and Kutz (2019); Peherstorfer *et al.* (2016)].

### 6.3.2 *Stiffness and mass matrices, and discrete Laplacian*

The stiffness and mass matrices in the surface finite element method are defined similarly as in finite element method in Chapter 3, by facewise integrals of P1 basis functions:

$$S_{ij} = \sum_{T\in\mathcal{T}} \int_T \langle \nabla\phi_i, \nabla\phi_j\rangle \,\mathrm{d}A, \qquad M_{ij} = \sum_{T\in\mathcal{T}} \int_T \phi_i\,\phi_j \,\mathrm{d}A.$$

Local face contributions yield the classical cotangent weights:

**Lemma 6.1.** *For $T = (i, j, k)$ with angles $\alpha_i, \alpha_j, \alpha_k$ at $(i, j, k)$,*

$$\int_T \langle \nabla\phi_i, \nabla\phi_j\rangle \,\mathrm{d}A = -\tfrac{1}{2}\cot\alpha_k, \qquad \int_T \langle \nabla\phi_i, \nabla\phi_i\rangle \,\mathrm{d}A = \tfrac{1}{2}(\cot\alpha_j + \cot\alpha_k). \tag{6.8}$$

*Proof.* On $T = (i, j, k)$, with face area $A_T$ and unit normal $n_T$ of the face, per-face gradients of nodal functions are given in (6.3). Thus we have

$$\int_T \nabla\phi_i\cdot\nabla\phi_j \,\mathrm{d}A = \frac{1}{4A_T^2}\int_T (n_T\times e_{kj})\cdot(n_T\times e_{ik})\,\mathrm{d}A = \frac{A_T}{4A_T^2}\,(e_{kj}\cdot e_{ik}^{\perp}) = -\tfrac{1}{2}\cot\alpha_k,$$

where $\alpha_k$ is the angle at vertex $k$. □

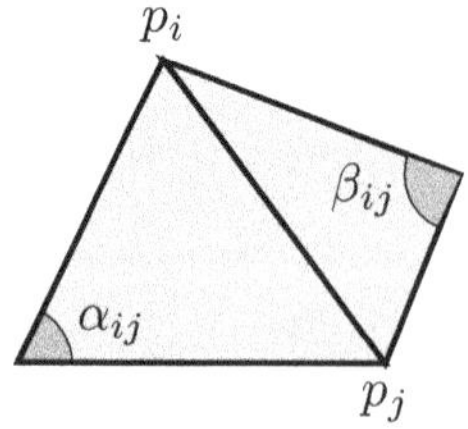


Fig. 6.4 The two angles for an interior edge connecting $p_i$ and $p_j$.

Summing over all faces $T$ gives the expression of stiffness and mass matrices.

**Proposition 6.1.** *For an interior edge $(i, j)$ with opposite angles $\alpha_{ij}, \beta_{ij}$ (see Figure 6.4),*

$$S_{ij} = -\frac{1}{2}\big(\cot \alpha_{ij} + \cot \beta_{ij}\big), \qquad S_{ii} = -\sum_{j \in \mathcal{R}_i} S_{ij}.$$

*On boundary edges, only the single adjacent angle contributes. For the mass matrix,*

$$M_{ii} = \frac{1}{6} \sum_{T \in \mathcal{T}_i} A_T, \qquad M_{ij} = \frac{1}{12} \sum_{T \supset (i,j)} A_T \ (i \neq j).$$

Since $\phi_i$ has compact support on the first ring of $p_i$, $M$ and $S$ are sparse matrices with off diagonal entries $M_{ij} = 0$ and $S_{ij} = 0$ when $p_j$ is not in the first ring of $p_i$. Moreover, $M$ and $S$ are symmetric matrices, with $M$ being positive definite. The matrix $S$ is positive semi-definite on closed meshes of $n$-vertices, with rank$(S) = n - 1$ and its null space is spanned by the constant one vector.

The following pseudo-code illustrates an implementation of assembling the stiffness and mass matrices from integrals over each triangle:

```
for each triangle T = (i,j,k):
    eij = p[j]-p[i]; eik = p[k]-p[i]
    n  = normalize(cross(eij, eik))
    AT = 0.5 * norm(cross(eij, eik))
    # gradients of hat functions
    grad_phi[i] = cross(n, p[k]-p[j]) / (2*AT)
    grad_phi[j] = cross(n, p[i]-p[k]) / (2*AT)
    grad_phi[k] = cross(n, p[j]-p[i]) / (2*AT)
    # local stiffness and mass
    for a,b in {(i,j,k)}:
        S[a,b] += AT * dot(grad_phi[a], grad_phi[b])
        M[a,b] += AT * (1/12 if a!=b else 2/12)
```

Notice that the mass matrix here is not diagonal, similar to the mass matrix defined without using quadrature in Chapter 3. On the other hand, a diagonal mass matrix is often convenient, e.g., the mass matrix becomes diagonal when approximating integrals by proper quadrature in Chapter 3. Here a diagonal alternative for the mass matrix is the *lumped mass* matrix, which is a diagonal matrix with diagonal entries derived by summing each row of $M$:

$$M_{ii}^{\text{lump}} = \sum_j M_{ij} = \frac{1}{3} \sum_{T \ni p_i} A_T,$$

which is useful for explicit time stepping and as a vertex-area weight. The simplest quadrature on a triangle $T = (i, j, k)$ for a function $f$ with vertex values $f_i, f_j, f_k$ is

$$\int_T f \, \mathrm{d}A \approx \frac{1}{3}(f_i + f_j + f_k)A_T, \tag{6.9}$$

which is exact for integrating linear polynomials but not quadratic polynomials. The *lumped mass* matrix can be equivalently obtained by approximating the integrals $\int_T \phi_i \phi_j \, \mathrm{d}A$ by this simple quadrature, which gives

$$\int_T \phi_i \phi_j \, \mathrm{d}A \approx \begin{cases} \frac{1}{3} A_T, & i = j \\ 0, & i \neq j \end{cases}.$$

Finally, to derive the discrete Laplacian operator, which is often called *Cotangent Laplacian* in this context of surface finite element method, we first notice that (6.2), (6.4) and (6.6) altogether imply the following discrete integration by parts:

$$\sum_i A_i \, (\nabla^{\mathrm{vtx}}_{\mathcal{M}_\mathrm{h}} \cdot X)_i \, u_i \;=\; -\sum_{T \in \mathcal{T}} \int_T \langle X, \nabla u_h \rangle \, \mathrm{d}A, \quad \forall u_h \in V_h.$$

By setting $X = \nabla v_h$ for any $v_h \in V_h$, it gives

$$\sum_i A_i \, (-\nabla^{\mathrm{vtx}}_{\mathcal{M}_\mathrm{h}} \cdot \nabla v_h)_i \, u_i \;=\; \sum_{T \in \mathcal{T}} \int_T \langle \nabla v_h, \nabla u_h \rangle \, \mathrm{d}A.$$

Plugging in $u_h = \sum_i u_i \phi_i$ and $v_h = \sum_j v_j \phi_j$, with $A_i = M^{\mathrm{lump}}_{ii}$, we get

$$\sum_i M^{\mathrm{lump}}_{ii} \, (\sum_j [-\nabla^{\mathrm{vtx}}_{\mathcal{M}_\mathrm{h}} \cdot \nabla \phi_j] v_j)_i \, u_i \;=\; \sum_i \sum_j \Big( \sum_{T \in \mathcal{T}} \int_T \langle \nabla \phi_j, \nabla \phi_i \rangle \, \mathrm{d}A \Big) u_i v_j.$$

Define two column vectors $\mathbf{u} = \begin{bmatrix} u_1 & \cdots & u_n \end{bmatrix}^T$ and $\mathbf{v} = \begin{bmatrix} v_1 & \cdots & v_n \end{bmatrix}^T$, and let $L$ be the matrix with entries $L_{ij} = [-\nabla^{\mathrm{vtx}}_{\mathcal{M}_\mathrm{h}} \cdot \nabla \phi_j]_i$, which is the matrix representation of the linear operator $-\nabla^{\mathrm{vtx}}_{\mathcal{M}_\mathrm{h}} \cdot \nabla$ under the basis $\{\phi_i\}$. Then similar to derivations in Chapter 3, we get the following equivalent matrix vector form,

$$\mathbf{u}^T M^{\mathrm{lump}} \, L \, \mathbf{v} = \mathbf{u}^T S \mathbf{v},$$

which implies

$$M^{\mathrm{lump}} \, L \, \mathbf{v} = S \mathbf{v}.$$

Thus we obtain the matrix for $-\nabla^{\mathrm{vtx}}_{\mathcal{M}_\mathrm{h}} \cdot \nabla$ as

$$L = [M^{\mathrm{lump}}]^{-1} S,$$

which is also known as the *cotangent Laplacian*, and it is consistent with FEM/DEC constructions.

**Remark 6.2** (Vertex Laplacian in stencil form). *Let $\mathcal{R}_i$ be the first ring of vertex $i$, using vertex areas $A_i$ and edge weights $w_{ij} = \frac{1}{2}(\cot\alpha_{ij} + \cot\beta_{ij})$, define the following discrete Laplacian operator as shown in Figure 6.5,*

$$(\Delta_h u)_i \ := \ \frac{1}{A_i}\sum_{j\in\mathcal{R}_i} w_{ij}\,(u_j - u_i),$$

*with the obvious single-angle modification on boundary edges. Since $M^{\text{lump}}$ is a diagonal matrix with diagonal entries $M^{\text{lump}}_{ii} = A_i$ and the stiffness matrix satisfies $S_{ii} = \sum_{j\in\mathcal{R}_i} w_{ij}$ and $S_{ij} = -w_{ij}$ for $i \neq j$, which is implied by Proposition 6.1. It is straightforward to verify that $-\Delta_h$ is exactly the same as $-\nabla^{\text{vtx}}_{\mathcal{M}_h}\cdot\nabla$ in the following sense*

$$-(\Delta_h u)_i = (L\mathbf{u})_i.$$

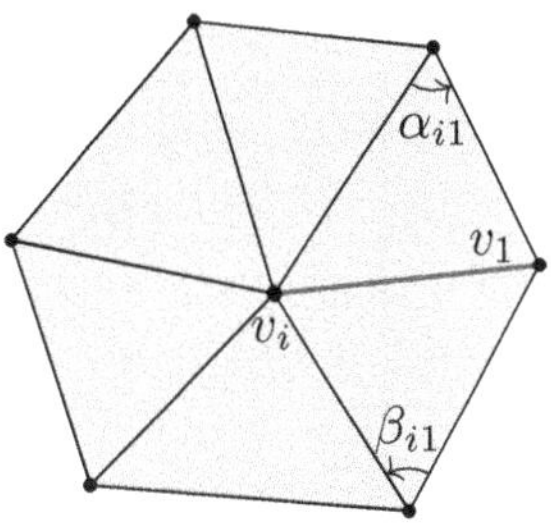


Weight for the edge $(v_i, v_1)$:
$w_{i1} = \frac{1}{2}(\cot\alpha_{i1} + \cot\beta_{i1})$

Discrete Laplacian (umbrella form):
$$(\Delta u)_i = \frac{1}{2A_i}\sum_{j\in\mathcal{R}_i}(\cot\alpha_{ij} + \cot\beta_{ij})\,(u_j - u_i)$$

Fig. 6.5 A vertex $v_i$ and its first ring $\mathcal{R}_i$ consisting of 6 vertices. Both the stiffness matrix and the discrete Laplace–Beltrami operator can be expressed by the cotangent formula using the two angles opposite each edge.

**Remark 6.3** (Normalization variants). *The following variants of discrete Laplacian matrices via graph-style normalizations are also common choices:*

$$L_{\text{rw}} = M^{-1}S \quad (\textit{random-walk}), \qquad L_{\text{sym}} = M^{-1/2}S\,M^{-1/2} \quad (\textit{symmetric}).$$

*see [Chung (1997)] for spectral graph theory and [Crane (2018)] for geometry-processing practice.*

**Remark 6.4.** *In this section, the triangular mesh is for a surface in $\mathbb{R}^3$, while the triangular mesh briefly discussed in Section 3.8 is a triangular mesh in $\mathbb{R}^2$. To have an easier understanding of the scheme derived in this section, we may consider it on a special triangular mesh for a flat surface on $x - y$ plane in $\mathbb{R}^3$, then all discussion in this section for approximating* *of Laplacian on a surface should be consistent with the discussion of approximation to Laplacian in $\mathbb{R}^2$ in Section 3.8. In particular, notice that the stiffness matrix $S$ and the lumped mass matrix in this section have the exactly the same formulae as those in* (3.20).

### 6.3.3 * *Random walk in two dimensions*

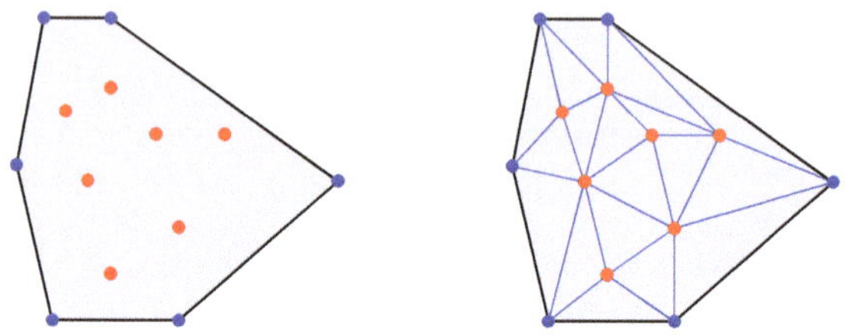

Fig. 6.6 Left: some scattered red points in a 2D polygonal domain. Right: random walk along edges.

To conclude this section, we demonstrate the usefulness of the formulae in (3.20) by discussing how the heat equation can be derived from random walks in two dimensions. In Chapter 1, we have seen how the heat equation $u_t = u_{xx}$ can be derived from the random walk, which is related to the discrete Laplacian in one dimension. How can such a derivation be extended to two dimensions? Consider a random walk of a particle in two dimensions as illustrated in Figure 6.6. The first question here is how we can define the transition probability of the random walk for going from one red node to another along the edge, which is not an easy question if we want to derive the heat equation.

One answer is to consider defining transition probability from the discrete Laplacian in (3.20). Let $M$ and $S$ be mass and stiffness matrices in (3.20), which are the lumped mass and stiffness matrices in this section but on a flat triangular mesh. Let $D$ be diagonal part of $L = M^{-1}S$, then $P = 2I - D^{-1}L$, where $I$ is identity, has entries with the same sparsity pattern as $S$,

$$P_{ij} = \begin{cases} 1, & i = j \\ \frac{1}{D_{ii}} \frac{\cot \alpha_{ij} + \cot \beta_{ij}}{2A_i}, & j \in \mathcal{R}_i \end{cases},$$

where

$$D_{ii} = \sum_{j \in \mathcal{R}_i} \frac{\cot \alpha_{ij} + \cot \beta_{ij}}{2A_i},$$

$A_i = \frac{1}{3}\sum_{T\in\mathcal{T}_i} A_T$ and the two angles $\alpha_{ij}, \beta_{ij}$ are defined as in Figure 3.4 (or Figure 6.4 with two triangles on a plane).

Assume there are in total $N$ nodes. In order to ensure $P_{ij} \geq 0$, we need $\cot\alpha_{ij} + \cot\beta_{ij} \geq 0$, which holds if and only if $\alpha_{ij} + \beta_{ij} \leq \pi$. Such an angle constraint condition $\alpha_{ij} + \beta_{ij} \leq \pi$ can be ensured in a Delaunay triangulation. So if we assume the red nodes in Figure 6.6 are distributed nicely so that the triangular mesh in Figure 6.6 is a Delaunay triangulation, then $P_{ij}$ defined above is indeed a random walk transition probability because we have $P_{ij} \geq 0$, and also $\sum_{j\in\mathcal{R}_i} P_{ij} = 1$ from the construction of $P$.

Finally, consider a particle locating at some red node and at time $t = 0$, and assume it moves from $p_i$ to another node $p_j$ for $j \in \mathcal{R}_i$ under probability $P_{ij}$ at each time $n\Delta t$. If $f(j,n)$ denotes the probability that the particle at location $p_j$ and time $t = n\Delta t$, then it satisfies

$$f(i,n+1) = \sum_{j\in\mathcal{R}_i} P_{ij} f(j,n),$$

thus

$$f(i,n+1) - f(i,n) = \sum_{j\in\mathcal{R}_i} P_{ij} f(j,n) - f(i,n) = \sum_{j=1}^{N} P_{ij} f(j,n) - 2f(i,n).$$

Multiplying it by $D_{ii}$, we get

$$D_{ii}[f(i,n+1) - f(i,n)] = \sum_{j=1}^{N} D_{ii}P_{ij} f(j,n) - 2D_{ii} f(i,n) = -\sum_{j=1}^{N} L_{ij} f(j,n). \tag{6.10}$$

If $D_{ii}$ is the same for all $i$, which is possible if red nodes are structured so that all the triangles are equilateral of the same size, then we can assume a time step $\Delta t = \frac{1}{D_{ii}}$ to obtain

$$\frac{1}{\Delta t}[f(i,n+1) - f(i,n)] = -\sum_{j=1}^{N} L_{ij} f(j,n).$$

If $h$ denotes the edge length in a uniform equilateral triangular mesh, then the nonzero entries in the discrete Laplacian matrix $L$ are of the order $\mathcal{O}(h^{-2})$, similar to the discrete Laplacian matrices in Part I. Thus the assumption of the time step gives a similar time step and mesh size ratio $\Delta t = \mathcal{O}(h^2)$ as in the 1D random walk. Suppose $p_i = (x_i, y_i)$ and there is a smooth function $u(x,y,t)$ such that $f(i,n) = u(x_i, y_i, n\Delta t)$, i.e., the point values of $u(x,y,t)$ at the node $p_i$ give the discrete probability $f(i,n)$. For fixed $i$ and $n$, let $(x, y) = (x_i, y_i)$ and $n\Delta t = t$, then we obtain

$$\frac{u(x_i, y_i, t + \Delta t) - u(x_i, y_i, t)}{\Delta t} = -\sum_{j=1}^{N} L_{ij} u(x_j, y_j, t).$$

Since $L$ is the discrete Laplacian matrix derived from P1 finite element method in scheme (3.20) in Chapter 3, we know that $Lu = -u_{xx} - u_{yy} + \mathcal{O}(h^2)$. If taking the limit $\Delta t \to 0$ and $h \to 0$ with the restriction $\Delta t = D_{ii}$ satisfied for structured vertices forming uniform equilateral triangular mesh, then the limit gives the heat equation $u_t = u_{xx} + u_{yy}$.

Obviously, to make this whole argument work for deriving $u_t = \Delta u$ from 2D random walks, we need to consider a very special set up, in which blue nodes in Figure 6.6 must form a hexagon and edges form a uniform equilateral triangular mesh. On the other hand, if only assuming edges form a Delaunay triangular mesh, then we can see that the limit of the right hand side (6.10) still recovers the Laplacian operator. Moreover, we can see that this argument can also be extended to random walks on point clouds along a surface, if the angle condition can be satisfied, which is possible for surfaces like a sphere but rather restrictive for general surfaces.

## 6.4 Numerical PDEs on surfaces

Recall that $\mathcal{M}_h$ is a triangulated surface approximating a given smooth surface $\mathcal{M} \subset \mathbb{R}^3$. Once the stiffness matrix $S$ and mass matrix $M$ are assembled on a triangulated surface $\mathcal{M}_h$, a few classical operators on the smooth manifold become accessible in a discrete form. The matrix $M$ represents the $L^2$ inner product associated with the piecewise linear basis, while $S$ encodes the Dirichlet energy and thus serves as the discrete Laplace–Beltrami operator. With these two ingredients, many evolution and spectral problems reduce to sparse linear systems or generalized eigenvalue problems.

### 6.4.1 *Galerkin formulation for a surface Poisson problem*

Consider solving the following Poisson equation defined on a surface $\mathcal{M}$ with the homogeneous Neumann boundary conditions:

$$\begin{cases} -\Delta_{\mathcal{M}} u = f \\ \nabla_{\mathcal{M}} u \cdot \vec{n}|_{\partial\mathcal{M}} = 0 \end{cases}. \tag{6.11}$$

We have defined $H^1(\Omega)$ function space in Chapter 3 for a flat domain $\Omega$. On a surface $\mathcal{M}$, the Sobolev function space $H^1(\mathcal{M})$ is similarly defined.

The continuous *bilinear forms* for $u, \varphi \in H^1(\mathcal{M})$ are defined as

$$m(u,\varphi) = \int_{\mathcal{M}} u\,\varphi\,\mathrm{dvol}_g, \qquad a(u,\varphi) = \int_{\mathcal{M}} \nabla_{\mathcal{M}} u \cdot \nabla_{\mathcal{M}} \varphi\,\mathrm{dvol}_g.$$

**Properties of the exact solutions.** We state a few facts of the exact solution to the surface Poisson equation.

First of all, for solving $-\Delta_{\mathcal{M}} u = f$ with no boundary conditions or with Neumann boundary conditions like (6.11), its weak or variational formulation can be derived by Green's identity (5.17). The *variational form* is to find $u \in H^1(\mathcal{M})$ satisfying

$$a(u,\varphi) = m(f,\varphi), \quad \forall \varphi \in H^1(\mathcal{M}).$$

Second, similar to the Euclidean case, the PDE $-\Delta_{\mathcal{M}} u = f$ is equivalent to the variational form. By plugging $\varphi \equiv 1$, we can see that the function $f$ must satisfy the compatibility condition $m(f,1) = \int_{\mathcal{M}} f\,\mathrm{dvol}_g = 0$ so that a solution can exist. Moreover, if $u$ is an exact solution, so is $u + c$ for any constant $c$. In particular, for any compatible $f \in L^2(\mathcal{M})$ with zero mean $\int_{\mathcal{M}} f\,\mathrm{dvol}_g = 0$, there is a unique exact solution $u$ satisfying

$$\int_{\mathcal{M}} u\,\mathrm{dvol}_g = 0.$$

Third, in Chapter 3, we have seen the *Elliptic Regularity* which states that the solution of (3.5) has second order derivative, i.e., the variational form (3.5) on a suitable Euclidean domain $\Omega$ has the property $f(x) \in L^2(\Omega) \Rightarrow u(x) \in H^2(\Omega)$. It also holds for a smooth surface, i.e., the variational form on a surface $\mathcal{M}$ satisfies

$$f(x) \in L^2(\mathcal{M}) \Rightarrow u(x) \in H^2(\mathcal{M}).$$

### 6.4.2 *Numerical scheme for Poisson equation*

Let $V_h$ be the finite element space consisting of all continuous piecewise linear functions defined on $\mathcal{M}_h$. Similar to what we see in Chapter 3, its variational form (a.k.a. the Galerkin weak form) on the triangulated mesh $\mathcal{M}_h$ is to seek $u_h \in V_h$ such that

$$a_h(u_h,\varphi_h) \;=\; m_h(f,\varphi_h), \qquad \forall\, \varphi_h \in V_h, \tag{6.12}$$

with the *discrete bilinear forms*

$$a_h(u_h,\varphi_h) = \sum_{T\in\mathcal{T}} \int_T \nabla u_h \cdot \nabla \varphi_h\,\mathrm{d}A, \quad m_h(f,\varphi_h) = \sum_{T\in\mathcal{T}} \int_T f\,\varphi_h\,\mathrm{d}A.$$

On each face, geometry is exact (planar), so all integrals reduce to triangle integrals. For simplicity, we assume $f$ is a continuous function. If we approximate the function $f$ by the linear interpolation at vertex values on each triangle, i.e., for the triangle $T = (i, j, k)$,

$$\int_T f\,\varphi_h\,\mathrm{d}A \approx \int_T (f_i\phi_i + f_j\phi_j + f_k\phi_k)\,\varphi_h\,\mathrm{d}A,$$

then the scheme has a matrix form $S\mathbf{u} = M\mathbf{f}$, where the stiffness matrix $S$ and the mass matrix $M$ are given in Section 6.3.2, and $\mathbf{f}$ is a vector storing point values of the function $f$ at vertices of the mesh $\mathcal{M}_h$. If we further approximate the integral $\int_T (f_i\phi_i + f_j\phi_j + f_k\phi_k)\,\varphi_h\,\mathrm{d}A$ by the simple quadrature rule (6.9), then the mass matrix $M$ reduces to the lumped mass matrix, which is diagonal.

To ensure that the linear system $S\mathbf{u} = M\mathbf{f}$ has at least one solution, we need to prepare the vector $\mathbf{f}$ such that it is in the column space of $M^{-1}S$, similar to what we did in Chapter 2 for Neumann boundary conditions. First, notice that $S$ satisfies $S\mathbf{1} = 0$ because

$$\sum_j S_{ij} = \sum_j \sum_{T\in\mathcal{T}} \int_T \langle \nabla\phi_i, \nabla\phi_j\rangle\,\mathrm{d}A = \sum_{T\in\mathcal{T}} \int_T \langle \nabla\phi_i, \nabla 1\rangle\,\mathrm{d}A = 0.$$

Second, the symmetry of $S$ gives $\mathbf{1}^T S = 0$. Thus, for the sake of deriving a consistent linear system, we only need to adjust or prepare $\mathbf{f}$ so that

$$\mathbf{1}^T M\mathbf{f} = 0,$$

which can be achieved by simply shifting nodal values of $f$ by order $\mathcal{O}(h^2)$, following similar discussions in Section 2.5.4. For simplicity, we assume that $\mathbf{f}$ has been prepared so that the discrete compatibility condition $\mathbf{1}^T M\mathbf{f} = 0$ is already satisfied, which is equivalent to $\sum_i \sum_{T\in\mathcal{T}} \int_T \phi_i\phi_j f_j \mathrm{d}A = 0$.

Similar to finite element method with Neumann boundary conditions in Chapter 3, it can be shown that the stiffness matrix $S$ of size $n\times n$ has rank $n-1$, and the discretized system does not have a unique solution, which is compatible with the fact that solutions of (6.11) are not unique. Recall that in Chapter 2 we have used eigenvalue decomposition of the discrete Laplacian matrix to invert the matrix, for which it is easy to define a zero sum solution, see Section 2.5.5. For the stiffness matrix $S$ on a surface, it is no longer possible to derive its eigenvalue decomposition. Nonetheless, to compute a zero average solution, we can consider solving the linear system $S\mathbf{u} = M\mathbf{f}$ with the zero average constraint $\mathbf{1}^T M\mathbf{u} = 0$, which is equivalent to the following constraint minimization

$$\min_{\mathbf{u}\in\mathbb{R}^n} \frac{1}{2}\mathbf{u}^T S\mathbf{u} - \mathbf{u}^T M\mathbf{f}, \quad \text{s.t.} \quad \mathbf{1}^T M\mathbf{u} = 0.$$

By introducing a scalar multiplier $\lambda$, the Lagrangian can be written as

$$\mathcal{L}(\mathbf{u},\lambda) = \frac{1}{2}\mathbf{u}^T S\mathbf{u} - \mathbf{u}^T M\mathbf{f} + \lambda \mathbf{1}^T M\mathbf{u}, \tag{6.13}$$

and its saddle point equation $\frac{\partial \mathcal{L}}{\partial \mathbf{u}} = \mathbf{0}$, $\frac{\partial \mathcal{L}}{\partial \lambda} = 0$ is

$$\begin{pmatrix} S & M\mathbf{1} \\ \mathbf{1}^T M & 0 \end{pmatrix} \begin{pmatrix} u \\ \lambda \end{pmatrix} = \begin{pmatrix} M\mathbf{f} \\ 0 \end{pmatrix}. \tag{6.14}$$

For solving the symmetric linear system (6.14), one can use the conjugate gradient (CG) method with incomplete Cholesky or algebraic multigrid (AMG) as a preconditioner, see Appendix B.4 for the preconditioned CG method. Multigrid exploits the mesh hierarchy to obtain resolution-independent convergence. On shape-regular meshes, the stiffness matrix condition number grows as $\mathcal{O}(h^{-2})$. The condition number of discrete Laplacian matrix in Euclidean space is also $\mathcal{O}(h^{-2})$, see Appendix B.3.

### 6.4.3 *Heat equation and the eigenvalue problem*

There are a few other equations that can be solved on surfaces, with the stiffness and mass matrices. We consider two quick examples.

**Heat equation.** For the surface heat equation $u_t = \Delta_{\mathcal{M}} u$, the standard Galerkin semi-discretization yields

$$M\dot{\mathbf{u}} + S\mathbf{u} = 0.$$

Implicit time integrators are commonly used due to robustness on irregular meshes. Backward Euler with step size $\Delta t > 0$ leads to

$$(M + \Delta t\, S)\, \mathbf{u}^{n+1} = M\, \mathbf{u}^n,$$

which is an unconditionally stable scheme. This formulation underlies many geometry-processing applications involving diffusion, smoothing, or geodesic computation. The matrix $M + \Delta t\, S$ is an invertible matrix, and the linear system can be solved by iterative methods like the preconditioned CG method in B.4.

**Eigenvalue problem.** The computation of Laplace–Beltrami spectra reduces to solving for unknown $\lambda \in \mathbb{R}$ and $\mathbf{u} \in \mathbb{R}^n$ satisfying

$$S\,\mathbf{u} = \lambda\, M\,\mathbf{u},$$

which is a generalized symmetric positive semi-definite eigenvalue problem. The resulting eigenvectors $\mathbf{u}$ approximate Laplace–Beltrami eigenfunctions which provide intrinsic coordinates and multiscale bases on the surface, and are central to spectral geometry, diffusion geometry, and the functional maps framework [Reuter *et al.* (2006); Botsch *et al.* (2010)]. They serve as building blocks in shape analysis, deformation processing, and the construction of reduced-order models.

## 6.5 *Error estimates of the surface finite element method

We now discuss the convergence properties of the surface finite element method (FEM) on a triangulated mesh $\mathcal{M}_h$ for solving the Poisson equation on a smooth surface $\mathcal{M}$. The objective is to quantify the discretization error under mesh refinement and to understand how geometric approximation of the surface influences the overall accuracy. For simplicity, we only focus on the key arguments, most of which are similar to those in classical Euclidean FEM convergence analysis in Chapter 3. On the other hand, the key distinction from Chapter 3 is that the surface $\mathcal{M}$ is approximated by a triangular mesh $\mathcal{M}_h$, thus the analysis must account for the discrepancy between the discrete triangulated surface $\mathcal{M}_h$ and its smooth counterpart $\mathcal{M}$. The discussion in this section follows the classical reference [Dziuk and Elliott (2013)], in which all the technical details can be found.

Recall that a triangulated surface is represented by a set of vertices and triangles $(\mathcal{P}, \mathcal{T})$, where $\mathcal{P}$ denotes the collection of mesh vertices and $\mathcal{T}$ is the set of triangular faces. For each triangle $T \in \mathcal{T}$, let $h(T)$ denote its diameter, and define the global mesh size by

$$h = \max_{T\in\mathcal{T}} h(T).$$

The union of all triangular faces constitutes a piecewise-flat surface, denoted by $\mathcal{M}_h = \bigcup_{T\in\mathcal{T}} T$. This discrete surface $\mathcal{M}_h$ serves as an approximation to a smooth, compact surface $\mathcal{M} \subset \mathbb{R}^3$.

### 6.5.1 *Projection, lift, and interpolation*

We introduce a few technical tools for the numerical analysis.

**Projection and lift.** To build a relation between $\mathcal{M}_h$ and $\mathcal{M}$, the following concepts are needed. First, for any $x \notin \mathcal{M}$, we define the projection onto $\mathcal{M}$ as

$$\mathrm{Proj}_{\mathcal{M}}(x) = \arg\min_{p\in\mathcal{M}} \|x - p\|. \tag{6.15}$$

Such a projection may not be uniquely defined, e.g., projection of the origin onto the unit sphere centered at the origin. On the other hand, by our intuition, $\mathrm{Proj}_{\mathcal{M}}(x)$ is uniquely defined if $x$ is very close to a smooth surface $\mathcal{M}$, which motivates the following definition [Niyogi *et al.* (2008)].

**Definition 6.5** (Tubular neighborhood and closest-point projection)**.** *Let* $\mathcal{M} \subset \mathbb{R}^n$ *be a* $C^2$ *embedded manifold. A radius* $\delta > 0$ *is called a* tubular radius *if every* $x \in U_\delta := \{x \in \mathbb{R}^n : \mathrm{Dist}(x, \mathcal{M}) < \delta\}$ *has a unique projection point* $Proj_{\mathcal{M}}(x) \in \mathcal{M}$. *The map* $Proj_{\mathcal{M}} : U_\delta \to \mathcal{M}$ *is the* closest-point projection. *The supremum of such* $\delta$ *is called the* reach *of* $\mathcal{M}$.

A tubular neighborhood $U_\delta$ always exists whenever $\mathcal{M}$ is compact and $C^2$. We assume that a sufficiently small mesh size $h$ is chosen so that $\mathcal{M}_h \subset U_\delta$ and the map

$$\mathrm{Proj}_{\mathcal{M}} : \mathcal{M}_h \to \mathcal{M}$$

is a bijection between $\mathcal{M}_h$ and $\mathcal{M}$. So with small enough mesh size, the projection $\mathrm{Proj}_{\mathcal{M}}$ provides a way for us to compare continuous and discrete surfaces.

Recall that $V_h = \mathrm{Span}\{\phi_i\}$ denotes the standard P1 finite element space, consisting of continuous piecewise linear functions defined on the mesh $\mathcal{M}_h$. We can *lift* a function in $V_h$ to a function defined on $\mathcal{M}$ as follows.

**Definition 6.6** (Lift of a function and lifted space)**.** *For any* $\phi_h \in V_h$*, it is a function* $\phi_h : \mathcal{M}_h \to \mathbb{R}$. *We define its* lift $\phi_h^\ell : \mathcal{M} \to \mathbb{R}$ *as*

$$\phi_h^\ell(p) = \phi_h(x), \quad \text{where } x \in \mathcal{M}_h \text{ satisfies } Proj_{\mathcal{M}}(x) = p.$$

In other words, the lifted function is defined by the inverse map of (6.15), i.e.,

$$\phi_h^\ell(p) = \phi_h[\mathrm{Proj}_{\mathcal{M}}^{-1}(p)].$$

The lifting operator permits an interpretation of the discrete problem as a perturbation of the continuous variational formulation posed on the exact surface $\mathcal{M}$. The corresponding *lifted finite element space* is $V_h^\ell = \{\phi_h^\ell \mid \phi_h \in V_h\}$. In Chapter 3, we have defined the Sobolev space $H^1(\Omega)$, and we have seen that the continuous FEM method on $\Omega \subset \mathbb{R}^2$ is *conforming* since the continuous piecewise polynomial space is a subset of the Sobolev space $H^1(\Omega)$. An important fact, for which we do not elaborate more, is that the lifted finite element space belongs to $H^1$ function space on the surface:

$$V_h^\ell = \{\phi_h^\ell \mid \phi_h \in V_h\} \subset H^1(\mathcal{M}).$$

**Estimates for interpolation, norms and bilinear forms.** In Section 3.2, for $\Omega \subset \mathbb{R}^d$, we have seen that $f \in H^2(\Omega)$ implies continuity of the function $f$ if the dimension satisfies $d \leq 3$. For a smooth surface ($d = 2$), similar results also hold, i.e., $f \in H^2(\mathcal{M})$ implies $f$ is continuous, thus its nodal values $f_j$ at vertices in $\mathcal{M}_h$ are well defined. With its nodal values, the linear interpolation is given as

$$I_h f = \sum_j f_j \phi_j \in V_h.$$

We can further define the lifted interpolation $I_h^\ell f$ as the lift of the piecewise polynomial function $I_h f$, and we have

$$I_h^\ell f \in V_h^\ell.$$

The best-approximation property of the discrete space is captured by the following interpolation bound. Detailed proofs can be found in [Dziuk and Elliott (2013), Sec. 4].

**Lemma 6.2** (Interpolation estimate on surfaces). *For any $f \in H^2(\mathcal{M})$, its lifted interpolant $I_h^\ell f \in V_h^\ell$ satisfies*

$$\|f - I_h^\ell f\|_{L^2(\mathcal{M})} + h\,\|\nabla_{\mathcal{M}}(f - I_h^\ell f)\|_{L^2(\mathcal{M})} \leq C h^2 \left(\|\nabla^2_{\mathcal{M}} f\|_{L^2(\mathcal{M})} + h \|\nabla_{\mathcal{M}} f\|_{L^2(\mathcal{M})}\right).$$

Because integration and differentiation are performed on different surfaces, geometric consistency must be quantified. The next result shows that norms of lifted and discrete functions differ only by $\mathcal{O}(h^2)$. Consequently, $\mathcal{M}_h$ is a second-order geometric approximation of $\mathcal{M}$.

**Lemma 6.3** (Norm equivalences under projection). *For any $\phi_h \in V_h$ with lift $\phi_h^\ell \in V_h^\ell$, for sufficiently small mesh size $h$, there is a constant $c$ depending only on curvature bounds of $\mathcal{M}$ so that the following estimates for norms and seminorms hold,*

$$(1 - ch^2)\,\|\phi_h^\ell\|^2_{L^2(\mathcal{M})} \leq \|\phi_h\|^2_{L^2(\mathcal{M}_h)} \leq (1 + ch^2)\,\|\phi_h^\ell\|^2_{L^2(\mathcal{M})}, \tag{6.16}$$

$$(1 - ch^2)\,|\phi_h^\ell|^2_{H^1(\mathcal{M})} \leq |\phi_h|^2_{H^1(\mathcal{M}_h)} \leq (1 + ch^2)\,|\phi_h^\ell|^2_{H^1(\mathcal{M})}. \tag{6.17}$$

**Lemma 6.4** (Bilinear form consistency). *For any $u_h, \varphi_h \in V_h$ with lifts $u_h^\ell, \varphi_h^\ell \in V_h^\ell$,*

$$|m(u_h^\ell, \varphi_h^\ell) - m_h(u_h, \varphi_h)| \leq C h^2\, \|u_h^\ell\|_{L^2(\mathcal{M})}\, \|\varphi_h^\ell\|_{L^2(\mathcal{M})}, \tag{6.18}$$

$$|a(u_h^\ell, \varphi_h^\ell) - a_h(u_h, \varphi_h)| \leq C h^2\, \|\nabla_{\mathcal{M}} u_h^\ell\|_{L^2(\mathcal{M})}\, \|\nabla_{\mathcal{M}} \varphi_h^\ell\|_{L^2(\mathcal{M})}. \tag{6.19}$$

### 6.5.2 Error estimates for the mean zero solution

Recall that we have discussed how the discrete compatibility condition can be enforced for a continuous function $f$. For a given compatible function $f \in L^2(\mathcal{M})$ with zero mean, it may not be continuous, but a piecewise polynomial $f_h \in V_h$ can be found as its $L^2$-norm projection onto $V_h$, defined by,

$$m_h(f_h, \varphi_h) = m(f, \varphi_h^\ell), \quad \forall \varphi_h \in V_h.$$

Notice that $\varphi_h \equiv 1$ has the lift $\varphi_h^\ell \equiv 1$, thus such a $L^2$-norm projection satisfies the discrete compatibility condition

$$m_h(f_h, 1) = \sum_{T \in \mathcal{T}} \int_T f_h \, \mathrm{d}A = 0.$$

By abusing notation, we let $f_h$ denote either the $L^2$-norm projection or the piecewise linear interpolation of shifted nodal values of $f$, such that $f_h \in V_h$ satisfies the discrete compatibility condition. For either case, with some additional assumptions such as smoothness of $f$, it can be shown that its lift $f_h^\ell$ is a second order accurate approximation:

$$\|f - f_h^\ell\|_{L^2(\mathcal{M})} \le c_f h^2,$$

where $c_f$ is a constant depending on the function $f$.

For the surface FEM scheme of seeking $u_h \in V_h$ satisfying

$$a_h(u_h, \varphi_h) = m_h(f_h, \varphi_h), \quad \forall \varphi_h \in V_h,$$

there is a unique zero mean FEM solution $U_h \in V_h$ satisfying

$$m_h(U_h, 1) = \sum_{T \in \mathcal{T}} \int_T U_h \, \mathrm{d}A = 0.$$

Now, we are ready to state the error estimates in $H^1$ seminorm and $L^2$-norm, with a sketchy proof.

**Theorem 6.1.** *Let $V_h$ be the finite element space on $\mathcal{M}_h$ and $V_h^\ell$ its lifted version on $\mathcal{M}$. Define $u_h := U_h^\ell$. Under all assumptions in this chapter, for sufficiently small $h \le h_0$,*

$$\|\nabla_{\mathcal{M}}(u - u_h)\|_{L^2(\mathcal{M})} \le c\,h\,\|f\|_{L^2(\mathcal{M})} + c\,\|f - f_h^\ell\|_{L^2(\mathcal{M})} \le Ch, \tag{6.20}$$

$$\|u - u_h\|_{L^2(\mathcal{M})} \le c\,h^2\,\|f\|_{L^2(\mathcal{M})} + c\,\|f - f_h^\ell\|_{L^2(\mathcal{M})} \le Ch^2. \tag{6.21}$$

*Proof.* From Lemma 6.4, for any $\phi_h \in V_h$ and its lift $\phi_h^\ell \in V_h^\ell \subset H^2(\mathcal{M})$,

$$\begin{aligned} a_h(\phi_h, \phi_h) &\geq a(\phi_h^\ell, \phi_h^\ell) - |a(\phi_h^\ell, \phi_h^\ell) - a_h(\phi_h, \phi_h)| \\ &\geq \|\nabla_{\mathcal{M}} \phi_h^\ell\|_{L^2(\mathcal{M})}^2 - ch^2 \|\nabla_{\mathcal{M}} \phi_h^\ell\|_{L^2(\mathcal{M})}^2 \end{aligned}$$

Therefore, for sufficiently small $h \leq h_0$, the discrete bilinear form $a_h$ is coercive:

$$a_h(\phi_h, \phi_h) \geq \tfrac{1}{2} \|\nabla_{\mathcal{M}} \phi_h^\ell\|_{L^2(\mathcal{M})}^2.$$

With the continuous and discrete weak forms

$$\begin{aligned} a(u, \varphi) &= m(f, \varphi), \quad \text{for any} \quad \varphi \in H^1(\mathcal{M}) \\ a_h(U_h, \phi_h) &= m_h(f_h, \phi_h), \quad \text{for any} \quad \phi_h \in V_h, \end{aligned}$$

Lemma 6.4 yields,

$$\begin{aligned} \tfrac{1}{2}\|\nabla_{\mathcal{M}}(u_h - \phi_h^\ell)\|_{L^2(\mathcal{M})}^2 &\leq a_h(U_h - \phi_h, U_h - \phi_h) \\ &= a(u - \phi_h^\ell, u_h - \phi_h^\ell) \\ &\quad + \left(a(\phi_h^\ell, u_h - \phi_h^\ell) - a_h(\phi_h, U_h - \phi_h)\right) \\ &\quad - \left(m(f, u_h - \phi_h^\ell - D) - m(f_h^\ell, u_h - \phi_h^\ell - D)\right) \\ &\quad - \left(m(f_h^\ell, u_h - \phi_h^\ell - D) - m_h(f_h, U_h - \phi_h - D)\right) \\ &\leq \|\nabla_{\mathcal{M}}(u - \phi_h^\ell)\|_{L^2(\mathcal{M})} \, \|\nabla_{\mathcal{M}}(u_h - \phi_h^\ell)\|_{L^2(\mathcal{M})} \\ &\quad + ch^2 \|\nabla_{\mathcal{M}} \phi_h^\ell\|_{L^2(\mathcal{M})} \, \|\nabla_{\mathcal{M}}(u_h - \phi_h^\ell)\|_{L^2(\mathcal{M})} \\ &\quad + \|f - f_h^\ell\|_{L^2(\mathcal{M})} \, \|u_h - \phi_h^\ell - D\|_{L^2(\mathcal{M})} \\ &\quad + ch^2 \|f_h^\ell\|_{L^2(\mathcal{M})} \, \|u_h - \phi_h^\ell - D\|_{L^2(\mathcal{M})}, \end{aligned} \tag{6.22}$$

where we have used the fact $m(f, D) = 0$ and $m_h(f_h, D) = 0$ for any constant $D \in \mathbb{R}$, since the means of $f$ on $\mathcal{M}$ and $f_h$ on $\mathcal{M}_h$ vanish.

If we choose

$$D = |\mathcal{M}|^{-1} \int_{\mathcal{M}} (u_h - \phi_h^\ell) \, \mathrm{dvol}_g,$$

then we can use *Poincaré inequality* (similar to Theorem 3.1):

$$\|u_h - \phi_h^\ell - D\|_{L^2(\mathcal{M})} \leq \tilde{c} \, \|\nabla_{\mathcal{M}}(u_h - \phi_h^\ell)\|_{L^2(\mathcal{M})},$$

which gives

$$\begin{aligned} \tfrac{1}{2}\|\nabla_{\mathcal{M}}(u_h - \phi_h^\ell)\|_{L^2(\mathcal{M})} \leq& \|\nabla_{\mathcal{M}}(u - \phi_h^\ell)\|_{L^2(\mathcal{M})} + ch^2 \|\nabla_{\mathcal{M}} \phi_h^\ell\|_{L^2(\mathcal{M})} \\ &+ c\|f - f_h^\ell\|_{L^2(\mathcal{M})} + ch^2 \|f\|_{L^2(\mathcal{M})}. \end{aligned}$$

Choosing $\phi_h^\ell = I_h^\ell u$ and invoking the interpolation estimates of Lemma 6.2,

$$\tfrac{1}{2}\|\nabla_{\mathcal{M}}(u_h - I_h^\ell u)\|_{L^2(\mathcal{M})} \le (1+ch^2)\|\nabla_{\mathcal{M}}(u - I_h^\ell u)\|_{L^2(\mathcal{M})} + ch^2\|\nabla_{\mathcal{M}}u\|_{L^2(\mathcal{M})} + c\|f - f_h^\ell\|_{L^2(\mathcal{M})} + ch^2\|f\|_{L^2(\mathcal{M})}.$$

Hence, using Lemma 6.2 again, we get

$$\|\nabla_{\mathcal{M}}(u_h - I_h^\ell u)\|_{L^2(\mathcal{M})} \le ch\|u\|_{H^2(\mathcal{M})} + c\|f - f_h^\ell\|_{L^2(\mathcal{M})} + ch^2\|f\|_{L^2(\mathcal{M})}.$$

The elliptic regularity $\|u\|_{H^2(\mathcal{M})} \le c\|f\|_{L^2(\mathcal{M})}$ gives

$$\|\nabla_{\mathcal{M}}(u - u_h)\|_{L^2(\mathcal{M})} \le ch\|f\|_{L^2(\mathcal{M})} + c\|f - f_h^\ell\|_{L^2(\mathcal{M})},$$

which proves (6.20).

The $L^2$-norm error estimate follows the *Aubin-Nitsche Lemma* in Chapter 3. Define the average of $u - u_h$ as

$$\overline{u - u_h} = \tfrac{1}{|\mathcal{M}|}\int_{\mathcal{M}} (u - u_h)\mathrm{dvol}_g,$$

then consider a dual problem to seek $v \in H^1(\mathcal{M})$ satisfying $m(v,1) = 0$ and

$$a(v,\varphi) = m(u - u_h - \overline{u - u_h}, \varphi), \quad \forall \varphi \in H^1(\mathcal{M}).$$

Then taking the test function $\varphi = u - u_h$ gives

$$a(u - u_h, v) = \|u - u_h\|_{L^2(\mathcal{M})}^2 - \tfrac{1}{|\mathcal{M}|}\left(\int_{\mathcal{M}} (u - u_h)\,\mathrm{dvol}_g\right)^2. \tag{6.23}$$

Using Lemma 6.4 and setting $\phi_h^\ell = I_h^\ell v \in V_h^\ell$,

$$\begin{aligned} a(u - u_h, v) &= a(u - u_h, v - \phi_h^\ell) + a(u - u_h, \phi_h^\ell) \\ &= a(u - u_h, v - \phi_h^\ell) + m(f, \phi_h^\ell) - m_h(f_h, \phi_h) \\ &\quad - (a(u_h, \phi_h^\ell) - a_h(U_h, \phi_h)) \\ &\le \|\nabla_{\mathcal{M}}(u - u_h)\|_{L^2(\mathcal{M})}\,\|\nabla_{\mathcal{M}}(v - \phi_h^\ell)\|_{L^2(\mathcal{M})} \\ &\quad + c(\|f - f_h^\ell\|_{L^2(\mathcal{M})} + h^2\|f_h^\ell\|_{L^2(\mathcal{M})})\,\|\nabla_{\mathcal{M}}\phi_h^\ell\|_{L^2(\mathcal{M})} \\ &\quad + ch^2\|\nabla_{\mathcal{M}}u_h\|_{L^2(\mathcal{M})}\,\|\nabla_{\mathcal{M}}\phi_h^\ell\|_{L^2(\mathcal{M})} \\ &\le c\big(h^2\|f\|_{L^2(\mathcal{M})} + \|f - f_h^\ell\|_{L^2(\mathcal{M})}\big)\|v\|_{H^2(\mathcal{M})} \\ &\le c\big(h^2\|f\|_{L^2(\mathcal{M})} + \|f - f_h^\ell\|_{L^2(\mathcal{M})}\big)\|u - u_h\|_{L^2(\mathcal{M})}, \end{aligned}$$

where we have used the elliptic regularity for the dual problem in the last step. For the mean term on the right of (6.23),

$$\begin{aligned} \tfrac{1}{|\mathcal{M}|}\left(\int_{\mathcal{M}} (u - u_h)\,\mathrm{dvol}_g\right)^2 &= c\,m(u - u_h, 1)^2 \\ &= c\,m(u_h, 1)^2 \\ &= c\,(m(u_h,1) - m_h(U_h,1))^2 \\ &\le ch^4\big(\|u\|_{L^2(\mathcal{M})}^2 + \|u - u_h\|_{L^2(\mathcal{M})}^2\big). \end{aligned}$$

Therefore,

$$
\begin{aligned}
(1 - ch^4)\|u - u_h\|^2_{L^2(\mathcal{M})} \leq & ch^4\|f\|^2_{L^2(\mathcal{M})} \\
& + c\big(h^2\|f\|_{L^2(\mathcal{M})} + \|f - f_h^\ell\|_{L^2(\mathcal{M})}\big)\|u - u_h\|_{L^2(\mathcal{M})},
\end{aligned}
$$

which for $h \leq h_0$ yields (6.21). □

## Chapter 7

# Laplacian on Point Cloud Representation of Manifolds

With the rapid advancement of sensing technologies and computational power, processing and analyzing geometric and scientific data in three or more dimensions has become a ubiquitous task across science, engineering, and data-driven modeling. In 3D acquisition and modeling, geometric objects such as surfaces are most naturally represented as *point clouds*, which are collections of discrete samples in space. Although many applications ultimately construct triangular meshes that encode both vertex positions and connectivity, the raw measurements obtained from scanners, simulations, or experiments are typically unstructured points. In higher-dimensional settings, such as data analysis, machine learning, or molecular dynamics, constructing a global mesh or structured grid is often infeasible. Hence, point clouds have emerged as a flexible and general representation for complex geometric data in high dimensions.

In many practical scenarios, these data points are not arbitrary but tend to concentrate near smooth low-dimensional structures. This motivates the *manifold assumption*: a data set can be viewed as samples from a smooth manifold $\mathcal{M}$ embedded in a possibly high-dimensional ambient space [Little *et al.* (2009b); Levina and Bickel (2004); Lin and Zha (2008)]. The manifold perspective provides a rigorous framework for capturing the intrinsic geometry of data, enabling the definition of tangent spaces, normal vectors, curvature, and geodesic distances. However, unlike signals or images defined on regular Euclidean grids, manifold-structured data lack global coordinates or connectivity, which makes it significantly more challenging to perform basic geometric and analytic operations, such as differentiation and integration.

On the other hand, a natural and unifying way to describe intrinsic geometry is through the language of geometric *PDEs on manifolds.*

Differential operators such as the gradient, divergence, and Laplace–Beltrami (LB) operator encode essential geometric and physical information. For example, the LB operator governs diffusion, heat flow, and harmonic functions, and its eigenfunctions provide intrinsic spectral coordinates that reveal the manifold's global structure [Belkin and Niyogi (2002); Coifman and Lafon (2006); Levy (2006); Sun *et al.* (2009); Belkin *et al.* (2009); Bronstein *et al.* (2010); Lai *et al.* (2010); Raviv *et al.* (2011)]. From this viewpoint, solving PDEs on manifolds not only enables geometric modeling and physical simulation but also bridges geometry, analysis, and data representation.

To transfer this continuous framework to discrete point cloud data, we need to construct numerical schemes that approximate intrinsic differential operators *directly from the samples*, without relying on global meshing or parameterization. Such a scheme allows geometric analysis, diffusion, and spectral computations to be performed intrinsically, using only the point cloud itself. The goal of this chapter is to present a systematic and unified framework for *solving PDEs on manifolds represented by point clouds.*

Unlike the classical finite element methods in Chapter 3 and Chapter 6, the analysis of the numerical method on point clouds in this chapter is underdeveloped, thus we only introduce the computational method, without discussing its convergence.

**Further reading:** This chapter is based on recent work of solving PDEs on point clouds discussed in [Lai *et al.* (2013); Liang *et al.* (2012); Liang and Zhao (2013)], where two systematic methods are introduced to approximate differential operators intrinsically at each point from a local approximation of the manifold and its metric by moving least squares or local mesh reconstruction through nearby neighbors. These methods achieve high accuracy and are capable of approximating general differential operators (possible non-diffusion type) on point clouds sampling manifolds with low intrinsic dimensions and possible high co-dimensions.

## 7.1 Moving least square method for solving PDEs on manifolds represented by point clouds

PDEs on smooth manifolds play a central role in geometry processing, physics, and data analysis. In many applications, however, the manifold is available only as a finite set of sampled points, without mesh connectivity or global parameterization. In this section, we present a general meshless framework for *solving PDEs directly on point clouds*, following [Liang *et al.* (2012); Liang and Zhao (2013)]. The key idea is to reconstruct the local geometry and function by *moving least squares (MLS)* in an intrinsic coordinate system derived from local principal component analysis (PCA). Once local metrics are obtained, differential operators such as the gradient, divergence, and Laplace–Beltrami operator can be consistently approximated and assembled into discrete systems for PDE solvers.

### 7.1.1 *Point cloud representation and local geometry*

Let's begin by describing how a smooth manifold can be represented by a finite set of discrete samples, or a *point cloud*, and how local geometric quantities can be recovered from this representation. Let the point cloud be

$$\mathcal{P} = \{\boldsymbol{p}_i \in \mathbb{R}^D \mid i = 1, \ldots, n\},$$

which consists of $n$ sample points in the ambient Euclidean space $\mathbb{R}^D$. We assume that these samples are drawn from a smooth $d$–dimensional Riemannian manifold $\mathcal{M} \subset \mathbb{R}^D$. Unlike a triangulated mesh, which encodes both vertex positions and connectivity information, a point cloud contains only positional data. Therefore, all geometric information, such as tangent directions, curvature, and local parametrizations, must be reconstructed locally from the distribution of nearby points.

For each point $\boldsymbol{p}_i \in \mathcal{P}$, we consider a neighborhood defined by its $K$–nearest neighbors:

$$\mathcal{P}(i) = \{\boldsymbol{p}_k \mid k \in \mathcal{I}(i)\}, \qquad |\mathcal{I}(i)| = K.$$

This local patch $\mathcal{P}(i)$ can be regarded as a discrete sampling of a small portion of the manifold around $\boldsymbol{p}_i$. The neighborhood size $K$ plays an important role: it must be sufficiently large to capture local geometric features, yet not too large so that the patch remains approximately flat in the intrinsic sense.

**Local PCA and tangent–normal decomposition.** The first step toward recovering local geometry is to estimate the *tangent space* and its orthogonal complement, the *normal space*, at each sample point using PCA. Intuitively, the tangent space describes the locally linear structure of the manifold, while the normal space captures how the manifold bends in the ambient space.

To this end, we compute the local barycenter and covariance matrix of the neighboring points:

$$\boldsymbol{c}_i = \frac{1}{K} \sum_{k \in \mathcal{I}(i)} \boldsymbol{p}_k, \qquad \mathrm{Cov}_i = \sum_{k \in \mathcal{I}(i)} (\boldsymbol{p}_k - \boldsymbol{c}_i)(\boldsymbol{p}_k - \boldsymbol{c}_i)^\top.$$

The covariance matrix $\mathrm{Cov}_i$ characterizes the spread of the neighborhood around $\boldsymbol{p}_i$. By performing eigen-decomposition

$$\mathrm{Cov}_i\, e_i^\alpha = \lambda_i^\alpha e_i^\alpha, \qquad \lambda_i^1 \geq \lambda_i^2 \geq \cdots \geq \lambda_i^D \geq 0,$$

we obtain an orthonormal basis $\{e_i^\alpha\}_{\alpha=1}^D$ of $\mathbb{R}^D$ together with the corresponding eigenvalues $\{\lambda_i^\alpha\}$, which quantify the variance of the local data along each direction.

For a well-sampled $d$–dimensional manifold, the first $d$ eigenvalues $\lambda_i^1, \ldots, \lambda_i^d$ are significantly larger than the remaining ones. The jump between $\lambda_i^d$ and $\lambda_i^{d+1}$ thus reveals the intrinsic dimension $d$. Accordingly, the tangent and normal spaces at $\boldsymbol{p}_i$ are defined as

$$\mathsf{T}_i = \mathrm{Span}\{e_i^1, \ldots, e_i^d\}, \qquad \mathsf{N}_i = \mathrm{Span}\{e_i^{d+1}, \ldots, e_i^D\},$$

so that the ambient space decomposes as $\mathbb{R}^D = \mathsf{T}_i \oplus \mathsf{N}_i$. In practice, the eigenvectors corresponding to the largest $d$ eigenvalues provide a reliable estimate of the local tangent directions. For noisy or non-uniform data, multiscale singular value decomposition (SVD) techniques can be used to improve robustness and determine the intrinsic dimension adaptively [Little *et al.* (2009a)].

**Local coordinates.** Having obtained an orthogonal basis $\{e_i^1, \ldots, e_i^D\}$ around $\boldsymbol{p}_i$, we can now introduce a local coordinate system in which the geometry of the manifold and functions defined on it can be described conveniently. Each neighboring point $\boldsymbol{p}_k \in \mathcal{P}(i)$ can now be expressed relative to $\boldsymbol{p}_i$ in this local basis:

$$(\boldsymbol{u}_k, \boldsymbol{v}_k) = \Big(\{\langle \boldsymbol{p}_k - \boldsymbol{p}_i, e_i^\alpha\rangle\}_{\alpha=1}^d, \{\langle \boldsymbol{p}_k - \boldsymbol{p}_i, e_i^\beta\rangle\}_{\beta=d+1}^D\Big).$$

Here $\boldsymbol{u}_k \in \mathbb{R}^d$ represents the *tangent coordinates* of $\boldsymbol{p}_k$, obtained by projecting onto the estimated tangent space $\mathsf{T}_i$, while $\boldsymbol{v}_k \in \mathbb{R}^{D-d}$ represents the *normal displacement* of $\boldsymbol{p}_k$ from this tangent plane.

This coordinate system provides an intrinsic local chart around $\boldsymbol{p}_i$. If the manifold is sufficiently smooth and the sampling is dense, the neighborhood $\mathcal{P}(i)$ can be viewed as a small graph of the form

$$\boldsymbol{v} = \boldsymbol{Z}_i(\boldsymbol{u}),$$

where $\boldsymbol{Z}_i : \mathbb{R}^d \to \mathbb{R}^{D-d}$ describes the local shape of the manifold as a function over its tangent plane. Such a local parametrization serves as the foundation for approximating geometric quantities and defining intrinsic differential operators directly on point clouds, as discussed in the following sections.

### 7.1.2 *Moving least squares approximation of the manifold*

Once a local coordinate system around each sample point is established, we can reconstruct the manifold geometry through a local approximation. Intuitively, near any point $\boldsymbol{p}_i$, a smooth manifold $\mathcal{M}$ can be well approximated by a small patch that behaves like a curved surface lying close to its tangent plane. To capture this local geometry, we model the manifold near $\boldsymbol{p}_i$ as a smooth graph over the tangent space $\mathsf{T}_i$:

$$(\boldsymbol{u}, \boldsymbol{Z}_i(\boldsymbol{u})) \subset \mathbb{R}^D,$$

where the function $\boldsymbol{Z}_i$ describes how the manifold deviates from its tangent plane along the normal directions.

More precisely, we write

$$\boldsymbol{Z}_i : \mathbb{R}^d \to \mathbb{R}^{D-d}, \qquad \boldsymbol{Z}_i(\boldsymbol{u}) = \big(Z_i^1(\boldsymbol{u}), \ldots, Z_i^{D-d}(\boldsymbol{u})\big),$$

where each component $Z_i^\mu$ represents the coordinate of the manifold in the $\mu$-th direction. To ensure sufficient smoothness for computing curvature and differential operators, we approximate each $Z_i^\mu$ by a polynomial of degree two:

$$Z_i^\mu(\boldsymbol{u}) = \sum_{|\boldsymbol{\xi}|\leq 2} c_{\boldsymbol{\xi}}^\mu \boldsymbol{u}^{\boldsymbol{\xi}},$$

where $\boldsymbol{u}^{\boldsymbol{\xi}} = u_1^{\xi_1} \cdots u_d^{\xi_d}$ and the coefficients $\{c_{\boldsymbol{\xi}}^\mu\}$ encode the local shape of the manifold.

**Moving Least Squares formulation.** The unknown coefficients are determined by fitting the local patch to the neighboring samples using the *moving least squares (MLS)* principle. At each $\boldsymbol{p}_i$, we solve the weighted least squares problem

$$\min_{\{c_{\boldsymbol{\xi}}^\mu\}} \sum_{k\in\mathcal{I}(i)} w(\|\boldsymbol{p}_k - \boldsymbol{p}_i\|)\, \|\boldsymbol{Z}_i(\boldsymbol{u}_k) - \boldsymbol{v}_k\|^2, \tag{7.1}$$

where $(\boldsymbol{u}_k, \boldsymbol{v}_k)$ are the local coordinates of $\boldsymbol{p}_k$ introduced earlier, and $w(\cdot) > 0$ is a weighting function that decays with distance from $\boldsymbol{p}_i$. The weights assign higher importance to points closer to $\boldsymbol{p}_i$, reflecting the fact that local geometry should be primarily determined by nearby samples.

This local optimization effectively finds a polynomial $\boldsymbol{Z}_i(\boldsymbol{u})$ that best approximates how neighboring points deviate from the tangent plane at $\boldsymbol{p}_i$. Since the problem is small and involves only local neighbors, it can be solved independently for each point and efficiently parallelized across the data set.

**Choice of weight functions.** The choice of the weighting function $w(d)$ influences both stability and locality of the approximation. Commonly used forms include

$$\begin{aligned}
&\text{Inverse-square: } w(d) = \frac{1}{d^2+\varepsilon^2},\\
&\text{Wendland:} \qquad w(d) = \left(1-\frac{d}{L}\right)^4\left(4\frac{d}{L}+1\right),\\
&\text{Equal-weight: } w(d) = \begin{cases} 1, & d=0,\\ 1/K, & d\neq 0,\end{cases}
\end{aligned}$$

where $d = \|\boldsymbol{p}_k - \boldsymbol{p}_i\|$ denotes the Euclidean distance in the ambient space. The inverse-square and Wendland kernels yield smooth distance-weighted fits and work well for uniformly sampled data. The equal-weight scheme, proposed in [Liang *et al.* (2012)], provides greater robustness for non-uniform or irregular sampling and produces stable results even when neighborhood densities vary significantly.

**Computation of the local metric tensor.** Once the local manifold patch $\boldsymbol{Z}_i(\boldsymbol{u})$ is determined, we can compute geometric quantities intrinsic to the manifold. A key quantity is the *metric tensor*, which characterizes how distances and angles are measured on the surface. For the local parametrization

$$\boldsymbol{x}(\boldsymbol{u}) = (u_1, \ldots, u_d, Z_i^1(\boldsymbol{u}), \ldots, Z_i^{D-d}(\boldsymbol{u})) \in \mathbb{R}^D, \tag{7.2}$$

the tangent vectors of the manifold at $\boldsymbol{u}$ are given by

$$\frac{\partial \boldsymbol{x}}{\partial u_s} = \boldsymbol{e}_i^s + \sum_{\mu=1}^{D-d} \frac{\partial Z_i^\mu}{\partial u_s} \boldsymbol{e}_i^{d+\mu}, \qquad s = 1, \ldots, d, \tag{7.3}$$

where $\{\boldsymbol{e}_i^1, \ldots, \boldsymbol{e}_i^D\}$ is the local orthonormal basis obtained from PCA. The induced Riemannian metric on the manifold is then computed as

$$g_{st}(\boldsymbol{u}) = \left\langle \frac{\partial \boldsymbol{x}}{\partial u_s}, \frac{\partial \boldsymbol{x}}{\partial u_t} \right\rangle = \delta_{st} + \sum_{\mu=1}^{D-d} \frac{\partial Z_i^\mu}{\partial u_s} \frac{\partial Z_i^\mu}{\partial u_t}, \qquad s,t = 1, \ldots, d. \tag{7.4}$$

In matrix form, this can be written as

$$G_i(\boldsymbol{u}) = I_d + (J_i(\boldsymbol{u}))^\top J_i(\boldsymbol{u}),$$

where $J_i(\boldsymbol{u}) \in \mathbb{R}^{(D-d)\times d}$ is the Jacobian matrix of $\boldsymbol{Z}_i(\boldsymbol{u})$. Thus, the local metric tensor $G_i$ measures how the manifold locally at the point $\boldsymbol{p}_i$ stretches or bends relative to its tangent plane.

Evaluating $G_i$ and its derivatives provides all the geometric ingredients needed to define intrinsic differential operators such as the gradient, divergence, and Laplace–Beltrami operator. The local coefficients of these operators are directly determined by $G_i$ and its inverse $G_i^{-1}$, linking the moving least squares approximation to the computation of intrinsic PDEs on point clouds.

The moving least squares approach offers several appealing properties:

- It is *local*: each polynomial fit depends only on a small neighborhood $\mathcal{P}(i)$, allowing the method to adapt to spatially varying curvature and sampling density.
- It is *intrinsic*: by operating in the tangent coordinate system, the approximation captures the manifold's geometry rather than that of the surrounding Euclidean space.
- It is *algebraic and general*: the formulation is purely based on least-squares fitting and extends naturally to manifolds of arbitrary dimension and co-dimension.

From a geometric viewpoint, the MLS polynomial $\boldsymbol{Z}_i(\boldsymbol{u})$ describes a smooth local chart that approximates the manifold near $\boldsymbol{p}_i$. Its Jacobian determines the local tangent vectors, and the induced metric tensor encodes how the manifold is curved in the ambient space. This local parametrization and metric serve as the foundation for constructing intrinsic differential operators and, ultimately, for solving PDEs directly on point clouds, as will be discussed in the next section.

### 7.1.3 *Local polynomial approximation of functions*

Once the local manifold geometry has been reconstructed, we turn to the approximation of scalar functions or vector fields defined on the manifold. Suppose a smooth function $f : \mathcal{M} \to \mathbb{R}$ (or a vector field $\boldsymbol{V} : \mathcal{M} \to \mathbb{R}^D$) is known only through its sampled values on the point cloud $\mathcal{P}$. To evaluate its differential quantities—such as gradients, divergences, or Laplacians—we first need a smooth local representation of $f$ near each sample point $\boldsymbol{p}_i$. This can be achieved by fitting a local polynomial model of $f$ in the same intrinsic coordinates used to approximate the manifold.

**Local polynomial model.** Given the local coordinates $\boldsymbol{u} = \{u_1, \cdots, u_d\}$ obtained from the tangent basis at $\boldsymbol{p}_i$, we represent the restriction of $f$ to the neighborhood $\mathcal{P}(i)$ by a polynomial expansion:

$$f_i(\boldsymbol{u}) = \sum_{|\boldsymbol{\xi}|\leq 2} b_{\boldsymbol{\xi}}\, \boldsymbol{u}^{\boldsymbol{\xi}}, \tag{7.5}$$

where $\boldsymbol{u}^{\boldsymbol{\xi}} = u_1^{\xi_1} \cdots u_d^{\xi_d}$ is the standard multi-index notation and the coefficients $\{b_{\boldsymbol{\xi}}\}$ encode the local variation of $f$ up to second order. The constant, linear, and quadratic terms of this polynomial respectively capture the local value, slope, and curvature of the function on the manifold patch.

To determine the coefficients $\{b_{\boldsymbol{\xi}}\}$, we again adopt a weighted least-squares formulation analogous to that used for approximating the manifold itself. Specifically, the coefficients minimize the objective

$$\min_{\{b_{\boldsymbol{\xi}}\}} \sum_{k\in\mathcal{I}(i)} w(\|\boldsymbol{p}_k - \boldsymbol{p}_i\|) \left(f_i(\boldsymbol{u}_k) - f(\boldsymbol{p}_k)\right)^2, \tag{7.6}$$

where $w(\cdot)$ is a positive, distance-based weight function. This optimization enforces that the polynomial $f_i(\boldsymbol{u})$ interpolates or approximates the observed function values $f(\boldsymbol{p}_k)$ within the neighborhood $\mathcal{P}(i)$, with stronger influence from points closer to $\boldsymbol{p}_i$. Since the problem is local and involves only a small number of neighbors, it can be efficiently solved using standard least-squares techniques, and the process can be repeated independently for all points.

This moving least squares procedure provides a smooth local surrogate of $f$ on the tangent coordinate system of $\mathcal{M}$. The fitted polynomial $f_i(\boldsymbol{u})$ serves as a local Taylor approximation to the true function, capturing its first-order and second-order variations intrinsic to the manifold. As a result:

- The zeroth-order term $b_{\mathbf{0}}$ approximates the function value $f(\boldsymbol{p}_i)$;
- The first-order coefficients $\{\partial_{u_s} f_i(\mathbf{0})\}_{s=1}^d$ provide the local gradient components of $f$ in the intrinsic directions;
- The second-order coefficients yield local Hessian information, which becomes essential in computing the Laplace–Beltrami operator and curvature-related quantities.

Thus, once the coefficients $\{b_{\boldsymbol{\xi}}\}$ are obtained, the local derivatives of $f$ with respect to $\boldsymbol{u}$ are immediately available through simple differentiation of the polynomial $f_i(\boldsymbol{u})$. In practice, evaluating these derivatives requires no additional fitting step: all partial derivatives can be computed analytically from the polynomial coefficients. Similarly, we can also have the same interpolation for vector fields defined on the point cloud.

**Geometric consistency.** It is important to emphasize that the above construction is *intrinsic*: the coordinates $\boldsymbol{u}$ are defined with respect to the local tangent space estimated from the point cloud, and the weighting function $w$ depends only on distances in the ambient space. This ensures that the polynomial approximation reflects the geometry of the manifold rather than that of the surrounding Euclidean space. Furthermore, since both the manifold reconstruction $\boldsymbol{Z}_i(\boldsymbol{u})$ and the function approximation $f_i(\boldsymbol{u})$ are expressed in the same coordinate system, they can be seamlessly combined to compute geometric differential operators such as the intrinsic gradient, divergence, and Laplace–Beltrami operator in the subsequent computation. In addition, these intrinsic differential operators are coordinate invariant by definition, thus the resulting numerical discretizations are geometrically consistent.

### 7.1.4 *Discrete differential operators on point clouds*

Once the local manifold geometry and function approximation are constructed, we are ready to approximate differential operators intrinsically defined on the manifold. This subsection bridges local geometric reconstruction and numerical computation by deriving the metric tensor, formulating differential operators in local coordinates, and discretizing them into sparse linear systems defined directly on the point cloud.

In the point cloud setting, each point $\boldsymbol{p}_i$ admits a local coordinate system $\boldsymbol{x}_i(\boldsymbol{u}) = (\boldsymbol{u}, \boldsymbol{Z}_i(\boldsymbol{u}))$ obtained from the MLS manifold reconstruction, together with a local MLS polynomial approximation $f_i(\boldsymbol{u})$ of a scalar function $f$ supported on the neighborhood $\mathcal{I}(i)$. Both the first derivatives $\partial f/\partial u_s$ and the metric tensor components $g_{st} = \langle \partial_{u_s}\boldsymbol{x}_i, \partial_{u_t}\boldsymbol{x}_i\rangle$ are directly available from the polynomial coefficients of the MLS fit in the aforementioned discussion. Consequently, all intrinsic differential quantities (tangent vectors, metric tensor, gradient, divergence, and Laplace–Beltrami) can be evaluated locally without any explicit mesh or global parametrization of the manifold.

**Linearity and matrix assembly.** A key observation is that the MLS differentiation procedure is *linear* in the nodal function values $\mathbf{f} = \{f(\boldsymbol{p}_k)\}_{k=1}^{n}$. More precisely, for each point $\boldsymbol{p}_i$, the MLS local fit takes the form

$$f_i(\boldsymbol{u}) = \sum_{k\in\mathcal{I}(i)} f(\boldsymbol{p}_k)\,\psi_{ik}(\boldsymbol{u}),$$

where the basis functions $\psi_{ik}$ depend only on the geometry of the neighborhood and the MLS weights. Differentiating the local polynomial gives

$$\left.\frac{\partial f}{\partial u_s}\right|_{\boldsymbol{p}_i} = \sum_{k\in\mathcal{I}(i)} f(\boldsymbol{p}_k)\left.\frac{\partial \psi_{ik}}{\partial u_s}\right|_{\boldsymbol{p}_i}, \qquad s = 1, \ldots, d.$$

Thus, each derivative at $\boldsymbol{p}_i$ is a weighted linear combination of neighboring values $f(\boldsymbol{p}_k)$, and the coefficients are known analytically from the MLS fit.

**Gradient matrix G.** The intrinsic gradient in local coordinates is

$$\nabla_{\mathcal{M}} f = \sum_{s,t=1}^{d} g^{st}\frac{\partial f}{\partial u_t}\,\partial_{u_s}\boldsymbol{x}_i.$$

Note that $\partial_{u_s}\boldsymbol{x}_i(\boldsymbol{p}_i)$ is a tangent vector at $\boldsymbol{p}_i$ represented as a vector in the ambient space $\mathbb{R}^D$ provided by (7.3), the resulting discrete gradient at $\boldsymbol{p}_i$ takes the form

$$(\nabla_{\mathcal{M}} f)(\boldsymbol{p}_i) = \sum_{k\in\mathcal{I}(i)} \mathrm{G}_{ik}\, f(\boldsymbol{p}_k), \qquad \mathrm{G}_{ik} \in \mathbb{R}^D.$$

Stacking these coefficients for all $i$ produces a sparse third-order tensor

$$\mathrm{G} \in \mathbb{R}^{n\times n\times D},$$

whose slice $\mathrm{G}_{i,:,:}$ acts on the vector $\mathbf{f}$ to produce the intrinsic gradient at $\boldsymbol{p}_i$. The sparsity pattern is localized: only entries with $k \in \mathcal{I}(i)$ are nonzero.

**Laplace–Beltrami matrix L.** The Laplace–Beltrami operator in local coordinates is

$$\Delta_{\mathcal{M}} f = \frac{1}{\sqrt{\det g}}\sum_{s,t=1}^{d} \partial_{u_s}\left(\sqrt{\det g}\, g^{st}\frac{\partial f}{\partial u_t}\right).$$

All quantities inside the derivative are known polynomials from the MLS manifold and function fits, so differentiating once more yields

$$(\Delta_{\mathcal{M}} f)(\boldsymbol{p}_i) = \sum_{k\in\mathcal{I}(i)} \mathrm{L}_{ik}\, f(\boldsymbol{p}_k), \qquad L_{ik} \in \mathbb{R}.$$

Assembling the coefficients $L_{ik}$ gives a sparse matrix

$$\mathrm{L} \in \mathbb{R}^{n\times n}$$

whose $i$-th row represents a local MLS-based stencil approximating $\Delta_{\mathcal{M}} f$ at $\boldsymbol{p}_i$. The structure of G and L mirror that of classical finite-difference discretizations but is entirely meshless and intrinsic to the manifold.

**Remark 7.1.** *To ensure numerical stability and physical consistency, particularly for diffusion type PDEs, one can further refine the discrete Laplacian weights $\{L_{ij}\}$ through a constrained quadratic optimization problem that enforces desirable matrix properties [Liang and Zhao (2013)]. Specifically, for each neighborhood $\mathcal{I}(i)$, the coefficients $\{L_{ij}\}_{j\in\mathcal{I}(i)}$ are chosen to minimize a quadratic form subject to linear constraints ensuring:*

$$L_{ii} < 0, \qquad L_{ij} \geq 0 \ (j \neq i), \qquad \sum_j L_{ij} = 0.$$

*These conditions guarantee that the discrete operator $L$ satisfies a discrete maximum principle, similar to discussions in Remark 2.12 for the M-matrix structure of the classical finite difference in Part I. The resulting matrix is diagonally dominant and behaves like an M-matrix (see Theorem 2.1), thereby ensuring monotonicity and stability even for non-uniform or noisy point clouds.*

### 7.1.5 *Solving PDEs on point clouds*

Once the intrinsic differential operators have been discretized, a wide range of partial differential equations defined on the manifold $\mathcal{M}$ can be solved directly on the point cloud $\mathcal{P}$, without the need for any global mesh or parameterization. The resulting framework is entirely meshless: geometric information is reconstructed locally, differential operators are assembled into sparse matrices, and standard numerical solvers can be applied on these discrete systems. We illustrate examples on how the discrete operators derived above enable the solution of various PDEs on point clouds.

**Elliptic problems and spectral analysis.** A fundamental example is the Laplace–Beltrami eigenvalue problem

$$-\Delta_{\mathcal{M}}\phi = \lambda\phi,$$

which arises in numerous applications including geometric processing, shape analysis, manifold learning, and spectral clustering. Discretizing the operator using the MLS-based Laplacian $L$ leads to the finite-dimensional eigenproblem

$$-\mathrm{L}\Phi = \lambda\Phi,$$

where $\Phi = [\phi(\boldsymbol{p}_1), \ldots, \phi(\boldsymbol{p}_n)]^\top$ collects the discrete values of the eigenfunction at the sample points. Since L is sparse and can be diagonally dominant under constrained optimization, standard sparse eigensolvers can be employed efficiently.

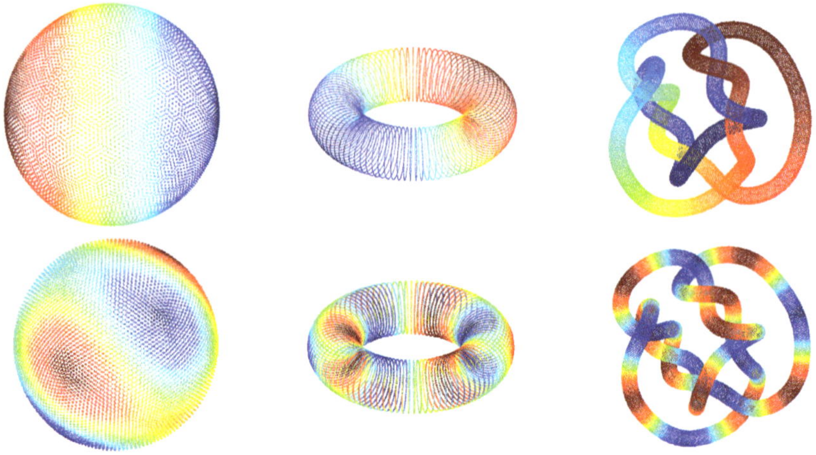

Fig. 7.1 Laplace-Beltrami eigenfunctions are color-coded on point clouds sampled from closed surfaces [Liang *et al.* (2012)]

The computed eigenpairs $(\lambda_j, \Phi_j)$ approximate the continuous spectrum of the Laplace–Beltrami operator on $\mathcal{M}$. The eigenfunctions $\Phi_j$ provide intrinsic coordinates on the manifold and reveal geometric structures such as symmetries, clusters, and diffusion geometry. They also serve as a spectral basis for expanding smooth functions or defining diffusion distances on $\mathcal{M}$, enabling spectral techniques to be applied directly to unstructured point clouds. Figure 7.1 illustrates a few examples of computing Laplace-Beltrami eigenfunctions on point clouds.

**Diffusion and heat flow.** Another important class of PDEs on manifolds is the diffusion (or heat) equation

$$\frac{\partial u}{\partial t} = \Delta_{\mathcal{M}} u,$$

which describes the evolution of scalar fields under intrinsic diffusion. Using the discrete Laplace–Beltrami operator L, the semi-discrete form becomes

$$\frac{dU(t)}{dt} = \mathrm{L}U(t),$$

where $U(t) = [u(\boldsymbol{p}_1, t), \ldots, u(\boldsymbol{p}_n, t)]^\top$. This linear system can be integrated in time by standard schemes. For instance, the standard implicit Crank–Nicolson time discretization scheme can be applied:

$$\left(I - \tfrac{\Delta t}{2}\mathrm{L}\right)U^{n+1} = \left(I + \tfrac{\Delta t}{2}\mathrm{L}\right)U^{n}.$$

It is well known that the Crank–Nicolson method is unconditionally stable and second order accurate in time, making it preferable for irregularly sampled point clouds and long-time diffusion processes. See Figure 7.2 for an example for solving the heat equation on Stanford bunny.

Fig. 7.2 Solving the heat equation on Stanford bunny [Liang and Zhao (2013)] The numerical solution is displayed at time $t = 0.1$, $0.2$, $0.4$ and $0.8$. The black dot is the heat source. A triangular mesh is used for better visualization, although solution is computed directly on point cloud.

**Reaction-Diffusion equation:** Similarly, the method extends naturally to reaction–diffusion systems on point clouds. Figure 7.3 shows the FitzHugh–Nagumo model generating a spiral wave on a point cloud representation of the unit sphere:

$$\frac{\partial u_s}{\partial t} = (a - u_s)(u_s - 1)u_s - v_s + \nu \Delta_{\mathcal{M}} u_s,$$

$$\frac{\partial v_s}{\partial t} = \epsilon(\beta u_s - v_s).$$

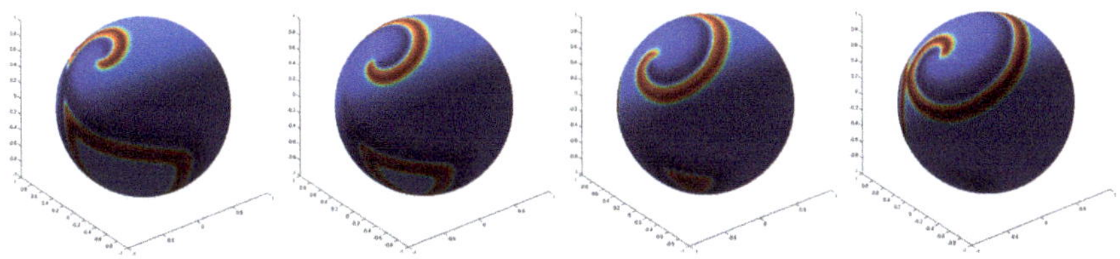

Fig. 7.3 Fitzhugh-Nagumo equation evolving on a sphere [Liang and Zhao (2013)] The excitation variable $u_s$ is displayed at time $t = 400$, $450$, $500$ and $550$. Triangular mesh is used for better visualization, although solution is computed based on a point cloud.

## 7.2 Local mesh method

In this section, we will discuss a different method for solving PDEs and variational problems directly on point clouds without requiring a global mesh. Global mesh construction is often difficult or impractical for raw point-sampled data, so the proposed approach relies instead on *local* mesh structures built around each point. At every point, a local tangent space is estimated via PCA, the neighboring points are projected onto this tangent plane, and a small triangulated patch is constructed. These overlapping local meshes supply the geometric framework needed to approximate intrinsic differential operators: local finite element formulas yield discrete mass and stiffness matrices, which are then assembled into global operators defined over the entire point cloud. This local mesh method enables a new way of numerically solving PDEs and variational models on point clouds. Examples include computing Laplace–Beltrami eigenfunctions and solving the Eikonal equation for distance-map computation and geodesic tracing, illustrating that accurate geometric and PDE-based computations can be achieved without constructing a global mesh.

### 7.2.1 *Local mesh construction and differential operators*

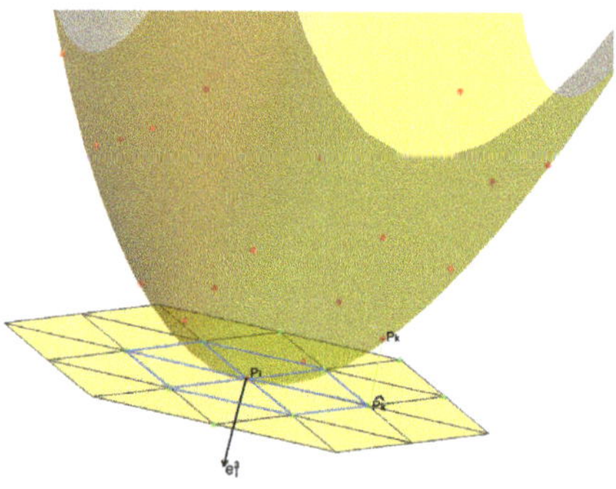

Fig. 7.4 Local mesh: Red stars mark the KNN (K-nearest neighbors) of $\boldsymbol{p}_i$, green stars mark the projection of red stars on the tangent plane at $\boldsymbol{p}_i$, and blue triangles color-code the connectivity of the first ring of $\boldsymbol{p}_i$.

The *local mesh method* proposed in [Lai *et al.* (2013)] provides an intrinsic framework for approximating differential operators directly on point clouds sampled from smooth manifolds. Unlike mesh-based methods, which require a global triangulation, the local mesh method constructs a small

Delaunay triangulation within each estimated tangent plane, thus preserving the manifold's local geometry while remaining coordinate invariant.

For simplicity, we assume that the point cloud $\mathcal{P} = \{\boldsymbol{p}_i \in \mathbb{R}^3\}_{i=1}^N$ is sampled from a smooth two-dimensional manifold $\mathcal{M} \subset \mathbb{R}^3$. The same idea extends naturally to point clouds sampled from manifolds of higher intrinsic dimension or higher codimension. Further details can be found in [Lai *et al.* (2013)]. At each point $\boldsymbol{p}_i$, we estimate the tangent plane $\mathcal{T}_i$ by local PCA on its $K$-nearest neighbors $\mathcal{N}(i)$. Denote the orthonormal frame by $\{\boldsymbol{e}_i^1, \boldsymbol{e}_i^2, \boldsymbol{e}_i^3\}$, with $\boldsymbol{e}_i^{1,2}$ spanning $\mathcal{T}_i$ and $\boldsymbol{e}_i^3$ as the normal vector. Each neighbor $\boldsymbol{p}_k \in \mathcal{N}(i)$ is projected onto $\mathcal{T}_i$ as

$$\hat{\boldsymbol{p}}_k = \mathrm{Proj}_{\mathcal{T}_i}(\boldsymbol{p}_k) = \boldsymbol{p}_k - \boldsymbol{p}_i - \langle \boldsymbol{p}_k - \boldsymbol{p}_i, \boldsymbol{e}_i^3 \rangle \boldsymbol{e}_i^3, \qquad k \in \mathcal{N}(i). \tag{7.7}$$

A standard Delaunay triangulation on $\{\hat{\boldsymbol{p}}_k\}$ defines a local mesh around $\boldsymbol{p}_i$. Let $\mathcal{R}(i) = \{T_i^1, \ldots, T_i^{\ell_i}\}$ denote all triangles incident to $\boldsymbol{p}_i$ (the first ring), and $\mathcal{V}(i)$ be the set of neighboring vertices. We call $\mathcal{C}_i = \{\boldsymbol{p}_i; \mathcal{V}(i), \mathcal{R}(i)\}$ the *local connectivity* of $\boldsymbol{p}_i$.

Within this first-ring structure, geometric differential operators are approximated intrinsically in local coordinates. For any scalar function $f$ and vector field $\boldsymbol{V}$ defined on $\mathcal{P}$, the gradient and divergence are computed by area-weighted averaging over the adjacent triangles:

$$\nabla_{\mathcal{P}} f(\boldsymbol{p}_i) = \frac{\sum_{T \in \mathcal{R}(i)} \mathrm{Area}(T) \nabla_T f(\boldsymbol{p}_i)}{\sum_{T \in \mathcal{R}(i)} \mathrm{Area}(T)},$$

$$\mathrm{div}_{\mathcal{P}} \boldsymbol{V}(\boldsymbol{p}_i) = \frac{\sum_{T \in \mathcal{R}(i)} \mathrm{Area}(T) \mathrm{div}_T \boldsymbol{V}(\boldsymbol{p}_i)}{\sum_{T \in \mathcal{R}(i)} \mathrm{Area}(T)}. \tag{7.8}$$

Furthermore, the Laplace–Beltrami operator is approximated by the cotangent formula

$$\Delta_{\mathcal{P}} f(\boldsymbol{p}_i) \approx \sum_{j \in \mathcal{V}(i)} \omega_{ij}(\boldsymbol{p}_i) \big( f(\boldsymbol{p}_j) - f(\boldsymbol{p}_i) \big), \tag{7.9}$$

where $\omega_{ij}(\boldsymbol{p}_i) = \frac{1}{2}\big(\cot \alpha_{ij}(\boldsymbol{p}_i) + \cot \beta_{ij}(\boldsymbol{p}_i)\big)$, and $\alpha_{ij}(\boldsymbol{p}_i)$, $\beta_{ij}(\boldsymbol{p}_i)$ are the two angles opposite to the edge $\overline{\boldsymbol{p}_i \boldsymbol{p}_j}$ in the local triangulation, as shown in Figure 6.4.

**Geometric consistency.** The above construction is *intrinsic*: the local coordinates are defined with respect to $\mathcal{T}_i$, and the weighting functions depend only on intrinsic geometric quantities such as edge lengths and triangle areas. As a result, the computed gradient, divergence, and Laplacian are coordinate invariant and geometrically consistent with the underlying manifold geometry.

### 7.2.2 Finite element analogue

The local mesh structure naturally induces a finite element analogue on the point cloud. This enables the computation of elliptic operators such as the Laplace–Beltrami eigensystem

$$-\Delta_{\mathcal{M}}\phi = \lambda\phi$$

through the construction of discrete mass and stiffness matrices defined using the local connectivity $\mathcal{C} = \{\mathcal{C}_i\}_{i=1}^N$.

**Mass matrix.** Let $\{\eta_j\}_{j=1}^N$ be a set of nodal "hat" functions on $\mathcal{P}$, defined as the piecewise-linear interpolation over $\mathcal{C}_i$ satisfying $\eta_j(\boldsymbol{p}_i) = \delta_{ij}$. We approximate their inner products as

$$\langle \eta_i, \eta_j \rangle = \begin{cases} \dfrac{1}{6}\displaystyle\sum_{k=1}^{\ell_i} \text{Area}(T_i^k), & i = j, \\ \dfrac{1}{12}\left[\text{Area}(T_i^{i_1}) + \text{Area}(T_i^{i_2})\right], & \text{if } \boldsymbol{p}_i\boldsymbol{p}_j \text{ is an edge shared by } T_i^{i_1}, T_i^{i_2}, \\ 0, & \text{otherwise.} \end{cases} \tag{7.10}$$

The symmetric mass matrix $\mathbb{M} = (m_{ij})$ is then defined by

$$m_{ij} = \int_{\mathcal{P}} \eta_i\eta_j \approx \frac{1}{2}\big(\langle \eta_i, \eta_j \rangle + \langle \eta_j, \eta_i \rangle\big). \tag{7.11}$$

**Stiffness matrix.** The stiffness matrix $\mathbb{S} = (S_{ij})$ is constructed by approximating $\int_{\mathcal{P}} \nabla_{\mathcal{P}}\eta_i \cdot \nabla_{\mathcal{P}}\eta_j$. Using the first-ring structure, we define

$$A_{ij} = \sum_{T\in\mathcal{R}(i)} \int_T \nabla_T\eta_i \cdot \nabla_T\eta_j = -\frac{1}{2}\left[\cot\alpha_{ij}(\boldsymbol{p}_i) + \cot\beta_{ij}(\boldsymbol{p}_i)\right], \qquad i \neq j. \tag{7.12}$$

Since local triangulations $\mathcal{R}(i)$ and $\mathcal{R}(j)$ may not coincide, $A_{ij}$ and $A_{ji}$ need not be equal. To ensure symmetry and numerical stability, two symmetrizations are commonly used:

$$S_{ij} = \begin{cases} \dfrac{1}{2}(A_{ij} + A_{ji}), & i \neq j, \\ -\displaystyle\sum_{k\neq i} S_{ik}, & i = j, \end{cases} \tag{7.13}$$

or the max/min symmetrization

$$S_{ij} = \begin{cases} \max(A_{ij}, A_{ji}), & A_{ij}, A_{ji} \le 0, \\ \min(A_{ij}, A_{ji}), & A_{ij}, A_{ji} \ge 0, \\ \min(A_{ij}, A_{ji}), & A_{ij} \cdot A_{ji} < 0, \\ -\sum_{k \neq i} S_{ik}, & i = j. \end{cases} \tag{7.14}$$

Both definitions guarantee the consistency condition $S\mathbf{1} = 0$, ensuring that constant functions are eigenfunctions with zero eigenvalue.

### 7.2.3 *Solving equations on point clouds*

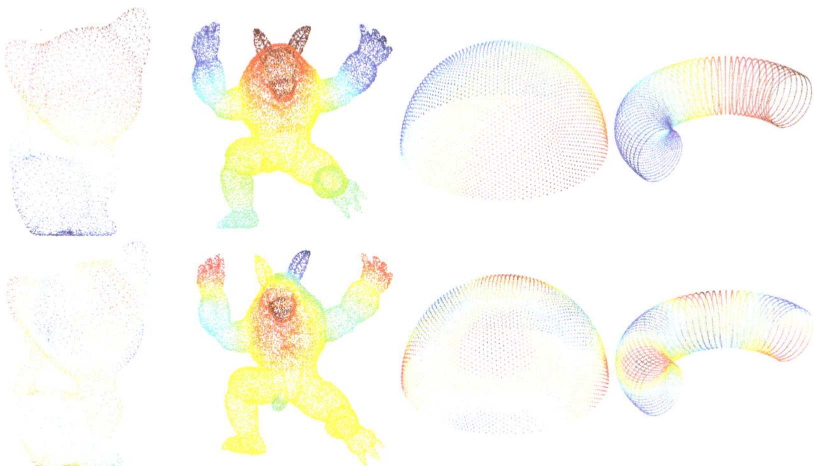

Fig. 7.5 Laplace-Beltrami eigenfunctions are color-coded on point clouds sampled from open and closed surfaces using local mesh method [Lai *et al.* (2013)]

**Laplace-Beltrami eigenvalue problem** The solutions to the Laplace–Beltrami (LB) eigenvalue problem can be computed from the following generalized matrix eigen-problem:

$$\begin{cases} \mathbb{S}v = \lambda \mathbb{M}v, \text{ where } v = (v_1, \cdots, v_N)^T, \\ \phi = \displaystyle\sum_{i=1}^{N} v_i \eta_i. \end{cases} \tag{7.15}$$

Note that the stiffness matrix $\mathbb{S}$ and the mass matrix $\mathbb{M}$ are symmetric and sparse matrices of size $N \times N$. Thus, all eigenvalues and eigenfunctions computed from the above method are real. This approach is not limited to LB eigenproblems and can be extended to other PDEs such as the Poisson equation, heat equation, and reaction–diffusion systems on point clouds.

**Eikonal equation on manifolds.** Such a local mesh method can also be adapted to solve non-diffusion type PDEs such as the Eikonal equation

$$\begin{cases} |\nabla_{\mathcal{M}} d(x)| = 1, \\ d(x) = 0, \quad x \in \Gamma \subset \mathcal{M}, \end{cases} \tag{7.16}$$

which defines the geodesic distance map $d(x)$ to a given set $\Gamma$ on a manifold $\mathcal{M}$. In particular, setting $\Gamma = \{p\}$ yields the distance map to a single point $p$. Since the first-ring local mesh structure provides a natural triangulation, we can directly apply the monotone upwind discretization for triangular meshes [Kimmel and Sethian (1998); Qian *et al.* (2007)], which is equivalent to approximating the dynamic programming principle locally on the first ring.

Efficient solvers such as the *fast marching method* [Sethian (1996); Kimmel and Sethian (1998)] or the *fast sweeping method* [Zhao (2005); Qian *et al.* (2007)] can then be applied using only the local mesh connectivity. Thus, the local mesh method can be seamlessly integrated with these schemes to solve the Eikonal equation directly on point clouds. This yields more accurate distance maps than earlier approaches such as [Mémoli and Sapiro (2005)], which approximate the Eikonal equation in a tubular neighborhood of $\mathcal{P} \subset \mathbb{R}^D$ rather than intrinsically on the manifold.

Fig. 7.6 Distance functions obtained from the Eikonal equation are color-coded on the point clouds, and geodesics are illustrated by black curves. Results are based on the local mesh method [Lai *et al.* (2013)].

## 7.3 Solving PDEs on incomplete-distance data

Incomplete inter-point distances arise naturally in applications such as protein structure determination from NMR spectroscopy [Crippen and Havel (1988); Berger *et al.* (1999); Leung and Toh (2009); Fang and Toh (2013)] and sensor network localization [Ji and Zha (2004); Biswas *et al.* (2006)]. In this data model, the input is not a set of embedded points but an *incomplete* distance matrix on an unknown point set $\{\boldsymbol{p}_i\}$ sampled from a manifold $\mathcal{M}$. While distance geometry traditionally focuses on recovering global coordinates from these distances, there has been less exploration of solving *geometric PDEs* directly from (incomplete) distances. A local strategy was proposed in [Lai and Li (2018)] to reconstruct only *local* coordinates sufficient to approximate intrinsic differential operators pointwise, then assemble numerical solvers for PDEs without global embedding.

Two key observations motivate this approach. First, geometric differential operators (gradient, divergence, Laplace–Beltrami) are *local* and defined pointwise from the manifold metric; second, these operators are invariant under rigid motions and local coordinate choices. Thus, it suffices to perform *local* coordinate reconstruction around each point from available distances, approximate operators intrinsically in the local chart, and build PDE solvers from these local pieces.

### 7.3.1 *Problem setup and local coordinate reconstruction*

We consider the setting where the discretization of $\mathcal{M}$ is given by an unknown point set $\{\boldsymbol{p}_1, \ldots, \boldsymbol{p}_n\} \subset \mathcal{M} \subset \mathbb{R}^D$, but the coordinates of these points are not provided. Instead, only a subset of their pairwise Euclidean distances is known:

$$D_{ij} = \|\boldsymbol{p}_i - \boldsymbol{p}_j\|, \qquad (i,j) \in \Omega,$$

where $\Omega \subset \{(i,j) \mid 1 \leq j < i \leq n\}$ is the index set of known measurements. If the full distance matrix were available, classical multidimensional scaling (MDS) would allow exact coordinate reconstruction. Here, however, only an *incomplete* distance matrix is given, and yet the objective is to approximate intrinsic differential operators and solve PDEs (e.g. the Laplace–Beltrami eigenproblem, Poisson, heat and Eikonal equations) directly from the incomplete distances. To this end, coordinates need to be reconstructed first. A computational framework for reconstructing local coordinates proceeds as follows.

*Step 1: KNN search.*

For finding KNN (K-nearest neighbors) for each point $\boldsymbol{p}_i$, choose a neighborhood $\mathcal{I}(i)$ of size $K$ using the available distances (or side information if provided), and form the local incomplete distance matrix $D \in \mathbb{R}^{K\times K}$ over $\mathcal{I}(i)$.

*Step 2: Local coordinate reconstruction (low-rank completion).*

Suppose $\mathcal{I}(i)$ contains points $\boldsymbol{p}_1, \cdots, \boldsymbol{p}_K$. For $\mathcal{I}(i)$, the Gram matrix $B \in \mathbb{R}^{K\times K}$ is defined as $B_{ij} = \langle \boldsymbol{p}_i, \boldsymbol{p}_j\rangle$. The Gram matrix satisfies $B = P^\top P$ where columns of $P$ are coordinates of $\boldsymbol{p}_i$. The classical multi-dimensional scaling (MDS) relation between a Gram matrix $B \succeq 0$ and squared distances is

$$D_{ij} = B_{ii} + B_{jj} - 2B_{ij}.$$

Our goal here is to reconstruct the Gram matrix using given data, which are the entries of distance matrix in a subset $\Omega \subset \{(i,j) \mid 1 \le j < i \le K\}$. In other words, the given data for $B$ can be formulated as

$$B_{ii} + B_{jj} - 2B_{ij} = D_{ij}, \quad (i,j) \in \Omega, \tag{7.17}$$

which is a linear constraint about unknown Gram matrix $B$.

For enforcing translation elimination, we can impose $B\mathbf{1} = \mathbf{0}$, which is equivalent to $\sum_i \boldsymbol{p}_i = \mathbf{0}$.

Notice that $B = PP^\top$ implies that $B$ is real symmetric positive semi-definite and the rank of $B$ should not exceed the dimension of ambient Euclidean space.

For reconstructing the unknown Gram matrix $B$, we seek a low-rank matrix $B$ satisfying the properties above:

$$\min_{B\in\mathcal{S}} \operatorname{Rank}(B) \quad \text{s.t.} \quad B \succeq 0, \;\; \mathcal{R}_\Omega(B) = D_\Omega, \;\; B\mathbf{1} = \mathbf{0}, \tag{7.18}$$

where $\mathcal{S} = \{B \in \mathbb{R}^{K\times K} \mid B = B^\top\}$ denotes symmetric matrices, and $\mathcal{R}_\Omega(B) = D_\Omega$ denotes the linear constraint (7.17), i.e., $D_\Omega$ is the restriction of $D$ on the index set $\Omega$, and $\mathcal{R}_\Omega(B)(i,j) = B_{ii} + B_{jj} - 2B_{ij}$ for $(i,j) \in \Omega$.

Following matrix completion principles [Candès and Recht (2008)], we relax rank to the matrix nuclear norm. See Appendix A.3 for the definition of the nuclear norm of a matrix. Let $\zeta : \mathcal{S} \to \mathbb{R}^{K(K+1)/2}$ be the canonical vectorization with inverse $\iota$, and define

$$\mathcal{A} : \mathbb{R}^{K(K+1)/2} \to \mathbb{R}^{|\Omega|} \times \mathbb{R}^K, \qquad \hat{B} \mapsto \big(\mathcal{R}_\Omega(\iota(\hat{B})),\; \iota(\hat{B})\mathbf{1}\big),$$

with target $\widetilde{D}_\Omega = (D_\Omega, \mathbf{0})$. Since $B \succeq 0$, we have $\|B\|_* = \operatorname{Tr}(B)$, thus it becomes a SDP (semi-definite programming)

$$\min_{\hat{B}\in\mathbb{R}^{K(K+1)/2}} \operatorname{Tr}(\iota(\hat{B})) \quad \text{s.t.} \quad \mathcal{A}\hat{B} = \widetilde{D}_\Omega, \;\; \iota(\hat{B}) \succeq 0. \tag{7.19}$$

**Algorithm 1** ADMM for local Gram matrix completion (7.19)

**Init:** $\hat{C}^0 = 0,\ H_1^0 = 0,\ H_2^0 = 0$; choose $\mu_1, \mu_2 > 0$ and tolerance $\varepsilon$.
**while** $\big|E(k) - E(k-1)\big|/E(k) \geq \varepsilon$ **do**
  Solve (7.21) for $\hat{B}^{k+1}$.
  $\hat{C}^{k+1} = \zeta \circ \mathcal{T}_E \circ \iota(\hat{B}^{k+1} + H_2^k)$.
  $H_1^{k+1} = H_1^k + \mathcal{A}\hat{B}^{k+1} - \widetilde{D}_\Omega$;
  $H_2^{k+1} = H_2^k + \hat{B}^{k+1} - \hat{C}^{k+1}$.
  $E(k+1) = \mathrm{Tr}(\iota(\hat{B}^{k+1})) + \frac{\mu_1}{2}\|\mathcal{A}\hat{B}^{k+1} - \widetilde{D}_\Omega\|_2^2 + \frac{\mu_2}{2}\|\hat{B}^{k+1} - \hat{C}^{k+1}\|_2^2$.
**end while**
**Return:** $B = \iota(\hat{B})$ and its MDS coordinates on $\mathcal{I}(i)$.

Theoretical guarantees differ from standard *restricted isometry property* settings [Recht *et al.* (2010)], but under coherence-type conditions, exact local completion can still be proved by dual-basis analysis [Tasissa and Lai (2018)]. After solving (7.19), local coordinates can be recovered by the eigendecomposition of $B$, since the Gram matrix satisfies $B = PP^\top$.

Interior-point SDP solvers can become expensive when constraints scale as $\mathcal{O}(K^2)$. Instead, we may consider an ADMM splitting (see Appendix C) by introducing an auxiliary $\hat{C} = \hat{B}$:

$$\min_{\hat{B},\hat{C}} \ \mathrm{Tr}(\iota(\hat{B})) \quad \text{s.t.} \quad \mathcal{A}\hat{B} = \widetilde{D}_\Omega, \quad \iota(\hat{C}) \succeq 0, \quad \hat{B} = \hat{C}. \tag{7.20}$$

A practical implementation is summarized in Algorithm 1, in which the updates are

$$(\mu_1 \mathcal{A}^*\mathcal{A} + \mu_2)\hat{B}^{k+1} = \mu_1 \mathcal{A}^*(\widetilde{D}_\Omega - H_1^k) + \mu_2(\hat{C}^k - H_2^k) - \zeta(I_K), \tag{7.21}$$

$$\hat{C}^{k+1} = \zeta \circ \mathcal{T}_E \circ \iota(\hat{B}^{k+1} + H_2^k), \tag{7.22}$$

where $\mathcal{T}_E(X) = V\,\mathrm{diag}(\max\{\Lambda_{ii}, 0\})\,V^\top$ is eigenvalue hard-thresholding of a symmetric matrix $X = V\Lambda V^\top$.

### 7.3.2 *Operator approximation from local reconstructions*

Once local coordinates for each neighborhood $\mathcal{I}(i)$ are obtained, intrinsic differential operators at $\boldsymbol{p}_i$ can be approximated by either:

- **MLS method** (Section 7.1): fit local polynomials in tangent coordinates and recover $\nabla_\mathcal{M}$, $\mathrm{div}_\mathcal{M}$, and $\Delta_\mathcal{M}$ from the fitted jets;
- **Local mesh method** (Section 7.2): build a Delaunay triangulation in the local chart and use area/cotangent formulae to assemble per-ring gradient/divergence and the Laplacian.

Both are *intrinsic* (coordinate invariant) and use only local reconstructions, and no global embedding is required. Row-by-row, we assemble discrete operators and key components in PDE solvers such as FEM-like mass and stiffness matrices. By combining local low-rank distance completion with MLS or local mesh discretizations, we obtain an intrinsic, meshless framework for solving PDEs directly from incomplete distance data—circumventing global embedding while retaining accuracy, robustness, and scalability [Lai and Li (2018)].

### 7.3.3 *Numerical examples*

Consider points $\{\boldsymbol{p}_1, \ldots, \boldsymbol{p}_n\}$ sampled uniformly distributed on a manifold $\mathcal{M} \subset \mathbb{R}^D$. Let $D$ denote the complete pairwise Euclidean distance matrix. To simulate incomplete measurements, select a random subset $\Omega_\gamma \subset \{(i,j) \mid 1 \le j < i \le n\}$ containing $|\Omega_\gamma| = \lceil \gamma n(n-1)/2 \rceil$ entries, and restrict to the $K$ nearest neighbors of each point:

$$D_{\gamma,K} = \{D(i,j) \mid j \in \mathcal{I}_K(i),\ (i,j) \in \Omega_\gamma,\ i = 1, \ldots, n\}.$$

This subsampled matrix $D_{\gamma,K}$ typically retains only a fraction $\gamma K/n$ of all distances, which can be extremely sparse when $K \ll n$. Local coordinate reconstruction (Step 1) and differential operator approximation (Step 2) are then performed based solely on this incomplete distance information.

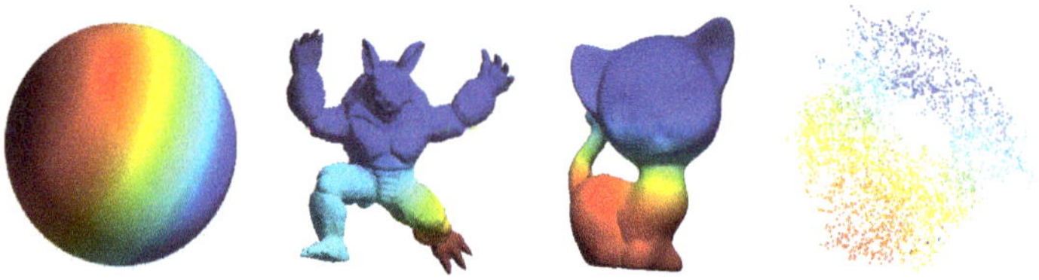

Fig. 7.7 The first nontrivial LB eigenfunctions are color-coded on the manifolds based on incomplete distance. From left to right: The unit sphere with $D_{50\%,30}$, armadillo with $D_{50\%,30}$, kitten with $D_{50\%,30}$ and a protein data with $D_{3\%,\infty}$ [Lai and Li (2018)].

**Laplace–Beltrami eigenvalue problem** The reconstructed local coordinates provide the necessary structure to assemble the discrete Laplace–Beltrami operator and associated mass matrix. The resulting generalized eigenvalue problem,

$$\mathbb{S}v = \lambda \mathbb{M}v,$$

is used to approximate the intrinsic spectrum of the manifold directly from incomplete distance data. Numerical experiments examine the convergence of the computed eigenvalues and eigenfunctions with respect to the sampling parameters $(\gamma, K)$, as well as the method's robustness under Gaussian noise perturbations. Additional tests extend the formulation to manifolds of higher intrinsic or ambient dimension, confirming that accurate spectral approximations can be achieved even from a small fraction of pairwise distances.

Table 7.1 Comparison of computational time (minutes) of the LB eigenvalue based on local/global reconstruction methods.

| number of points | | | | |
|---|---|---|---|---|
| 1002 | 1962 | 4002 | 7842 | 16002 |
| $\gamma = 100\%$, $K = 6$ | | | | |
| 0.26 | 0.51 | 1.01 | 2.03 | 4.05 |
| $\gamma = 80\%$, $K = 9$ | | | | |
| 2.28 | 5.60 | 11.17 | 22.28 | 45.02 |
| $\gamma = 50\%$, $K = 18$ | | | | |
| 4.03 | 8.09 | 16.14 | 32.44 | 64.71 |
| $\gamma = 30\%$, $K = 30$ | | | | |
| 15.13 | 30.19 | 60.42 | 120.95 | 241.63 |
| **global** reconstruction using 3% distance ($K = 6$ for MLS) | | | | |
| 2.09 | 9.86 | 40.13 | 154.40 | 597.06 |

The local coordinate reconstruction is performed independently for each neighborhood, leading to significant advantages in both computational efficiency and memory usage. In contrast to global reconstruction approaches that recover all coordinates before approximating differential operators, the local strategy reconstructs only the $K$ nearest neighbors of each point. The most expensive step of the augmented Lagrangian solver is the eigenvalue hard-thresholding operation. For an incomplete distance matrix of size $n$, the cost of global reconstruction scales as $\mathcal{O}(n^2 m)$ when computing the largest $m$ eigenvalues. The local strategy reduces this to $\mathcal{O}(nK^2 m)$, since only $K$-neighborhoods are processed. When $n$ is large and $K \leq 30$, the method achieves a speedup on the order of $n/K$ and naturally supports parallel execution across neighborhoods.

The memory requirement is also reduced from $\mathcal{O}(n^2)$ for the dense global matrix to $\mathcal{O}(nK^2)$ for local reconstruction, avoiding the prohibitive storage demands of global completion. Empirical results in Table 7.1 confirm that the local approach substantially accelerates PDE computation. For example, in computing the first 100 Laplace–Beltrami eigenvalues with $n = 16002$ and $K = 6$, the local reconstruction is approximately 150 times faster than the global method.

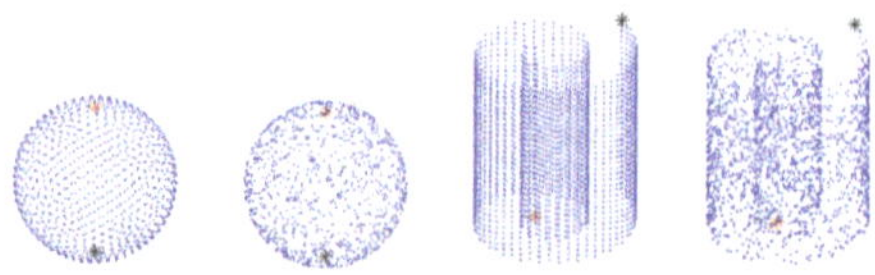

| | | 1002 | 1962 | 4002 | 7842 | 16002 |
|---|---|---|---|---|---|---|
| | | **Uniform sampling on $S^2$** | | | | |
| DK | $E_{av}$ | 0.0348 | 0.0285 | 0.0248 | 0.0232 | 0.0224 |
| | $E_{se}$ | 0.00862 | 0.00861 | 0.00830 | 0.01064 | 0.01150 |
| FMD | $E_{av}$ | 0.0113 | 0.0095 | 0.0080 | 0.0064 | 0.0022 |
| | $E_{se}$ | 0.00810 | 0.00589 | 0.00411 | 0.00288 | 0.00216 |
| | | **Non-Uniform sampling on $S^2$** | | | | |
| DK | $E_{av}$ | 0.0363 | 0.0344 | 0.0319 | 0.0305 | 0.0294 |
| | $E_{se}$ | 0.01121 | 0.01609 | 0.01838 | 0.01639 | 0.01995 |
| FMD | $E_{av}$ | 0.0200 | 0.0163 | 0.0141 | 0.0124 | 0.0088 |
| | $E_{se}$ | 0.01202 | 0.00879 | 0.00374 | 0.00174 | 0.00277 |
| | | **Uniform sampling on a Swiss roll** | | | | |
| DK | $E_{av}$ | 0.0119 | 0.0156 | 0.0198 | 0.0200 | 0.0203 |
| | $E_{se}$ | 0.01310 | 0.02124 | 0.02456 | 0.02431 | 0.02600 |
| FMD | $E_{av}$ | 0.0065 | 0.0044 | 0.0052 | 0.0033 | 0.0022 |
| | $E_{se}$ | 0.00313 | 0.00164 | 0.00113 | 0.00078 | 0.00062 |
| | | **Non-Uniform sampling on a Swiss roll** | | | | |
| DK | $E_{av}$ | 0.0400 | 0.0105 | 0.0114 | 0.0151 | 0.0180 |
| | $E_{se}$ | 0.01661 | 0.01578 | 0.01457 | 0.01659 | 0.01865 |
| FMD | $E_{av}$ | 0.0138 | 0.0053 | 0.0047 | 0.0029 | 0.0073 |
| | $E_{se}$ | 0.00475 | 0.00519 | 0.00309 | 0.00517 | 0.00725 |

Fig. 7.8 Left to right images: Incomplete distance data $D_{60\%,20}$ from 1002 points sampled uniformly and non-uniformly on the unit sphere with fixed north pole (red star) and south pole (blue star) for calculating geodesic distance, and $D_{60\%,20}$ from 1962 points sampled uniformly and non-uniformly on the Swiss roll with a fixed starting point (red star) and an ending point (blue star) for calculating geodesic distance. Table: Error comparison between Dijkstra (DK) and Fast Marching on distance data (FMD). Relative error of geodesic distances $E_{av}$ (averaging from starting to all the points) and $E_{se}$ (from the starting point to the ending point).

**Eikonal equation and geodesic accuracy.** The framework further extends to non-diffusive PDEs, most notably the Eikonal equation. By employing the local mesh method to reconstruct local coordinates from pairwise distances—referred to as fast marching on distance data (FMD) [Lai and Li (2018)]—the Eikonal equation can be solved directly on distance-sampled manifolds. We demonstrate its numerical accuracy in comparison with the classical Dijkstra algorithm. Figure 7.8 presents the relative errors of the computed geodesic distances from a fixed source to a target point and to all other points on both the unit sphere and the Swiss roll manifold, using incomplete distance data $D_{60\%,20}$. The results demonstrate that FMD yields substantially smaller errors than Dijkstra's method [Dijkstra (1959)]. Moreover, the relative errors decrease as the number of sampled points increases. For uniformly sampled data, the geodesic distance from the north pole to the south pole exhibits first-order convergence with respect to sampling density under the fast marching discretization, whereas Dijkstra's method does not exhibit comparable convergence behavior.

# Chapter 8

# Applications of Laplacian on Manifold-Structured Data

The Laplace–Beltrami operator serves as a unifying tool linking geometry, analysis, and computation on manifolds. While previous chapters have developed its theoretical foundations and numerical discretizations on meshes and point clouds, this chapter illustrates how these constructions enable a wide range of practical applications in geometry processing, scientific computing, and data analysis.

Many problems in physics, biology, and computer vision involve data or fields defined not in Euclidean space but on curved surfaces or high-dimensional manifolds. Examples include texture and signal processing on anatomical surfaces, analysis of molecular conformations, nonrigid registration of shapes, and extraction of geometric features from unstructured point samples. In these contexts, intrinsic operators such as the gradient, divergence, and Laplace–Beltrami operator provide the natural mathematical framework for computation that respects the underlying geometry.

The first part of this chapter presents variational formulations for image processing on surfaces, including denoising, deblurring, and segmentation. These models extend classical total-variation and Chan–Vese formulations to manifold domains and are efficiently solved through alternating direction methods of multipliers (ADMM). The second part is devoted to computational optimal transport on manifolds. The third part focuses on shape analysis and explores intrinsic approaches for parameterization, skeleton extraction, and nonrigid registration. These tasks make use of conformal mappings, Laplace–Beltrami eigenfunctions, and spectral embeddings to capture both local and global geometric structure. The final part discusses additional applications in scientific modeling, including the computation of committor functions for stochastic dynamics using local-mesh methods on point clouds.

Together, these examples demonstrate how intrinsic differential operators provide a unified computational language for geometry-aware modeling and analysis. They bridge discrete and continuous viewpoints, enabling robust algorithms that operate directly on unorganized data while preserving geometric and topological information.

## 8.1 Image processing on surfaces

Image processing on manifolds naturally arises when data are defined not on a flat Euclidean domain but on a curved surface embedded in $\mathbb{R}^3$, such as texture patterns on anatomical shapes, scalar fields on 3D meshes, or color distributions defined on point clouds. In this setting, intrinsic differential operators, most notably the *Laplace–Beltrami operator* $\Delta_{\mathcal{M}}$, replace the Euclidean Laplacian and encode the local geometry of the surface. Classical image processing tasks such as denoising, deblurring, and segmentation extend to manifold domains by formulating their *variational models* using intrinsic quantities proposed in [Lai and Chan (2011)], which provides the main content of this section. These formulations lead to elliptic or parabolic PDEs on $\mathcal{M}$, often involving Poisson-type equations with $\Delta_{\mathcal{M}}$ within optimization loops. An efficient computational framework for these problems is provided by the *Alternating Direction Method of Multipliers* (ADMM). See Appendix C.4 for a short introduction of ADMM. A general variational model on a surface $\mathcal{M}$ can be written as

$$\min_{\varphi\in\mathcal{S}} \mathcal{J}(\varphi)+\mathcal{H}(\varphi), \tag{8.1}$$

where $\mathcal{S}$ is a suitable function space such as $BV(\mathcal{M})$ (functions with a bounded variation), and $\mathcal{J}$, $\mathcal{H}$ are convex functionals. When $\mathcal{J}(\varphi) = TV(\varphi) = \int_{\mathcal{M}} |\nabla_{\mathcal{M}}\varphi|\,\mathrm{d}s$, one obtains the total-variation (TV) model on $\mathcal{M}$,

$$\min_{\varphi} \int_{\mathcal{M}} |\nabla_{\mathcal{M}}\varphi|\,\mathrm{d}s+\mathcal{H}(\varphi), \tag{8.2}$$

with $\mathcal{H}$ typically representing a fidelity term. The TV-norm denoising example that we see in Section 4.2 is a special case in which $\mathcal{M}$ reduces to a flat square. To decouple the nonsmooth TV term, introduce an auxiliary tangent vector field $\mathbb{W}\in\Gamma(T\mathcal{M})$ and rewrite the problem as

$$\min_{\varphi,\mathbb{W}\in\Gamma(T\mathcal{M})} \|\mathbb{W}\|_1+\mathcal{H}(\varphi) \quad \text{subject to} \quad \mathbb{W}=\nabla_{\mathcal{M}}\varphi. \tag{8.3}$$

The ADMM iteration for this constrained problem reads

$$\varphi^{k+1} = \arg\min_{\varphi} \ \mathcal{H}(\varphi) + \frac{\lambda}{2} \left\| \mathbb{W}^k - \nabla_{\mathcal{M}}\varphi - \mathbb{B}^k \right\|_2^2, \tag{8.4}$$

$$\mathbb{W}^{k+1} = \arg\min_{\mathbb{W}\in\Gamma(T\mathcal{M})} \ \|\mathbb{W}\|_1 + \frac{\lambda}{2} \left\| \mathbb{W} - \nabla_{\mathcal{M}}\varphi^{k+1} - \mathbb{B}^k \right\|_2^2, \tag{8.5}$$

$$\mathbb{B}^{k+1} = \mathbb{B}^k + \nabla_{\mathcal{M}}\varphi^{k+1} - \mathbb{W}^{k+1}. \tag{8.6}$$

The update (8.5) admits a vector soft-thresholding (shrinkage) form,

$$\mathbb{W}^{k+1} = \max\left\{ \left|\nabla_{\mathcal{M}}\varphi^{k+1} + \mathbb{B}^k\right| - \tfrac{1}{\lambda}, 0 \right\} \frac{\nabla_{\mathcal{M}}\varphi^{k+1} + \mathbb{B}^k}{\left|\nabla_{\mathcal{M}}\varphi^{k+1} + \mathbb{B}^k\right|}. \tag{8.7}$$

**(A) ROF Denoising on surfaces.** The ROF model [Rudin *et al.* (1992)] balances surface smoothness and data fidelity through the following minimization

$$\min_{\varphi\in BV(\mathcal{M})} E_1(\varphi) = \int_{\mathcal{M}} |\nabla_{\mathcal{M}}\varphi| \, \mathrm{d}s + \frac{\mu}{2} \int_{\mathcal{M}} (\varphi - I)^2 \, \mathrm{d}s, \tag{8.8}$$

where $I : \mathcal{M} \to \mathbb{R}$ denotes the observed image. Here $\mathcal{H}(\varphi) = \frac{\mu}{2}\|\varphi - I\|_2^2$, and the optimality condition of (8.4) leads to the Poisson equation

$$(\mu \, \mathrm{Id} - \lambda \, \Delta_{\mathcal{M}}) \, \varphi^{k+1} = \mu \, I + \lambda \, \mathrm{div}_{\mathcal{M}}\big(\mathbb{B}^k - \mathbb{W}^k\big). \tag{8.9}$$

Recall that we have seen how such a Poisson equation emerges in ADMM for ROF denoising in Section 4.2. Hence, ADMM for (8.8) can be implemented as

(1) Initialize $\mathbb{W}^0 = \mathbb{B}^0 = 0$;
(2) Solve $(\mu\mathrm{Id} - \lambda\Delta_{\mathcal{M}}) \, \varphi^{k+1} = \mu I + \lambda \, \mathrm{div}_{\mathcal{M}}(\mathbb{B}^k - \mathbb{W}^k)$;
(3) $\mathbb{W}^{k+1} = \mathrm{shrink}\big(\nabla_{\mathcal{M}}\varphi^{k+1} + \mathbb{B}^k, \, 1/\lambda\big)$;
(4) $\mathbb{B}^{k+1} = \mathbb{B}^k + \nabla_{\mathcal{M}}\varphi^{k+1} - \mathbb{W}^{k+1}$;
(5) Repeat until convergence.

Replacing $(\varphi - I)^2$ by $(K\varphi - I)^2$ with a linear blur operator $K$ on $\mathcal{M}$ yields the total-variation deblurring model

$$\min_{\varphi\in BV(\mathcal{M})} E_2(\varphi) = \int_{\mathcal{M}} |\nabla_{\mathcal{M}}\varphi| \, \mathrm{d}s + \frac{\mu}{2} \int_{\mathcal{M}} (K\varphi - I)^2 \, \mathrm{d}s, \tag{8.10}$$

while TV inpainting on $\mathcal{M}$ with missing region $D \subset \mathcal{M}$ is formulated as

$$\min_{\varphi\in BV(\mathcal{M})} E_3(\varphi) = \int_{\mathcal{M}} |\nabla_{\mathcal{M}}\varphi| \, \mathrm{d}s + \frac{\mu}{2} \int_{\mathcal{M}\setminus D} (\varphi - I)^2 \, \mathrm{d}s. \tag{8.11}$$

(a1) (a2) (a3)

Fig. 8.1 ROF denoising results of the image of a lady $I$ on a 25075 vertices cup surface. (a1) the clean image. (a2) the noisy image with Gaussian noise $\sigma$=40. (a3) the denoised result.

**(B) Chan–Vese segmentation and its convex variants.** The Chan–Vese (CV) model [Chan and Vese (2001)] represents the segmentation boundary on a surface $\mathcal{M}$ as the zero level set of $\varphi : \mathcal{M} \to \mathbb{R}$:

$$\min_{\varphi,c_1,c_2} \int_{\mathcal{M}} \left|\nabla_{\mathcal{M}} H(\varphi)\right| \mathrm{d}s + \mu \int_{\mathcal{M}} (c_1 - I)^2 H(\varphi)\, \mathrm{d}s + \mu \int_{\mathcal{M}} (c_2 - I)^2 (1 - H(\varphi))\, \mathrm{d}s, \tag{8.12}$$

where $H$ denotes the Heaviside function. Because this energy is nonconvex, a convexified variant (CCV) [Chan *et al.* (2006)] alternates between updating $\varphi$ and the mean intensities $(c_1, c_2)$:

$$(1)\ \varphi^{k+1} = \arg\min_{0\le\varphi\le1} \int_{\mathcal{M}} |\nabla_{\mathcal{M}}\varphi|\, \mathrm{d}s + \mu \int_{\mathcal{M}} \varphi\big((c_1^k - I)^2 - (c_2^k - I)^2\big)\, \mathrm{d}s;$$

$$(2)\ c_1^{k+1} = \int_{\Omega^+(\varphi^{k+1})} I\, \mathrm{d}s, \qquad c_2^{k+1} = \int_{\Omega^-(\varphi^{k+1})} I\, \mathrm{d}s.$$

Let $r^k = (c_1^k - I)^2 - (c_2^k - I)^2$, then we have $\mathcal{H}(\varphi) = \mu \int_{\mathcal{M}} \varphi r^k\, \mathrm{d}s$, so that the ADMM subproblem (8.4) becomes

$$\varphi^{k+1} = \arg\min_{0\le\varphi\le1} \frac{\lambda}{2} \left\|\mathbb{W}^k - \nabla_{\mathcal{M}}\varphi - \mathbb{B}^k\right\|_2^2 + \mu \int_{\mathcal{M}} \varphi r^k\, \mathrm{d}s. \tag{8.13}$$

Its solution involves a Poisson equation followed by a projection onto $[0, 1]$:

$$\begin{aligned} \text{solve:}\quad & \Delta_{\mathcal{M}}\varphi^{k+1} = \frac{\mu}{\lambda} r^k + \mathrm{div}_{\mathcal{M}}\big(\mathbb{W}^k - \mathbb{B}^k\big), \\ \text{project:}\quad & \varphi^{k+1}(p_i) \leftarrow \min\big\{\max\{\varphi^{k+1}(p_i), 0\},\, 1\big\}. \end{aligned} \tag{8.14}$$

The complete ADMM iteration for the CCV model proceeds as

(1) Initialize $\mathbb{W}^0 = \mathbb{B}^0 = 0$;
(2) Update $r^k = (c_1^k - I)^2 - (c_2^k - I)^2$;
(3) Solve $\Delta_{\mathcal{M}}\varphi^{k+1} = \frac{\mu}{\lambda} r^k + \mathrm{div}_{\mathcal{M}}(\mathbb{W}^k - \mathbb{B}^k)$, then project $0 \le \varphi^{k+1} \le 1$ as in (8.14);
(4) $\mathbb{W}^{k+1} = \mathrm{shrink}\big(\nabla_{\mathcal{M}}\varphi^{k+1} + \mathbb{B}^k,\ 1/\lambda\big)$;
(5) $\mathbb{B}^{k+1} = \mathbb{B}^k + \nabla_{\mathcal{M}}\varphi^{k+1} - \mathbb{W}^{k+1}$;
(6) $c_1^{k+1} = \int_{\Omega^+(\varphi^{k+1})} I \,\mathrm{d}s, \quad c_2^{k+1} = \int_{\Omega^-(\varphi^{k+1})} I \,\mathrm{d}s$;
(7) Repeat until convergence.

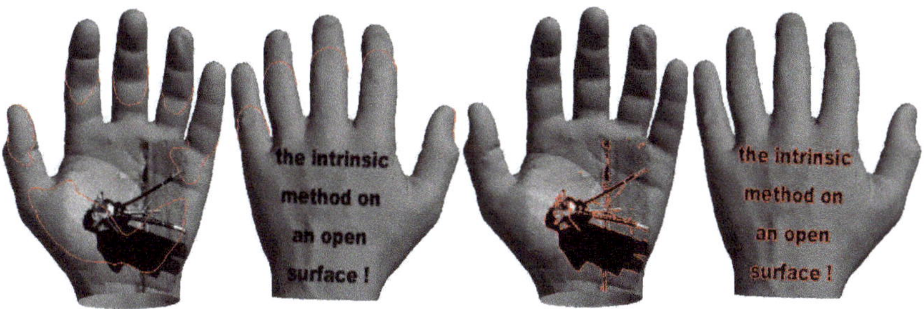

(a1) (a2) (b1) (b2)

Fig. 8.2 CCV segmentation results on a 58875 vertices a hand surface [Lai and Chan (2011)]. (a1), (a2) two views the image and the initial segmentation curve marked by the red contour. (b1), (b2) correspond to two views of the CCV segmentation result.

## 8.2 Computational optimal transport on manifolds

In Section 4.4, we have seen optimal transport (OT) on a flat domain. OT on manifolds provides a natural geometric framework to quantify distances between probability measures while respecting intrinsic curvature. When the underlying space $\mathcal{M}$ is a smooth compact Riemannian manifold with metric $g$, the dynamic formulation of OT generalizes the celebrated Benamou–Brenier formula from $\mathbb{R}^d$ to curved spaces [Lavenant *et al.* (2018)]. It can be further viewed as a special case of the potential mean-field game (MFG) functional studied in [Yu *et al.* (2023, 2024)], and solved efficiently by a projected (proximal) gradient descent scheme on triangular meshes. See Appendix C.2 for a quick introduction to the proximal gradient method.

**Dynamic formulation of OT on manifolds.** Let $\rho_0, \rho_1 \in \mathcal{P}(\mathcal{M})$ be two smooth probability densities. The $L^2$–Wasserstein distance on $(\mathcal{M}, g)$

can be written as the solution to a convex dynamic problem:

$$\begin{aligned} W_2^2(\rho_0, \rho_1) &= \inf_{\rho,\, m} \int_0^1 \int_{\mathcal{M}} \frac{\|m(x,t)\|_g^2}{\rho(x,t)} \,\mathrm{d}V_g(x)\,\mathrm{d}t, \\ \text{s.t. } &\partial_t \rho + \mathrm{Div}_{\mathcal{M}} m = 0,\ \rho(\cdot,0) = \rho_0,\ \rho(\cdot,1) = \rho_1. \end{aligned} \tag{8.15}$$

Here $m(x,t) \in T_x\mathcal{M}$ represents the probability flux, and $\|m\|_g^2 = g_x(m,m)$ is the squared norm under metric $g$. The continuity equation evolves densities along vector fields tangent to $\mathcal{M}$, ensuring mass conservation.

Introducing a potential $\phi : \mathcal{M} \times [0,1] \to \mathbb{R}$ as Lagrange multiplier for the constraint, the corresponding Lagrangian reads

$$\mathcal{A}(\rho, m, \phi) = \int_0^1 \int_{\mathcal{M}} \frac{\|m\|_g^2}{2\rho} + \rho\,\partial_t\phi + \langle m, \nabla_{\mathcal{M}}\phi\rangle_g \,\mathrm{d}V_g\,\mathrm{d}t. \tag{8.16}$$

Stationarity in $(\rho, m, \phi)$ yields the manifold Hamilton–Jacobi–continuity system

$$\begin{cases} -\partial_t\phi + \frac{1}{2}\|\nabla_{\mathcal{M}}\phi\|_g^2 = 0, \\ \partial_t\rho - \mathrm{Div}_{\mathcal{M}}(\rho\nabla_{\mathcal{M}}\phi) = 0, \end{cases} \qquad \rho(\cdot,0) = \rho_0, \quad \rho(\cdot,1) = \rho_1. \tag{8.17}$$

This is the canonical form of the dynamic OT equations on manifolds, and a special case of the potential MFG system [Yu *et al.* (2023)] with zero interaction cost.

**Variational formulation and convex functional.** For a general convex Lagrangian $L(x,p)$, the dynamic OT cost generalizes to

$$\begin{aligned} \min_{\rho,m} \mathcal{J}(\rho, m) &:= \int_0^1 \int_{\mathcal{M}} \rho(x,t)\, L\Big(x, \frac{m(x,t)}{\rho(x,t)}\Big) \,\mathrm{d}V_g(x)\,\mathrm{d}t, \\ &\text{s.t. } \partial_t\rho + \mathrm{Div}_{\mathcal{M}} m = 0. \end{aligned} \tag{8.18}$$

With $L(x,p) = \frac{1}{2}\|p\|_g^2$, the optimality condition recovers (8.17). When additional interaction or terminal terms $F(\rho)$, $F_T(\rho(1))$ are added, one obtains the full potential MFG functional [Yu *et al.* (2023)]:

$$\mathcal{Y}(\rho, m) = \int_0^1 \int_{\mathcal{M}} \rho\, L\Big(x, \frac{m}{\rho}\Big) \,\mathrm{d}V_g\,\mathrm{d}t + \int_0^1 F(\rho(\cdot,t))\,\mathrm{d}t + F_T(\rho(\cdot,1)).$$

**Projected (proximal) gradient descent on discrete manifolds.** Discretizing $(\mathcal{M}, g)$ by a triangular mesh $\mathcal{M}_h$ with vertices $\{V_i\}$, local basis $\{e_i\}$, and intrinsic gradient/divergence operators $(\nabla_{\mathcal{M}_h}, \mathrm{Div}_{\mathcal{M}_h})$, we denote by $P_i(t) \approx \rho(V_i, t)$ the discrete density and by $M_j(t) \approx m(T_j, t)$ the flux on triangle $T_j$. The semi-discrete dynamic OT problem becomes

$$\min_{P,M} \widetilde{\mathcal{J}}(P, M) = \sum_{k=1}^{n} \sum_{T_j} A_{T_j} \, P(T_j, t_k) \, L\Big(T_j, \frac{M(T_j, t_k)}{P(T_j, t_k)}\Big), \tag{8.19}$$

$$\text{s.t. } \partial_t P + \mathrm{Div}_{\mathcal{M}_h} M = 0, \;\; P(\cdot, 0) = P_0, \;\; P(\cdot, 1) = P_1. \tag{8.20}$$

To enforce the discrete continuity constraint, Yu–Lai–Li–Osher [Yu *et al.* (2023, 2024)] proposed a proximal gradient method:

$$(P^{\ell+\frac{1}{2}}, M^{\ell+\frac{1}{2}}) = (P^{\ell}, M^{\ell}) - \eta_\ell \nabla_{P,M} \widetilde{\mathcal{J}}(P^{\ell}, M^{\ell}), \tag{8.21}$$

$$(P^{\ell+1}, M^{\ell+1}) = \mathrm{proj}_{\mathcal{C}(P_0,P_1)}(P^{\ell+\frac{1}{2}}, M^{\ell+\frac{1}{2}}), \tag{8.22}$$

where $\mathcal{C}(P_0, P_1)$ is the discrete constraint set ensuring $\partial_t P + \mathrm{Div}_{\mathcal{M}_h} M = 0$. The projection step is equivalent to minimizing

$$\min_{(P,M) \in \mathcal{C}(P_0,P_1)} \frac{1}{2} \|P - P^{\ell+\frac{1}{2}}\|_V^2 + \frac{1}{2} \|M - M^{\ell+\frac{1}{2}}\|_T^2, \tag{8.23}$$

whose first-order optimality condition introduces a dual potential $\psi$ satisfying a discrete Poisson-type equation:

$$(-\partial_t \partial_t - \mathrm{Div}_{\mathcal{M}_h} \nabla_{\mathcal{M}_h}) \, \psi = \partial_t P^{\ell+\frac{1}{2}} + \mathrm{Div}_{\mathcal{M}_h} M^{\ell+\frac{1}{2}}. \tag{8.24}$$

Then the projected variables are updated via

$$P^{\ell+1} = P^{\ell+\frac{1}{2}} + \partial_t \psi, \qquad M^{\ell+1} = M^{\ell+\frac{1}{2}} + \nabla_{\mathcal{M}_h} \psi. \tag{8.25}$$

The linear operator in (8.24) can be pre-factored, and the iteration preserves both mass and positivity (with a safeguard $P^{\ell+1} \leftarrow \max(P^{\ell+1}, \varepsilon)$).

**Algorithm 2** Projected gradient method for OT on triangular meshes

Initialize $P^{(0)}, M^{(0)}$ satisfying $P(\cdot, 0) = P_0$ and $P(\cdot, 1) = P_1$.
**for** $\ell = 0, 1, 2, \ldots$ **do**
    Gradient step: $(P^{\ell+\frac{1}{2}}, M^{\ell+\frac{1}{2}}) = (P^{\ell}, M^{\ell}) - \eta_\ell \nabla_{P,M} \widetilde{\mathcal{J}}(P^{\ell}, M^{\ell})$.
    Projection: solve $(\partial_t \partial_t^* - \mathrm{Div}_{\mathcal{M}_h} \nabla_{\mathcal{M}_h}) \psi = \partial_t P^{\ell+\frac{1}{2}} + \mathrm{Div}_{\mathcal{M}_h} M^{\ell+\frac{1}{2}}$.
    Update: $P^{\ell+1} = P^{\ell+\frac{1}{2}} + \partial_t \psi$, $M^{\ell+1} = M^{\ell+\frac{1}{2}} + \nabla_{\mathcal{M}_h} \psi$.
    Safeguard: $P^{\ell+1} \leftarrow \max(P^{\ell+1}, \varepsilon)$.
**end for**

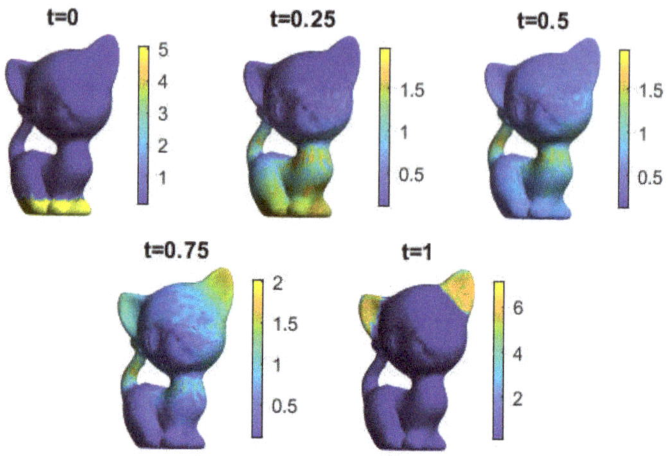


Fig. 8.3 Dynamical optimal transport on the kitten surface [Yu *et al.* (2023)].

## 8.3 Shape Analysis

Shape analysis seeks to extract, compare, and register geometric information from manifolds and surfaces in a coordinate-invariant manner. This section presents several representative methodologies built upon intrinsic operators such as the Laplace–Beltrami operator. The first part describes global conformal parameterization of genus zero surfaces, formulated as a constrained variational problem and efficiently solved by a manifold-based gradient descent method. The second part explains how spectral information from Laplace–Beltrami eigenfunctions can be used to derive skeletal structures that summarize global geometry. The final part develops intrinsic registration techniques for both nearly isometric and non-isometric shapes through Laplace–Beltrami eigenmaps.

### 8.3.1 *Conformal parameterization for Genus-0 surfaces*

Conformal parameterizations provide angle-preserving mappings from curved surfaces to simple parameter domains, enabling computations that respect intrinsic geometry. They are fundamental in computer vision, computer graphics, and medical imaging applications [Eck *et al.* (1995); Kanai *et al.* (1998); Levy *et al.* (2002); Alliez *et al.* (2002); Gu and Yau. (2002); Gu and Yau (2003); Gu *et al.* (2004); Springborn *et al.* (2008); Lui *et al.* (2008)]. For a compact, closed, genus-zero Riemannian surface $(\mathcal{M}, g)$, a global conformal parameterization onto the unit sphere $(S^2, g_0)$ corresponds, up to Möbius transformations, to a harmonic map $\vec{F} = (f_1, f_2, f_3) : \mathcal{M} \to S^2$ that minimizes the Dirichlet energy under a spherical constraint [Schoen and Yau (1994); Jost (2008)]:

$$\min_{\vec{F}=(f_1,f_2,f_3)} \quad \mathcal{E}(\vec{F}) = \frac{1}{2}\int_{\mathcal{M}} \left(\|\nabla_{\mathcal{M}} f_1\|^2 + \|\nabla_{\mathcal{M}} f_2\|^2 + \|\nabla_{\mathcal{M}} f_3\|^2\right) \mathrm{d}\mathcal{M},$$
$$\text{subject to} \quad \|\vec{F}(x)\|^2 = f_1^2(x) + f_2^2(x) + f_3^2(x) = 1, \quad \forall x \in \mathcal{M}. \tag{8.26}$$

Unlike optimization problems in Euclidean space, where the objective can be decreased along straight-line search directions such as the negative gradient, optimization on a curved manifold requires motion constrained to the manifold itself. A natural choice for such motion is the geodesic, the intrinsic analogue of a straight line and the path of shortest length between two points on the manifold. Alternatively, iterative projection methods descend along straight directions and then project the intermediate results back to the constraint manifold. Both strategies are widely studied in optimization on manifolds [Absil *et al.* (2008)]. Since at each $x \in \mathcal{M}$, the constraint $\|\vec{F}(x)\| = 1$ defines a unit sphere, an efficient scheme can be designed using a sphere-geodesic descent approach [Lai *et al.* (2014)], following the framework introduced in [Wen and Yin (2013)]. This method admits the use of acceleration techniques such as Barzilai–Borwein step selection and nonmonotone line search, while maintaining global convergence.

**Constraint-preserving update.** The constraint in (8.26) defines a manifold of unit-length vector fields, which can be regarded as a special case of the Stiefel manifold. The corresponding optimization can be viewed as a projected gradient method constrained to this manifold [Lai *et al.* (2014)]. The Lagrangian associated with (8.26) is

$$\mathcal{L}(\vec{F}, \lambda) = \mathcal{E}(\vec{F}) - \frac{1}{2}\int_{\mathcal{M}} \lambda(x)\left(\|\vec{F}(x)\|^2 - 1\right) \mathrm{d}\mathcal{M},$$

where $\lambda$ acts as a Lagrange multiplier enforcing the pointwise unit-norm condition. The first-order optimality conditions are

$$\begin{cases} H - \lambda\vec{F} = 0, \\ \|\vec{F}(x)\| = 1, \quad \forall x \in \mathcal{M}, \end{cases} \tag{8.27}$$

where $H = \nabla_{\vec{F}}\mathcal{E}(\vec{F}) = -(\Delta_{\mathcal{M}} f_1, \Delta_{\mathcal{M}} f_2, \Delta_{\mathcal{M}} f_3)$ is the Fréchet derivative of $\mathcal{E}(\vec{F})$, and $\Delta_{\mathcal{M}}$ is the Laplace–Beltrami operator. Applying the linear

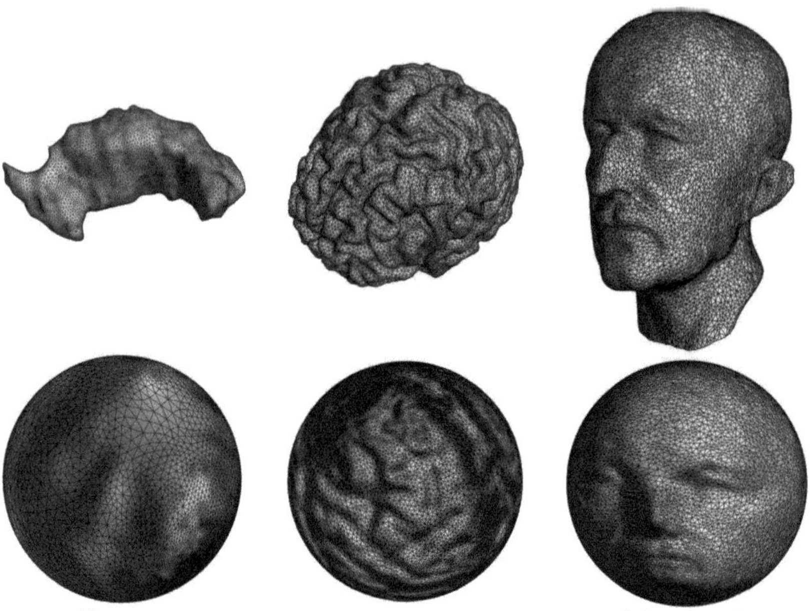


Fig. 8.4 First row: three input surfaces; Second row: the resulting conformal maps.

functional $\vec{F}^*$ to the first equation in (8.27) and using $\vec{F}^*\vec{F} = 1$ yields $\lambda = \vec{F}^*H = H^*\vec{F}$. Substitution into (8.27) gives

$$H - (\vec{F}^*H)\vec{F} = 0, \quad \text{equivalently,} \quad A\vec{F} = 0, \quad \text{with} \quad A := H\vec{F}^* - \vec{F}H^*.$$

The operator $A(x)$ is skew-symmetric for each $x \in \mathcal{M}$. Following [Wen and Yin (2013)], $A$ defines a tangential search direction that preserves the unit-sphere constraint.

In Euclidean space, a standard gradient descent update is $\vec{Y}(x) = \vec{F}(x) - \tau A(x)\vec{F}(x)$, where $\tau > 0$ is a step size. However, this update generally violates the unit-norm constraint. To ensure $\|\vec{Y}(x)\| = 1$, an implicit update of the form

$$\vec{Y}(x) = \vec{F}(x) - \frac{\tau}{2}A(x)\big(\vec{F}(x) + \vec{Y}(x)\big)$$

is adopted. Solving for $\vec{Y}(x)$ yields

$$\vec{Y}(x) = \big(I + \tfrac{\tau}{2}A(x)\big)^{-1}\big(I - \tfrac{\tau}{2}A(x)\big)\vec{F}(x). \tag{8.28}$$

By the Cayley transformation, the operator $\left(I + \frac{\tau}{2}A(x)\right)^{-1}\left(I - \frac{\tau}{2}A(x)\right)$ is orthogonal, guaranteeing $\|\vec{Y}(x)\| = \|\vec{F}(x)\| = 1$. The step size $\tau$ can be adjusted adaptively through a line search scheme to ensure sufficient decrease of the objective. For example, a nonmonotone curvilinear search strategy can be adopted. We refer more details in [Wen and Yin (2013); Lai *et al.* (2014)].

The method efficiently minimizes the harmonic energy under the spherical constraint and produces a globally consistent conformal parameterization for genus-zero surfaces.

It is worth to note that the procedure requires only a discretization of the Laplace–Beltrami operator. Hence, the same algorithmic framework applies to point clouds: using the corresponding discrete Laplace–Beltrami operator for point-sampled surfaces yields a fully analogous method for computing conformal parameterizations.

### 8.3.2 *Skeleton extraction from point clouds*

The first nontrivial eigenfunction of the Laplace–Beltrami operator encodes essential global geometric information of a surface and can be used to extract its topological structure. The approach introduced in [Shi *et al.* (2008); Liang *et al.* (2012)] constructs surface skeletons by forming the Reeb graph associated with this eigenfunction. Given a manifold $\mathcal{M}$ and its first nontrivial Laplace–Beltrami eigenfunction $\phi_1$, the Reeb graph $R(\phi_1)$ is defined as the quotient space under the equivalence relation $x \sim y$ if and only if $\phi_1(x) = \phi_1(y)$ for $x, y \in \mathcal{M}$. The graph $R(\phi_1)$ captures the connectivity among the level sets of $\phi_1$, providing a concise representation of the shape's global structure.

**Algorithm (level-set barycenters).**

(1) Compute $\phi_1$ using the point-cloud LB eigenproblem (see Sec. 7.3.2 for MLS discretizations).

(2) For a set of values $\{c_m\}$, extract approximate level contours $\{\phi_1 = c_m\}$: around each point $p_i$ with $\phi_1(p_i) \in [c_m - \delta, c_m + \delta]$, solve
$$\min_{p\in\mathbb{R}^3} \|p - p_i\|^2 + \beta\big(\Phi_1(p) - c_m\big)^2,$$
where $\Phi_1$ is the local polynomial fit of $\phi_1$ (MLS), $\delta$ small, $\beta$ large.

(3) For each level $c_m$, take the barycenter of the reconstructed contour as a skeleton node; connect nodes in order of $c_m$ (or by nearest-neighbor in parameter $c$) to form the skeleton curve graph.

This procedure produces intuitive "backbone" summaries (e.g., figure-eight, armadillo, knot), showing that a single intrinsic eigenfunction can drive high-level structure recovery from unorganized points (see Fig. 8.5).

Fig. 8.5 Skeletons computed from the first nontrivial LB eigenfunction on point clouds. Blue: samples; black: level contours of $\phi_1$; red: barycenters and skeleton connections [Liang *et al.* (2012)].

### 8.3.3 *Nonrigid manifolds registration via LB eigenmaps*

Intrinsic registration between two isometric or nearly isometric shapes is a central problem in geometry processing. For geometric shapes, the task is complicated by the absence of a global parameterization and by ambiguities arising from their embeddings in the ambient space $\mathbb{R}^D$. An effective strategy employs the Laplace–Beltrami eigenmap to construct intrinsic coordinates, resolves the eigenbasis ambiguity through an orthogonal alignment, and matches the resulting point distributions using a rotation invariant sliced Wasserstein distance.

**Preliminaries: LB eigensystem and functional maps.** For a closed Riemannian surface $(\mathcal{M}, g)$, the LB operator

$$\Delta_g \phi = \frac{1}{\sqrt{G}} \sum_{i=1}^{2} \frac{\partial}{\partial x_i}\Big(\sqrt{G} \sum_{j=1}^{2} g^{ij} \frac{\partial \phi}{\partial x_j}\Big), \qquad G = \det(g_{ij}), \tag{8.29}$$

is self-adjoint and elliptic, with spectrum $0 = \lambda_0 < \lambda_1 \le \lambda_2 \le \cdots$ and orthonormal eigenfunctions $\{\phi_i\}_{i\ge 0}$:

$$-\Delta_g \phi_i = \lambda_i \phi_i, \qquad \int_{\mathcal{M}} \phi_i \phi_j \,\mathrm{dvol}_g = \delta_{ij}.$$

The LB basis generalizes Fourier modes and is invariant under isometries. The resulting eigenpairs are intrinsic and stable under small metric perturbations [Chavel (1984); Jost (2008)]. Generic eigenfunctions are

Morse functions [Uhlenbeck (1976)], which support topological characterization through Morse theory [Milnor (1963)], and they provide global descriptors widely used in geometric comparison and retrieval [Rustamov (2007); Sun *et al.* (2009); Bronstein and Kokkinos (2010); Lai *et al.* (2010)]. Laplace–Beltrami based representations have become fundamental in surface analysis, medical shape correspondence, and manifold learning [Reuter (2010); Vallet and Lévy (2008); Raviv *et al.* (2011); Bronstein *et al.* (2010); Lai *et al.* (2010); Shi *et al.* (2012); Lai and Zhao (2017)].

**Nearly isometric registration** Let $(\mathcal{M}, g_{\mathcal{M}})$ and $(\mathcal{N}, g_{\mathcal{N}})$ be $d$-dimensional Riemannian manifolds sampled by point clouds. Denote their first $n$ nontrivial LB eigenpairs by $\Phi_n = \{(\lambda_k, \phi_k)\}_{k=1}^n$ and $\Psi_n = \{(\eta_k, \psi_k)\}_{k=1}^n$. The LB eigenmap $\iota_{\mathcal{M}}^n : \mathcal{M} \to \mathbb{R}^n$ is

$$\iota_{\mathcal{M}}^n(u) = \left(\lambda_1^{-d/4}\phi_1(u), \ldots, \lambda_n^{-d/4}\phi_n(u)\right), \tag{8.30}$$

$$\iota_{\mathcal{N}}^n(v) = \left(\eta_1^{-d/4}\psi_1(v), \ldots, \eta_n^{-d/4}\psi_n(v)\right). \tag{8.31}$$

which is intrinsic and stable to isometries [Bérard *et al.* (1994); Lai *et al.* (2010)]. However, eigenfunctions are defined up to sign and, within multiplicity blocks, up to orthogonal mixing; hence LB embeddings are only defined up to $R \in O(n)$.

Let $P \in \mathbb{R}^{\ell_P \times n}$ and $Q \in \mathbb{R}^{\ell_Q \times n}$ collect the row-wise embeddings of samples $\{u_i\}_{i=1}^{\ell_P} \subset \mathcal{M}$ and $\{v_j\}_{j=1}^{\ell_Q} \subset \mathcal{N}$, with discrete probability weights $\mu^P \in \mathbb{R}_+^{\ell_P}$ and $\mu^Q \in \mathbb{R}_+^{\ell_Q}$ (e.g., uniform). We align $(PR, \mu^P)$ and $(Q, \mu^Q)$ *in distribution* by optimal transport, while optimizing over $R \in O(n)$ to remove eigenbasis ambiguity [Lai and Zhao (2017)]:

$$\begin{aligned} \mathrm{RSWD}\big((P, \mu^P), (Q, \mu^Q)\big)^2 &= \min_{R \in O(n)} \int_{S^{n-1}} \min_{\sigma^{\theta,R} \in \mathrm{ADM}(\mu_{\theta,R}^P, \mu_\theta^Q)} \\ &\times \sum_{i,j} \sigma_{ij}^{\theta,R} \left(p_i R\theta^\top - q_j\theta^\top\right)^2 \mathrm{d}\theta, \end{aligned} \tag{8.32}$$

where $\mu_{\theta,R}^P$ and $\mu_\theta^Q$ are the pushforward 1D measures induced by projections $p_i \mapsto p_i R\theta^\top$, $q_j \mapsto q_j\theta^\top$. Each inner 1D OT is solved by sorting in $O(m \log m)$, giving scalability [Bonneel *et al.* (2015); Lai and Zhao (2017)]. The optimization over $R$ handles sign/multiplicity/ordering ambiguities in the LB basis. (Under a natural equivalence that identifies point sets up to permutations and $O(n)$, RSWD defines a distance; see [Lai and Zhao (2017)].)

In practice, approximate (8.32) by Monte Carlo directions $\{\theta_t\}_{t=1}^T$:

(1) **Initialize.** $R \leftarrow I_n$ (or from coarse scale; see multiscale below).
(2) **Project.** For each $t$, compute $a^{(t)} = PR\theta_t^\top \in \mathbb{R}^{\ell_P}$ and $b^{(t)} = Q\theta_t^\top \in \mathbb{R}^{\ell_Q}$.
(3) **1D OT.** Sort $a^{(t)}$ and $b^{(t)}$ and compute the optimal coupling $\sigma^{(t)}$ matching their cumulative masses (linear-time after sorting).
(4) **Update $R$.** Minimize

$$\min_{R\in O(n)} \frac{1}{T}\sum_{t=1}^{T}\sum_{i,j}\sigma_{ij}^{(t)}\left(p_i R\theta_t^\top - q_j\theta_t^\top\right)^2 \quad\Longleftrightarrow\quad \max_{R\in O(n)} \operatorname{tr}\Big(R^\top H\Big),$$

with $H = \frac{1}{T}\sum_t P^\top D^{(t)} Q\,\theta_t\theta_t^\top$ and $D^{(t)}$ a sparse matrix encoding $\sigma^{(t)}$ (derived from pairings after sorting). Solve by orthogonal Procrustes: $H = U\Sigma V^\top$, $R \leftarrow UV^\top$.
(5) **Repeat** 2–4 until convergence of the RSWD objective or of $R$.

LB eigenmaps are naturally multiscale: small $n$ captures coarse/global structure, larger $n$ adds detail. We thus register progressively:

$$n_1 < n_2 < \cdots < n_L, \qquad R_k \text{ initialized by } R_{k-1} \text{ (padded to } n_k).$$

This improves robustness, speed, and accuracy.

**Remarks.** (i) If shapes are nearly isometric, the optimal $R$ captures only sign/permutation within multiplicity blocks; RSWD reduces to standard functional-map alignment. (ii) Weights $\mu^P, \mu^Q$ can reflect sampling density or area elements (via local PCA/MLS). (iii) The method integrates seamlessly with additional descriptors by concatenating to the LB embedding prior to RSWD.

**Non-isometric registration via conformal basis alignment.** For large nonrigid deformations, classical spectral descriptors may misalign. A conformal deformation rescales the metric by a positive factor $g \mapsto \tilde{g} = w^2 g$. On genus-0 surfaces, conformal equivalence implies that alignment can be mediated by a scalar weight $w > 0$ on the target. The LB eigensystem on $(\mathcal{M}_2, \tilde{g})$ is equivalent to a *weighted* problem on $(\mathcal{M}_2, g_2)$:

$$-\Delta_{g_2}\psi = \lambda\, w^2\psi, \qquad \int_{\mathcal{M}_2} \psi_i\psi_j\, w^2\, \mathrm{dvol}_{g_2} = \delta_{ij}. \tag{8.33}$$

This suggests coupling a conformal weight with a weighted LB basis so that *feature coefficients* match across shapes [Schonsheck *et al.* (2021)].

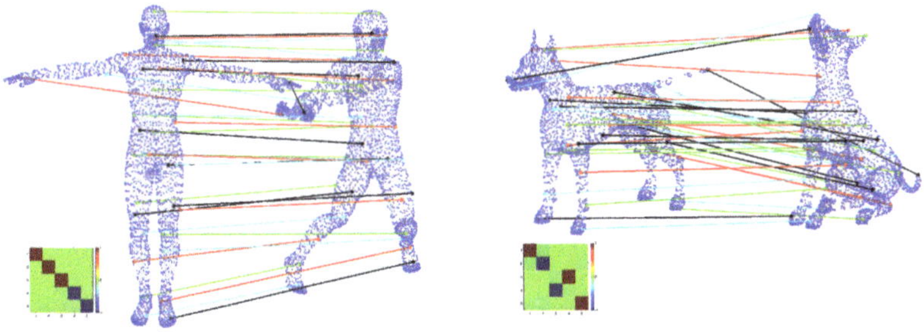

Fig. 8.6 Registration using the first 5 nontrivial LB modes ($n = 5$). We plot aligned embeddings and the resulting $5 \times 5$ orthogonal matrices $R^\star$; uniform weights $\mu^P = \mu^Q = 1/N$.

Let $\Phi = \{\phi_i\}_{i=1}^N$ be fixed on $\mathcal{M}_1$ and $\Psi = \{\psi_i\}_{i=1}^N$ be a weighted-orthonormal basis on $(\mathcal{M}_2, w^2 g_2)$. Given paired features $F = \{f_\ell\}_{\ell=1}^k$ and $G = \{g_\ell\}_{\ell=1}^k$, consider

$$\begin{aligned} \min_{w>0,\ \Psi} \quad & \tfrac{r_1}{2} \Big\| \underbrace{\Big(\int_{\mathcal{M}_1} f_\ell \phi_i \,\mathrm{dvol}_{g_1}\Big)}_{\langle F,\Phi\rangle_{g_1}} - \underbrace{\Big(\int_{\mathcal{M}_2} g_\ell \psi_i \, w^2 \,\mathrm{dvol}_{g_2}\Big)}_{\langle G,\Psi\rangle_{w^2 g_2}} \Big\|_F^2 \\ & + \tfrac{r_2}{2} \textstyle\sum_{i=1}^N \int_{\mathcal{M}_2} \|\nabla_{g_2} \psi_i\|^2 \,\mathrm{dvol}_{g_2} \\ \text{s.t.} \quad & \textstyle\int_{\mathcal{M}_2} \psi_i \psi_j \, w^2 \,\mathrm{dvol}_{g_2} = \delta_{ij}. \end{aligned} \tag{8.34}$$

The first term aligns feature *coefficients*; the second enforces near-harmonicity of $\Psi$ on $(\mathcal{M}_2, w^2 g_2)$.

**Optimization outline.** A practical scheme alternates [Schonsheck *et al.* (2021)]:

(1) *Basis update* ($w$ fixed): minimize (8.34) w.r.t. $\Psi$ under the weighted orthonormality constraint. Use a splitting method (e.g., ADMM): a quadratic "alignment + smoothing" subproblem for $\Psi$, followed by substituting $Q \leftarrow M^{1/2}(w)\Psi$ onto the weighted-orthonormal Stiefel set, where $M(w)$ is the $w^2$-weighted mass matrix.
(2) *Weight update* ($\Psi$ fixed): parameterize $w^2 = \exp(\eta)$ and solve a regularized least-squares problem for $\eta$ (optionally with Laplacian smoothing), then renormalize $M(w)$ and repeat.

This alternating procedure progressively reduces coefficient mismatch while preserving weighted orthonormality, yielding robust alignments under large conformal distortions.

After the final basis is obtained, a point-to-point correspondence between the two surfaces can be established by comparing the values of the corresponding basis functions. This procedure follows the principle of the functional map framework introduced in [Ovsjanikov *et al.* (2012)], applied here to the newly constructed basis. The correspondence is determined by performing a nearest-neighbor search (with $K = 1$) between the rows of $\Phi$ and $\Psi$, which involves comparing $n$ points in a $k$-dimensional space. This approach provides a more efficient and accurate correspondence than direct delta-function matching.

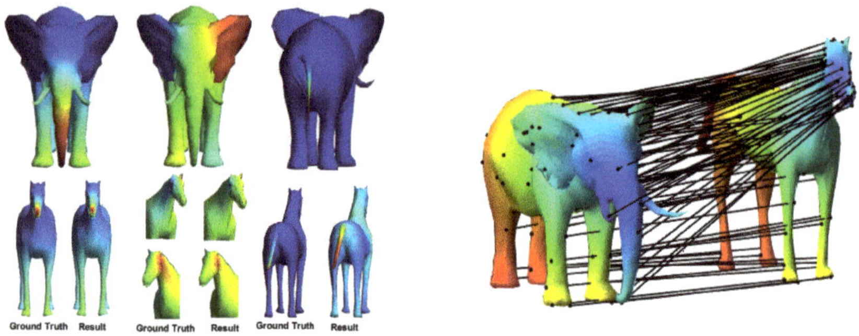


Fig. 8.7 Top left: 9th, 11th and 44th natural LB eigenfunctions on source. Bottom left: results and ground truth. Right: visualization of point-to-point map and texture transfer [Schonsheck *et al.* (2021)].

## 8.4 Committor problem on point clouds

The committor function describes the probability that a stochastic trajectory first reaches a prescribed product region before a reactant region. It provides a quantitative characterization of transition pathways in molecular and chemical systems. The local mesh method introduced in [Lai *et al.* (2013); Lai and Lu (2018)] can be adapted to compute the committor on point clouds using only sampled data, even when samples concentrate near transition manifolds.

**Model.** Consider the overdamped Langevin dynamics

$$\mathrm{d}X_t = -\nabla U(X_t)\,\mathrm{d}t + \sqrt{2\beta^{-1}}\,\mathrm{d}W_t, \tag{8.35}$$

where $U : \Omega \to \mathbb{R}$ is a potential, $\beta = (k_B T)^{-1}$ is the inverse temperature, and $W_t$ is a standard Wiener process. For disjoint reactant and product sets $A, B \subset \Omega$, the committor $q(x) = \mathbb{P}_x(\tau_B < \tau_A)$ satisfies

$$\begin{cases} -\beta^{-1}\Delta q + \nabla U \cdot \nabla q = 0, & x \in \Omega \setminus (A \cup B), \\ q = 0 \text{ on } A, \quad q = 1 \text{ on } B, \quad \nabla q \cdot \vec{n} = 0 \text{ on } \partial\Omega, \end{cases} \tag{8.36}$$

which is a drift–diffusion equation with mixed boundary conditions. Introducing the Gibbs weight $\rho(x) \propto e^{-\beta U(x)}$, the weak form becomes

$$\int_\Omega \nabla q \cdot \nabla \eta \, e^{-\beta U} \, \mathrm{d}x = 0, \qquad \text{for all test functions } \eta \text{ vanishing on } A \cup B. \tag{8.37}$$

Reactive quantities follow as

$$\nu_R = k_B T \int_\Omega |\nabla q|^2 \rho \, \mathrm{d}x, \qquad J_R = k_B T \, \rho \nabla q,$$

representing the reactive flux and current density, respectively.

Let $\{e_i\}_{i=1}^N$ denote the local hat basis functions constructed from one-ring triangulations around each point, as described in Section 7.3.2. The bilinear form in (8.37) gives the weighted stiffness entries

$$S_{ij}^{(\beta)} = \int_\Omega \nabla e_i \cdot \nabla e_j \, e^{-\beta U} \, \mathrm{d}x \approx \sum_T \Big( \int_T \nabla_T e_i \cdot \nabla_T e_j \mathrm{d}x \Big) \, w_T,$$

where $\nabla_T$ is the intrinsic gradient on the local triangle $T$, and $w_T$ is a quadrature of $e^{-\beta U}$ on $T$ (for instance, a vertex average). Using cotangent weights for $\int_T \nabla_T e_i \cdot \nabla_T e_j$ and scaling each triangle contribution by $w_T$ yields a sparse symmetric matrix with $S^{(\beta)}\mathbf{1} = 0$ on free nodes. Let the index sets of nodes correspond to $A$, $B$, and $C = \Omega \setminus (A \cup B)$. With $q = (q_A, q_B, q_C)^\top$, where $q_A \equiv 0$ and $q_B \equiv 1$, the discrete equations on the free set $C$ read

$$S^{(\beta)}(C, C) \, q_C = - \, S^{(\beta)}(C, B) \, \mathbf{1}. \tag{8.38}$$

The matrix $S^{(\beta)}(C, C)$ is symmetric positive definite and can be solved efficiently by preconditioned conjugate gradient method. Gradients of $q$ and subsequent computation of $\nu_R$ and $J_R$ reuse the local-mesh intrinsic gradients and employ a Gibbs-weighted mass matrix for integration.

**Illustration on the rugged Müller potential.** The classical Müller potential and its rugged variant serve as benchmarks for transition computations. On the domain $\Omega = [-1.5, 1] \times [-0.5, 2]$, define

$$U_{\mathrm{rm}}(x, y) = \textstyle\sum_{i=1}^4 D_i \, e^{a_i(x - X_i)^2 + b_i(x - X_i)(y - Y_i) + c_i(y - Y_i)^2} + \gamma \sin(2k\pi x) \sin(2k\pi y),$$

with parameters $\gamma = 9$, $k = 5$, and $(D_i, a_i, b_i, c_i, X_i, Y_i)$ following [Lai and Lu (2018)]. The reactant and product wells are defined as $A = \{U < -120, y > 0.75\}$ and $B = \{U < -82, y < 0.35\}$. Point clouds containing $2 \times 10^4$–$2 \times 10^5$ samples are generated from simulations of (8.35), resulting in higher sampling density near the minimum-energy transition path. The computed committor agrees closely with a finite-element reference $q_{\mathrm{FE}}$ (transition rate $\nu_R^{\mathrm{FE}} = 0.92960$), with relative errors

$$E_{\nu_R} = \frac{|\nu_R - \nu_R^{\mathrm{FE}}|}{\nu_R^{\mathrm{FE}}}, \qquad E_q = \frac{\|q - q_{\mathrm{FE}}\|_2}{\|q_{\mathrm{FE}}\|_2},$$

typically below one percent at moderate sample sizes.

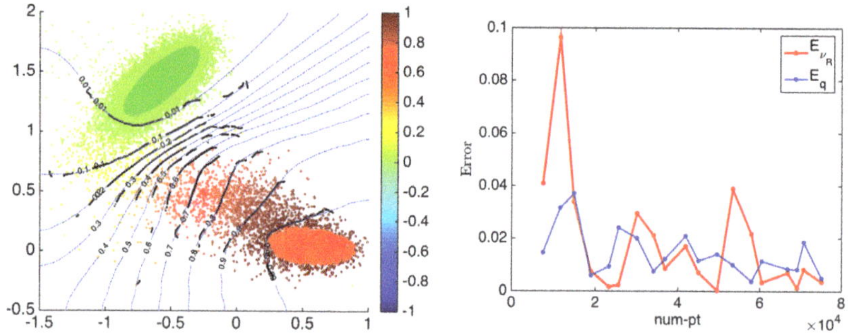


Fig. 8.8 Committor on the rugged Müller potential. Left is contour overlay: computed committor $q$ (black) and FEM solution $q_{\mathrm{FE}}$ (blue dashed) at a representative sample size. Right is convergence of relative errors $E_{\nu_R}$ and $E_q$ with respect to the number of sample points. The local-mesh solution on point clouds exhibits close agreement with the finite-element reference and demonstrates systematic convergence.

This formulation is entirely intrinsic and extends naturally to point clouds sampled from low-dimensional manifolds embedded in high-dimensional spaces. The Gibbs weighting introduces negligible additional cost, requiring only one scalar per local triangle, and Neumann conditions on $\partial\Omega$ are automatically incorporated through the weighted weak form (8.37).

## Appendix A

# Linear Algebra

For readers who are familiar with eigenvalues, the first appendix chapter is an introduction to singular values. For a matrix $A$, $A^\top$ denote the transpose, $A^*$ denote the conjugate transpose of $A$. A matrix $A \in \mathbb{C}^{n\times n}$ is called Hermitian if $A^* = A$.

## A.1 Eigenvalues and min-max principle

Basic facts from linear algebra:

- Any Hermitian matrix $A \in \mathbb{C}^{n\times n}$ has real eigenvalues $\lambda_1 \geq \lambda_2 \geq \cdots \geq \lambda_n$ with a complete set of orthonormal eigenvectors.
- Any real symmetric matrix $A \in \mathbb{R}^{n\times n}$ has real eigenvalues $\lambda_1 \geq \lambda_2 \geq \cdots \geq \lambda_n$ with a complete set of **real** orthonormal eigenvectors.

For a Hermitian matrix $A$, its *Rayleigh-Ritz quotient* is defined as

$$R_A(x) = \frac{x^*Ax}{x^*x}, \quad x \in \mathbb{C}^n.$$

**Theorem A.1** (Courant-Fischer-Weyl min-max principle). *Let $\lambda_1$ and $\lambda_n$ be the largest and the smallest eigenvalues of a Hermitian matrix $A$, then for any vector $x \in \mathbb{C}^n$,*

$$\lambda_n \leq \frac{x^*Ax}{x^*x} \leq \lambda_1.$$

*Proof.* Let $\{v_j \in \mathbb{C}^n : j = 1, \cdots, n\}$ be orthonormal eigenvectors of $A$ then they form a basis. Thus $x = \sum\limits_{j=1}^{n} a_j v_j$. Let $V$ be a matrix with columns as $v_j$ and $a$ be a column vector with entries $a_j$. Then $x = Va$ and $x^*x = a^*V^*Va = a^*a = \sum\limits_{j=1}^{n} |a_j|^2$. Let $\Lambda$ be a diagonal matrix with diagonal entries $\lambda_j$. We have $Av_j = \lambda_j v_j$ thus $Ax = \sum_{j=1}^{n} a_j A v_j = \sum_{j=1}^{n} a_j \lambda_j v_j = V\Lambda a$. Thus $x^*Ax = a^*V^*V\Lambda a = a^*\Lambda a = \sum_{j=1}^{n} \lambda_j |a_j|^2$. The min-max principle holds because

$$\lambda_n \sum_{j=1}^{n} |a_j|^2 \leq \sum_{j=1}^{n} \lambda_j |a_j|^2 \leq \lambda_1 \sum_{j=1}^{n} |a_j|^2.$$

□

## A.2 Singular values and SVD

For a matrix $A \in \mathbb{C}^{m\times n}$, $A^*A$ and $AA^*$ are both positive semi-definite (or definite) Hermitian matrices thus have real non-negative eigenvalues, denoted as $\lambda_i(A^*A)$ and $\lambda_i(AA^*)$, ordered by magnitudes.

The matrix $A$ has $l = \min\{m, n\}$ *singular values*, defined as

$$\sigma_i(A) = \sqrt{\lambda_i(A^*A)} = \sqrt{\lambda_i(AA^*)}.$$

The singular values are defined for a matrix $A$ of any shape, and they are always non-negative real numbers. In contrast, eigenvalues are defined for square matrices and are not necessarily real.

**Theorem A.2.** *Let $l \leq \min\{m, n\}$. Any matrix $A \in \mathbb{C}^{m\times n}$ of rank $k$ has a decomposition $A = U\Sigma V^*$, called* **singular value decomposition (SVD)**, *where $U$ of size $m \times l$ and $V$ of size $n \times l$ both have orthonormal columns, and $\Sigma$ of size $l \times l$ is a diagonal matrix with singular values of $A$. It also has a compact decomposition $A = U_1\Sigma_1 V_1^*$ (***compact SVD***) where $U$ of size $m \times k$ and $V$ of size $n \times k$ have orthonormal columns and $\Sigma_1$ of size $k \times k$ is diagonal matrix with nonzero singular values of $A$.*

*Proof.* Assume $n \leq m$, we consider the matrix $A^*A$ (if $n > m$, similar procedure for $AA^*$). The matrix $A^*A$ is positive semi-definite Hermitian thus has non-negative real eigenvalues with a complete set of orthonormal eigenvectors. Since $A^*A$ has the same rank as $A$ (why? good exercise to figure it out), $A^*A$ has $k$ nonzero eigenvalues. Let $D$ be a $k \times k$ diagonal matrix with all nonzero eigenvalues of $A^*A$ as diagonal entries, and $V$ be a $n \times n$ matrix with orthonormal eigenvectors as columns. Then

$$V^*A^*AV = \begin{bmatrix} D & 0 \\ 0 & 0 \end{bmatrix}.$$

Let $V = [V_1\, V_2]$ corresponding to nonzero and zero eigenvalues, then

$$\begin{bmatrix} V_1^* \\ V_2^* \end{bmatrix} A^*A \begin{bmatrix} V_1 & V_2 \end{bmatrix} = \begin{bmatrix} D & 0 \\ 0 & 0 \end{bmatrix}.$$

By multiplying matrices in the left hand side above, we get

$$V_1^*A^*AV_1 = D, \quad V_2^*A^*AV_2 = 0.$$

Recall $V = [V_1\, V_2]$ has orthonormal columns thus $VV^* = I$, which implies $V_1V_1^* + V_2V_2^* = I$.

Next, since $V_2$ consists of eigenvectors to zero eigenvalue of $A^*A$, we get $A^*AV_2 = 0$ thus $V_2^*A^*AV_2 = 0$. So we must have $AV_2 = 0$ because otherwise it contradicts with $V_2^*A^*AV_2 = 0$.

Let $U_1 = AV_1D^{-\frac{1}{2}}$ where $D^{\frac{1}{2}}$ is defined as taking square root for diagonal entries of $D$. Then

$$U_1D^{\frac{1}{2}}V_1^* = AV_1V_1^* = A(I - V_2V_2^*) = A - (AV_2)V_2^* = A.$$

The decomposition $A = U_1D^{\frac{1}{2}}V_1^*$ is exactly the compact SVD. Pick any $U_2$ of size $n \times (n-k)$ such that $U = [U_1\ U_2]$ is a unitary matrix and define $\Sigma$ of size $n \times n$ as

$$\Sigma = \begin{bmatrix} D^{\frac{1}{2}} & 0 \\ 0 & 0 \end{bmatrix},$$

then $A = U\Sigma V^*$ is the full SVD. □

From the proof above, we get the following facts:

- The columns of $V$ (right-singular vectors) are eigenvectors of $A^*A$.
- The columns of $U$ (left-singular vectors) are eigenvectors of $AA*$.
- A real matrix $A$ has real singular vectors.
- Let $u_i$ and $v_i$ be $i$-th columns of $U$ and $V$ corresponding $i$-th singular value $\sigma_i(A)$, then
$$Av_i = \sigma_i u_i, \quad A^*u_i = \sigma_i v_i.$$
- The rank of $A$ is also the number of nonzero singular values of $A$.
- The compact SVD of $A$ looks like this:
$$A = \boxed{U_1}\,\Sigma_1\,\boxed{\ V_1^*\ }$$
with $\Sigma_1 = diag\{\sigma_1, \cdots, \sigma_k\}$. It is a convention to order $\sigma_i$ in decreasing order: $\sigma_1 \geq \sigma_2 \geq \cdots \geq \sigma_k$.
- For a Hermitian (or real symmetric) positive semi-definite (PSD) matrix $A$ and its SVD $A = U\Sigma V^*$ we must have $U = V$, thus its SVD $A = U\Sigma U^*$ is also its eigenvalue decomposition. Therefore, singular values are also eigenvalues for PSD matrices.

## A.3 Matrix norms

For $x = \begin{bmatrix} x_1 \, x_2 \cdots x_n \end{bmatrix}^\top$:

- *2-norm*: $\|x\| = \sqrt{\sum_{j=1}^{n} |x|_j^2}$.
- *1-norm*: $\|x\|_1 = \sum_{j=1}^{n} |x|_j$.
- *$\infty$-norm*: $\|x\|_\infty = \max_j |x|_j$.

For a rank $k$ matrix $A = (a_{ij})$ of size $m \times n$, assume its SVD is $A = U\Sigma V$ with nonzero singular values $\sigma_1 \geq \sigma_2 \geq \cdots \geq \sigma_k$. Let $\boldsymbol{\sigma} = \begin{bmatrix} \sigma_1 \, \sigma_2 \cdots \sigma_k \end{bmatrix}^\top$. There are many norms of matrices. The following are a few important ones:

- *Spectral norm*: $\|A\|$ is defined as $\|A\| = \max_{x \in \mathbb{C}^n} \frac{\|Ax\|}{\|x\|}$ ($x \in \mathbb{R}^n$ for real matrices) and $\|A\|$ is equal to the largest singular value of $A$. By Courant-Fischer-Weyl min-max principle Theorem A.1,
$$\frac{\|Ax\|}{\|x\|} = \sqrt{\frac{\|Ax\|^2}{\|x\|^2}} = \sqrt{\frac{x^*A^*Ax}{x^*x}} \leq \sqrt{\lambda_1(A^*A)}.$$
By taking $x = v_1$, the eigenvector of $A^*A$ corresponding to $\lambda_1(A^*A)$, we get $\|A\| = \sqrt{\lambda_1(A^*A)} = \sigma_1$.
- *Frobenius norm*: $\|A\|_F = \sqrt{tr(A^*A)} = \sqrt{\sum_{i=1}^{m} \sum_{j=1}^{n} |a_{ij}|^2}$. We have $\|A\|_F = \|\boldsymbol{\sigma}\|$ because
$$\|A\|_F = \sqrt{tr(V^*\Sigma U^*U\Sigma V)} = \sqrt{tr(V^*\Sigma^2 V)} = \sqrt{tr(VV^*\Sigma^2)} = \sqrt{\textstyle\sum_j \sigma_j^2},$$
where we have used the property of trace function $tr(ABC) = tr(CAB)$ for three matrices $A, B, C$ of proper sizes.
- *Nuclear norm*: $\|A\|_* = \sigma_1 + \sigma_2 + \cdots + \sigma_k$. Then the nuclear norm of $A$ is simply $\|\boldsymbol{\sigma}\|_1$.
- *Matrix 1-norm*: $\|A\|_1 = \max_{x \in \mathbb{C}^n} \frac{\|Ax\|_1}{\|x\|_1}$ ($x \in \mathbb{R}^n$ for real matrices). Since $Ax$ is a linear combination of columns of $A$, therefore $\|Ax\|_1$ for $\|x\|_1 = 1$ is less than or equal to a convex combination of 1-norm of columns of $A$ thus $\|A\|_1 = \max_j \sum_{i=1}^{m} |a_{ij}|$.
- *Matrix $\infty$-norm*: $\|A\|_\infty = \max_{x \in \mathbb{C}^n} \frac{\|Ax\|_\infty}{\|x\|_\infty}$ ($x \in \mathbb{R}^n$ for real matrices). It can be proven that $\|A\|_\infty = \max_i \sum_{j=1}^{n} |a_{ij}|$.

Useful facts:

- For a matrix norm $|||A|||$ induced by vector norms such as spectral norm, 1-norm and $\infty$-norm, by definition we have

$$|||Ax||| \leq |||A||| \cdot |||x|||.$$

Since $|||ABx||| \leq |||A||| \cdot |||Bx||| \leq |||A||| \cdot |||B||| \cdot |||x|||$, we also have

$$|||AB||| \leq |||A||| \cdot |||B|||.$$

- For a matrix norm $|||A|||$ defined through singular values such as spectral norm, Frobenius norm and nuclear norm, it is invariant after unitary transformation: let $T$ and $S$ be unitary matrices, then $|||A||| = |||TAS|||$. Notice that $TAS = (TU)\Sigma(V^*S)$ is the SVD of $TAS$, so $TAS$ has the same singular values as $A$.

## A.4 Normal matrices

A matrix $A$ is normal if $A^*A = AA^*$. The following are equivalent:

- $A^*A = AA^*$.
- Its singular values and eigenvalues satisfy $\sigma_i(A) = |\lambda_i(A)|$.
- $A$ is diagonalizable by a unitary matrix: $A = V\Lambda V^*$ where $\Lambda$ is diagonal, $VV^* = I$ and $A = V\Lambda V^*$ is its eigenvalue decomposition.

In other words, a normal matrix $A$ has a complete set of orthonormal eigenvectors. For a normal matrix $A = V\Lambda V^*$, if $\Lambda$ has negative or complex diagonal entries, then $A = V\Lambda V^*$ is not SVD. Instead, its SVD can be derived as follows. Let $|\Lambda|$ be a diagonal matrix with diagonal entries $|\lambda_i|$. Let $S$ be a diagonal matrix with diagonal entries $s_i = \lambda_i/|\lambda_i|$. Then one particular SVD is given by $A = U|\Lambda|V^*$ with $U = VS$.

Three special cases of normal matrices include:

(1) All Hermitian matrices ($A^* = A$) are normal, and they have real eigenvalues.
(2) A matrix $A$ is skew-Hermitian if $A^* = -A$. Skew-Hermitian matrices are normal and always have zero or purely imaginary eigenvalues.
(3) The rotation matrix $A = \begin{pmatrix} \cos\theta & -\sin\theta \\ \sin\theta & \cos\theta \end{pmatrix}$ with $\theta \in (0, \frac{\pi}{2})$ is neither Hermitian nor Skew-Hermitian, but $A$ is normal.

**Remark A.1.** *A normal matrix is diagonalizable by a unitary matrix, which implies that eigenspaces of a normal matrix are orthogonal to one* *another. Thus a diagonalizable matrix can be written as* $A = PDP^{-1}$ *with a diagonal matrix* $D$*, which is not necessarily normal, since a diagonalizable matrix can have non-orthogonal eigenspaces, e.g.,*

$$A = \begin{pmatrix} 1 & 1 \\ 0 & 2 \end{pmatrix} \begin{pmatrix} 3 & 0 \\ 0 & 4 \end{pmatrix} \begin{pmatrix} 1 & 1 \\ 0 & 2 \end{pmatrix}^{-1}.$$

**Problem A.1.** *Prove the equivalence above about normal matrices by using Theorem A.2.*

## Appendix B

# Iterative Methods for Solving Linear Systems

In Chapter 2 and Chapter 3, we have used a simple direct inversion method for the big discrete Laplacian matrix $K_x \otimes I_y + I_x \otimes K_y$ in two dimensions and also discrete Laplacian in three dimensions, via their eigenvalue decompositions, which are however no longer available for general problems. For instance, if we solve an elliptic equation in the form of

$$-\nabla \cdot (a(x, y)\nabla u) = f,$$

then any non-constant coefficient $a(x, y)$ will destroy the tensor structure.

One example of variable coefficient problems is the Poisson equation $u_{xx} + u_{yy} = f$ on a disk. We can consider the finite difference method in the polar coordinates, under which the disk becomes a rectangle. The Poisson equation $-u_{xx} - u_{yy} = f$ in the polar coordinates becomes

$$-\frac{1}{r}\frac{\partial}{\partial r}\left(r\frac{\partial u}{\partial r}\right) - \frac{1}{r^2}\frac{\partial^2 u}{\partial \theta^2} = f(r, \theta).$$

In this appendix chapter, we introduce a few iterative methods for approximately solving these large linear systems.

## B.1 Linear iterative methods

In Chapter 2, we have derived a linear system $Au = f$, for approximating the 1D Poisson equation $-u_{xx} = f$ on the interval $[0, 1]$ with homogeneous Dirichlet boundary conditions:

$$\frac{-u_{j-1} + 2u_j - u_{j+1}}{h^2} = f_j, \tag{B.1}$$

or the 2D Poisson equation $-u_{xx} - u_{yy} = f$ on the square $[0, 1] \times [0, 1]$ with homogeneous Dirichlet boundary conditions (assuming $h = \Delta x = \Delta y$):

$$\frac{-u_{i-1,j} - u_{i,j-1} + 4u_{i,j} - u_{i+1,j} - u_{i,j+1}}{h^2} = f_{i,j}. \tag{B.2}$$

So the matrix $A = K$ for the 1D case, and $A = K \otimes I + I \otimes K$ for the 2D case with $\Delta x = \Delta y$ on a square domain, where $K$ is the same matrix as defined in Chapter 2.

Here we use (B.1) and (B.2) only as examples to understand iterative methods. In practice, for implementing (B.1) and (B.2), the direct inversion method via eigenvalue decompositions in Chapter 2 is a simple and acceptably efficient method for most applications on modern computing devices.

Suppose we split the matrix $A$ as $A = B - C$ for any nonsingular matrix $B$, then the exact solution $u$ to the linear system $Au = f$ satisfies:

$$Bu = Cu + f,$$

thus we use the following iterative methods,

$$Bu^{k+1} = Cu^k + f,$$

in which the limit of $u^k$ will be the solution if this iteration converges. For a convergent iteration, we may use $u^N$ for large enough $N$ as an approximate solution.

By subtracting the two equations above, the error $e^k = u^k - u$ satisfies,

$$e^{k+1} = B^{-1}Ce^k = Me^k,$$

where $M = B^{-1}C$ is called the iteration matrix. The spectral radius of a square matrix is the maximum of the absolute values of its eigenvalues. Thus $e^k = M^k e^0$, and the convergence is guaranteed if the spectral radius $\rho(M) < 1$ due to the following fact:

**Theorem B.3.** *A square matrix $M$ satisfies $\lim_{k\to\infty} M^k = 0$ if and only if its spectral radius $\rho(M) < 1$.*

If $M$ is a normal matrix, i.e, it can be unitarily diagonalized (see Appendix A.4), then

$$M = U \begin{pmatrix} \lambda_1 & & \\ & \ddots & \\ & & \lambda_n \end{pmatrix} U^* \Rightarrow M^k = U \begin{pmatrix} \lambda_1^k & & \\ & \ddots & \\ & & \lambda_n^k \end{pmatrix} U^*,$$

thus $\|M^k\| = \max_j |\lambda_j|^k = \rho(M)^k$ and

$$\|e^k\| \leq \|M^k\|\|e^0\| = \rho(M)^k\|e^0\|,$$

which implies that the convergence rate is $\rho(M)$.

If $M$ is not a normal matrix, then the asymptotic convergence rate is $\rho(M)$, i.e., there is a constant $C$ such that

$$\|e^k\| \leq C\rho(M)^k, \quad \text{for large enough } k,$$

due to the following fact:

**Theorem B.4** (Gelfand's formula). *Any square matrix $M$ satisfies*

$$\lim_{k\to\infty} \|M^k\|^{\frac{1}{k}} = \rho(M).$$

**Jacobi and weighted Jacobi iterations.** Let $D$ be a diagonal matrix denoting the diagonal part of the matrix $A$, $L$ be a lower triangular matrix denoting the lower triangular part of the matrix $-A$, $U$ be a upper triangular matrix denoting the upper triangular part of the matrix $-A$. In other words, assume $A = D - L - U$. The Jacobi iteration for solving $Au = f$ is defined as

$$Du^{k+1} = (L+U)u^k + f.$$

For the 1D Poisson scheme (B.1), the Jacobi iteration is equivalent to

$$-u_{j-1}^k + 2u_j^{k+1} - u_{j+1}^k = h^2 f_j.$$

It can be implemented as an iteration of

$$u_j^{k+1} = \frac{1}{2}u_{j-1}^k + \frac{1}{2}u_{j+1}^k + \frac{1}{2}h^2 f_j, \tag{B.3}$$

and the 2D case is

$$u_{i,j}^{k+1} = \frac{1}{4}u_{i,j-1}^k + \frac{1}{4}u_{i,j+1}^k + \frac{1}{4}u_{i-1,j}^k + \frac{1}{4}u_{i+1,j}^k + \frac{1}{4}h^2 f_{i,j}.$$

Now let us compute the spectral radius of the iteration matrix $M = D^{-1}(D-A)$ for the second order finite difference scheme on a $N\times N$ mesh solving the 2D Poisson equation with zero boundary conditions, in which $A = K\otimes I + I\otimes K$ and $D = \frac{4}{h^2} I\otimes I$. We have $M = D^{-1}(D-A) = I - \frac{h^2}{4}A$ thus the eigenvalues (see Chapter 2 for the eigenvalues of $A$) are

$$\lambda_{i,j}(M) = 1 - \frac{h^2}{4}\lambda_{i,j}(A) = \frac{1}{2}\cos\left(i\frac{\pi}{N+1}\right) + \frac{1}{2}\cos\left(j\frac{\pi}{N+1}\right).$$

Thus $\rho(M) = \cos\theta$ with $\theta = \frac{\pi}{N+1}$. Similarly for the 1D case $\lambda_j(M) = \cos\left(j\frac{\pi}{N+1}\right) = \cos(j\theta)$. In particular, the convergence rate for large $N$ is

$$\rho(M) = \cos\theta \approx 1 - \frac{1}{2}\theta^2 = 1 - \frac{1}{2}\frac{\pi^2}{(N+1)^2} \approx 1 - cN^{-2},$$

which means that the convergence is slower for a larger system.

The Jacobi iteration can be regarded as solving a time-dependent heat equation till the steady state. If using the forward Euler time discretization for the equation $u_t = u_{xx} + f$, we get

$$u_j^{n+1} = u_j^n + \frac{\Delta t}{\Delta x^2}(u_{j-1}^n - 2u_j^n + u_{j+1}^n) + \Delta t f_j,$$

which becomes (B.3) if $\Delta t = \frac{1}{2}\Delta x^2$.

**Remark B.2.** *The Lax-Richtmyer stability for using forward Euler approximating time-dependent PDE like the heat equation requires* $\Delta t \leq \frac{1}{2}\Delta x^2$, *i.e., the iteration above is stable only if* $\Delta t \leq \frac{1}{2}\Delta x^2$.

The weighted Jacobi iteration is to use the splitting $A = D/w - (D/w - A)$ where $w$ is a constant parameter/weight, which results in the iteration

$$D/wu^{k+1} = (D/w - A)u^k + f,$$

and its iteration matrix is $M = (D/w)^{-1}(D/w - A) = I - wD^{-1}A$. Notice that $w = 1$ is the original Jacobi iteration. For the 1D Poisson equation case, we have $\lambda_j(M) = 1 - \frac{1}{2}w\lambda_j(A) = 1 - w + w\cos(j\theta)$ with $\theta = \frac{\pi}{N+1}$. For the comparison of the two eigenvalues, see Figure B.1, where $|\lambda_j(M)|$ in the weighted Jacobi is smaller for large $j$ (meaning faster convergence for high frequencies) and $|\lambda_j(M)|$ in the weighted Jacobi is larger for smaller $j$ (meaning slower convergence for low frequencies). Here frequencies refer to the indices of the eigenvectors of $K$ matrix in Chapter 2 since its eigenvectors are point values of trigonometric functions of different frequencies.

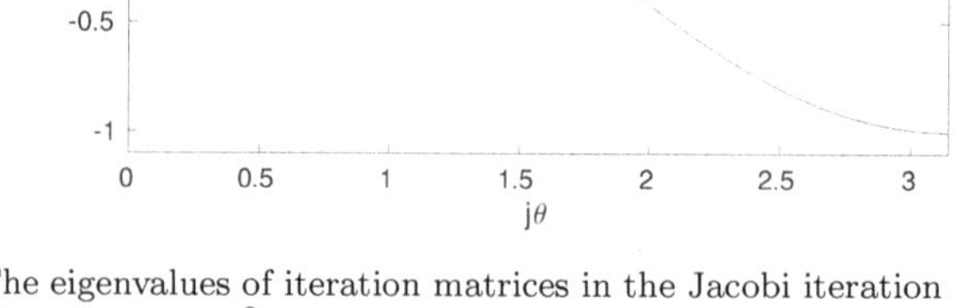


Fig. B.1 The eigenvalues of iteration matrices in the Jacobi iteration and the weighted Jacobi iteration with $w = \frac{2}{3}$.

**Gauss-Seidel iteration.** Suppose $A = D - L - U$ as before, then the Gauss-Seidel iteration is

$$(D - L)u^{k+1} = Uu^k + f,$$

where the system $(D-L)x = b$ can be easily solved by forward substitution since $D-L$ is lower triangular. For the 1D Poisson scheme (B.1), the Gauss-Seidel iteration is conceptually

$$-u_{j-1}^{k+1} + 2u_j^{k+1} - u_{j+1}^k = h^2 f_j,$$

and in 2D it is

$$-u_{i-1,j}^{k+1} - u_{i,j-1}^{k+1} + 4u_{i,j}^{k+1} - u_{i+1,j}^k - u_{i,j+1}^k = h^2 f_{i,j}.$$

To find the eigenvalues for $M = (D-L)^{-1}U$, assume $\lambda$ and $v$ form a eigenvalue-eigenvector pair for $M$, then $Mv = \lambda v$ thus

$$Uv = \lambda(D-L)v \Rightarrow \lambda Dv = \lambda Lv + Uv,$$

which in 2D is equivalent to

$$4\lambda v_{i,j} = \lambda v_{i-1,j} + \lambda v_{i,j-1} + v_{i+1,j} + v_{i,j+1}.$$

Consider a change of variable by introducing a vector $w$ so that $v_{i,j} = \lambda^{\frac{i+j}{2}} w_{i,j}$, then $w$ satisfies

$$4\lambda^{\frac{i+j+2}{2}} w_{i,j} = \lambda^{\frac{i+j+1}{2}} w_{i-1,j} + \lambda^{\frac{i+j+1}{2}} w_{i,j-1} + \lambda^{\frac{i+j+1}{2}} w_{i+1,j} + \lambda^{\frac{i+j+1}{2}} w_{i,j+1}.$$

Therefore the vector $w$ satisfies

$$4\lambda^{\frac{1}{2}} w_{i,j} = w_{i-1,j} + w_{i,j-1} + w_{i+1,} + w_{i,j+1},$$

whose matrix form is precisely $\lambda^{\frac{1}{2}} Dw = (L+U)w$, i.e., $D^{-1}(L+U)w = \lambda^{\frac{1}{2}} w$. Since $D^{-1}(L+U)$ is the iteration matrix in the Jacobi iteration, this means that $\lambda(M_{GS}) = \lambda(M_{Jacobi})^2$. Thus $\rho(M_{GS}) = \cos^2\theta$, which means that the Gauss-Seidel iteration is twice as fast as the Jacobi iteration.

The Gauss-Seidel iteration can also be regarded as a scheme solving a time-dependent equation. Consider the equation

$$u_t = u_{xx} - \varepsilon u_{xt} + f,$$

and the scheme

$$\frac{u_j^{n+1} - u_j^n}{\Delta t} = \frac{u_{j-1}^n - 2u_j^n + u_{j+1}^n}{\Delta x^2} - \varepsilon\frac{(u_j^{n+1} - u_{j-1}^{n+1}) - (u_j^n - u_{j-1}^n)}{\Delta x \Delta t} + f_j,$$

which is accurate around $t = (n+\frac{1}{2})\Delta t$ and $x = x_j$. If setting $\Delta t = \Delta x^2$ and $\varepsilon = \Delta x$, the scheme becomes the Gauss-Seidel iteration.

**Successive overrelaxation.** Another popular iteration method is successive overrelaxation (SOR), which is a combination of Jacobi and Gauss-Seidel. The matrix splitting is $A = \frac{1}{w}(D - wL) - \frac{1}{w}[(1-w)D + wU]$, thus

$$\frac{1}{w}(D - wL)u^{k+1} = \frac{1}{w}[(1-w)D + wU]u^k + f,$$

which is equivalent to

$$Du^{k+1} = Du^k + w(Lu^{k+1} + Uu^k - Du^k) + wf.$$

For the 2D Poisson scheme, this is

$$4u_{i,j}^{k+1} = 4u_{i,j}^k + w(u_{i,j-1}^{k+1} + u_{i-1,j}^{k+1} + u_{i,j+1}^k + u_{i+1,j}^k - 4u_{i,j}^k) + wh^2 f_{i,j}.$$

There are ways to choose $w$ to improve the spectral radius from $\rho(M) = 1 - cN^{-2}$ to $\rho(M) = 1 - cN^{-1}$.

## B.2 Steepest descent

Consider any linear system $Ax = b$ where $A$ is a real square matrix and $x$ and $b$ are real vectors of size $n$. It is equivalent to solving a minimization problem:

$$\min_{x \in \mathbb{R}^n} \frac{1}{2}\|Ax - b\|^2.$$

The method of gradient descent is the simplest first order method for solving a minimization problem $\min\limits_{x \in \mathbb{R}^n} f(x)$:

$$x_{k+1} = x_k - \alpha_k \nabla f(x_k),$$

where $\alpha_k > 0$ is the optimal step size to be chosen to guarantee convergence. The steepest descent method is to use the optimal step size $\alpha_k$ along the search direction $-\nabla f(x_k)$. Recall that the function ascent/descent the most along the positive/negative gradient: the directional derivative along a unit vector $\mathbf{u}$ is

$$D_{\mathbf{u}} f(x) = \nabla f(x) \cdot \mathbf{u} = \|\nabla f(x)\| \|\mathbf{u}\| \cos\theta,$$

where $\theta$ is the angle between two vectors $\nabla f(x)$ and $\mathbf{u}$, thus $D_{\mathbf{u}} f(x)$ attends its maximum (minimum) at $\theta = 0$ ($\theta = \pi$).

**Remark B.3.** *First order here refers to the fact that only the first order derivative $\nabla f(x)$ is used in the algorithm. A typical second order method is the Newton's method: for example, suppose $n = 1$, if the minimizer of $f(x)$ exists, it must be a critical point of $f(x)$, thus we can solve $f'(x) = 0$ instead, and the second order derivative $f''(x)$ is needed in Newton's method solving $f'(x) = 0$.*

To simplify the discussion, from now on, we assume $A$ is real symmetric and positive semi-definite. If the coefficient matrix cannot be made symmetric or positive definite after simple transformations, then we consider solving $A^\top Ax = A^\top b$ instead. For $Ax = b$, define the cost function

$$f(x) = \frac{1}{2}x^\top Ax - x^\top b,$$

and its gradient is

$$\nabla f(x) = Ax - b.$$

Since $\nabla^2 f(x) = A \geq 0$ is positive semi-definite, the function $f(x)$ is a convex function, which guarantees that the minimizer of $f(x)$ must be a solution to $\nabla f(x) = 0$.

In the steepest descent method $x_{k+1} = x_k - \alpha_k \nabla f(x_k)$, the step size $\alpha_k$ can be chosen so that $f(x_{k+1})$ is the smallest:

$$\alpha_k = \arg\min_{\alpha \geq 0} f(x_k - \alpha \nabla f(x_k)).$$

Let $\phi(\alpha) = f(x_k - \alpha \nabla f(x_k))$. Let $r_k$ denote $-\nabla f(x_k) = b - Ax_k$, which is also called *residue*. Then

$$\begin{aligned}
\phi(\alpha) &= f(x_k + \alpha r_k) \\
&= \frac{1}{2}x_k^\top Ax_k + \alpha r_k^\top Ax_k + \frac{1}{2}\alpha^2 r_k^\top Ar_k - x_k^\top b - \alpha r_k^\top b \\
&= \frac{1}{2}x_k^\top Ax_k - x_k^\top b + \alpha r_k^\top (Ax_k - b) + \frac{1}{2}\alpha^2 r_k^\top Ar_k \\
&= f(x_k) - \alpha r_k^\top r_k + \frac{1}{2}\alpha^2 r_k^\top Ar_k,
\end{aligned}$$

which is quadratic in $\alpha$ thus attains its minimum at the critical point. By setting $\phi'(\alpha) = 0$, we get

$$\alpha_k = \arg\min_{\alpha \geq 0} \phi(\alpha) = \frac{r_k^\top r_k}{r_k^\top Ar_k}.$$

The steepest descent method can be implemented as iterations of the following steps:

(1) $r_k = b - Ax_k$.
(2) $\alpha_k = \frac{r_k^\top r_k}{r_k^\top Ar_k}$.
(3) $x_{k+1} = x_k + \alpha_k r_k$.

We can use the eigenvectors $v_j$ and eigenvalues $\lambda_j$ of $A$ to understand the convergence of the steepest descent method. Define the error $e_k = x_k - x$, then $r_k = b - Ax_k = Ax - Ax_k = -Ae_k$ and we have:

$$e_{k+1} = e_k + \alpha_k r_k.$$

Since $A$ is real symmetric, we can choose orthonormal eigenvectors $v_j$ ($j = 1, \cdots, n$) which can span the whole space $\mathbb{R}^n$. The error can be expressed as linear combinations of $v_j$:

$$e_k = \sum_{j=1}^{n} a_j v_j.$$

We also have

$$\begin{aligned} r_k &= -Ae_k = -\sum_{j=1}^{n} a_j A v_j = -\sum_{j=1}^{n} a_j \lambda_j v_j, \\ e_k^\top A e_k &= (\sum_{j=1}^{n} a_j v_j^\top)(\sum_{j=1}^{n} a_j \lambda_j v_j) = \sum_{j=1}^{n} a_j^2 \lambda_j, \\ r_k^\top r_k &= \sum_{j=1}^{n} a_j^2 \lambda_j^2, \\ r_k^\top A r_k &= \sum_{j=1}^{n} a_j^2 \lambda_j^3. \end{aligned}$$

Assume $0 \le \lambda_1 \le \lambda_2 \le \cdots \le \lambda_n$, by the min-max principle in Appendix A.1,

$$\lambda_1 r_k^\top r_k \le r_k^\top A r_k \le \lambda_n r_k^\top r_k,$$

and also

$$\lambda_1 e_k^\top A e_k \le r_k^\top r_k \le \lambda_n e_k^\top A e_k.$$

Define the energy norm $\|e\|_A = (e^\top A e)^{\frac{1}{2}}$ which is easier to work with than the Euclidean norm. Then

$$
\begin{aligned}
\|e_{k+1}\|_A^2 &= e_{k+1}^\top A e_{k+1} \\
&= (e_k + \alpha_k r_k)^\top A (e_k + \alpha_k r_k) \\
&= e_k^\top A e_k + 2\alpha_k r_k^\top A e_k + \alpha_k^2 r_k^\top A r_k \\
&= e_k^\top A e_k - 2\frac{r_k^\top r_k}{r_k^\top A r_k} r_k^\top r_k + \left(\frac{r_k^\top r_k}{r_k^\top A r_k}\right)^2 r_k^\top A r_k \\
&= e_k^\top A e_k - \frac{(r_k^\top r_k)^2}{r_k^\top A r_k} \\
&= \|e_k\|_A^2 \left(1 - \frac{(r_k^\top r_k)^2}{r_k^\top A r_k e_k^\top A e_k}\right) \\
&= \|e_k\|_A^2 w,
\end{aligned}
$$

where

$$
w = 1 - \frac{r_k^\top r_k}{r_k^\top A r_k}\frac{r_k^\top r_k}{e_k^\top A e_k} \le 1 - \frac{1}{\lambda_n}\lambda_1 = 1 - \frac{\lambda_1}{\lambda_n}.
$$

**Definition B.1.** *The condition number of a matrix $A$ is defined as $\kappa(A) = \|A\| \cdot \|A^{-1}\|$. If $\sigma_1$ and $\sigma_m$ are the largest and the smallest singular values of a matrix $A$, then $\kappa(A) = \frac{\sigma_1}{\sigma_m}$.*

Since $A$ is a positive definite real symmetric matrix, its eigenvalues are also its singular values (see Appendix A), its condition number is $\kappa = \frac{\lambda_n}{\lambda_1}$, thus $w \le \frac{\kappa-1}{\kappa}$. Therefore,

$$
\|e_k\|_A \le \sqrt{\frac{\kappa-1}{\kappa}}\|e_{k-1}\|_A \le \left(\frac{\kappa-1}{\kappa}\right)^{\frac{k}{2}} \|e_0\|_A.
$$

Recall that $Ax = b$, therefore $x^\top Ax = x^\top b$ thus

$$
x^\top Ax = -2(\frac{1}{2}x^\top Ax - x^\top b) = -2f(x).
$$

For a symmetric $A$, we have

$$
\begin{aligned}
\|e_k\|_A^2 &= (x_k - x)^\top A(x_k - x) = x_k^\top A x_k - x^\top A x_k - x_k^\top A x + x^\top A x \\
&= x_k^\top A x_k - b^\top x_k - x_k^\top b + x^\top A x = 2f(x_k) - 2f(x).
\end{aligned}
$$

We can summarize the convergence rate of steepest descent for a linear system as the following theorem.

**Theorem B.5.** *Let $A$ be a positive definite matrix with eigenvalues $0 < \lambda_1 \le \cdots \le \lambda_n$. For the steepest descent method minimizing a quadratic* *function* $f(x) = \frac{1}{2}x^\top Ax - x^\top b$*, the convergence rate is linear (exponentially fast):*

$$\|e_{k+1}\|_A \leq \sqrt{\frac{\lambda_n - \lambda_1}{\lambda_n}}\|e_k\|_A,$$

*and*

$$f(x_{k+1}) - f(x_*) \leq \left(1 - \frac{\lambda_1}{\lambda_n}\right)[f(x_k) - f(x_*)].$$

**Remark B.4.** *The steepest descent has an exponential convergence rate* $\sqrt{1 - \frac{1}{\kappa}}$*, which implies very slow convergence for solving the Poisson equation on fine grids with small mesh size* $h$*.*

**Problem B.2.** *In MATLAB, we can use the functions numgrid and delsq to generate the classical second order discrete Laplacian with homogeneous Dirichlet b.c. on a few more general domains:*

```
% size of geometry, larger Length gives larger matrix
Length = 10;
% 'numgrid' generates uniform grid points for the interior
% of the specified shape.
% other possible shapes include L,S,N,C,D,A,H,B
DomainGrid = numgrid('L',Length); % L shape domain
% 'delsq' generate the second order discrete Laplacian
% defined on the grid points
A = delsq(DomainGrid);
n = size(A,1);
figure;
% visualize the domain and grid points
subplot(1,2,1), spy(DomainGrid), title('domain shape')
% visualize the sparsity pattern of the matrix A
subplot(1,2,2), spy(A), title('system matrix')
%% Generate a linear system Ax=b with a random vector b
b = rand(n,1);
x = A\b;
```

*For example, if we specify the domain as the square, then the obtained matrix above is exactly* $K \otimes I + I \otimes K$*. Use the script above to generate a linear system* $Ax = b$ *where* $A$ *is* $n \times n$ *and* $b$ *is generated randomly. You can use whatever shape you like and use a proper length such that* $n$ *is at least around one thousand. You can verify in MATLAB that the matrix* $A$ *above is indeed symmetric and positive definite:*

```
norm(A-A','fro') % If A is real symmetric, then this is zero
min(eig(A))      % The smallest eigenvalue should be positive
```

*The following can be used to compute the decomposition $A = D - L - U$:*

```
D=diag(A);  L=-tril(A)+diag(D);  U=diag(D)-L-A;
```

*Implement the following methods: Steepest Descent, Jacobi iteration, Weighted Jacobi iteration with $w = \frac{2}{3}$, Gauss-Seidel iteration, Successive overrelaxation (SOR) for which $w$ can be chosen as a number close to 2, e.g., use $w = 1.9$. Store relative error $\|x^k - x\|/\|x\|$ and relative residue $\|b - Ax^k\|/\|b\|$ for each iteration. Use zero initial guess thus the initial relative error and the initial relative residue are both one for each method. For all methods, set the iteration number as $n/2$ where $n$ is the size of the vector $b$. Generate a figure of semilogy plot of relative residue for these methods to compare them.*

## B.3 The conjugate gradient method

We still assume $A$ is real symmetric and $A > 0$. In practice, the steepest descent method is simple to apply, but very slow. The conjugate gradient method is a faster popular choice. The conjugate gradient method can be regarded as an acceleration of steepest descent:

$$x_{k+1} = x_k + \alpha_k(r_k + \gamma_k(x_k - x_{k-1})),$$

where $\alpha_k$ and $\gamma_k$ are parameters. This formula shows that the new change in position, $x_{k+1} - x_k$ , is a linear combination of the steepest descent direction and the previous change in position $x_k - x_{k-1}$. It can be rewritten as

$$x_{k+1} = x_k + \alpha_k p_k,$$

where the search direction $p_k$ is

$$p_k = r_k + \gamma_k(x_k - x_{k-1}) = r_k + \gamma_k\alpha_{k-1}p_{k-1} = r_k + \beta_{k-1}p_{k-1}.$$

These formulas can be summarized as

$$x_{k+1} = x_k + \alpha_k p_k \tag{B.4}$$

$$r_{k+1} = r_k - \alpha_k A p_k \tag{B.5}$$

$$p_{k+1} = r_{k+1} + \beta_k p_k. \tag{B.6}$$

We still need to determine the initial search direction $p_0$ and the parameters $\alpha$ and $\beta$. Suppose $p_k$ is known, then we seek $\alpha_k$ so that

$f(x_k+\alpha_k p_k)$ is the smallest. Minimizing the function $\phi(\alpha) = f(x_k+\alpha p_k) = f(x_k) - \alpha r_k^\top p_k + \frac{1}{2}\alpha^2 p_k^\top A p_k$ gives us the best $\alpha_k$:

$$\alpha_k = \frac{p_k^\top r_k}{p_k^\top A p_k}. \tag{B.7}$$

Using this optimal $\alpha_k$, we have

$$f(x_{k+1}) = f(x_k) - \frac{1}{2}\frac{(p_k^\top r_k)^2}{p_k^\top A p_k}.$$

From the formula above, we can see that $p_0 = r_0$ will guarantee $f(x_1) < f(x_0)$. Now we assume $p_0 = r_0$ which will imply other useful properties. By (B.5) and (B.7), we get

$$p_k^\top r_{k+1} = p_k^\top r_k - \alpha_k p_k^\top A p_k = 0.$$

Together with (B.6), we get

$$p_{k+1}^\top r_{k+1} = r_{k+1}^\top r_{k+1} + \beta_k p_k^\top r_{k+1} = r_{k+1}^\top r_{k+1}, \quad k \geq 0.$$

Notice that $p_0 = r_0$ ensures $p_k^\top r_k = r_k^\top r_k$ for $k \geq 0$. Thus (B.7) becomes

$$\alpha_k = \frac{r_k^\top r_k}{p_k^\top A p_k},$$

and we have

$$f(x_{k+1}) = f(x_k) - \frac{1}{2}\frac{(r_k^\top r_k)^2}{p_k^\top A p_k}.$$

Next we choose some $\beta$ to minimize $p_k^\top A p_k$ thus to minimize $f(x_{k+1})$. Since (B.6) implies

$$p_k^\top A p_k = r_k^\top A r_k + 2\beta_{k-1} r_k^\top A p_{k-1} + \beta_{k-1}^2 p_{k-1}^\top A p_{k-1},$$

the best choice of $\beta_{k-1}$ is,

$$\beta_{k-1} = -\frac{r_k^\top A p_{k-1}}{p_{k-1}^\top A p_{k-1}}, \quad k \geq 1,$$

or equivalently

$$\beta_k = -\frac{r_{k+1}^\top A p_k}{p_k^\top A p_k}, \quad k \geq 0,$$

which with (B.6) implies

$$p_{k+1}^\top A p_k = r_{k+1}^\top A p_k + \beta_k p_k^\top A p_k = 0. \tag{B.8}$$

The property $p_{k+1}^\top A p_k = 0$ means that the search direction $p_{k+1}$ is A-orthogonal to the previous one $p_k$. In other words, $p_{k+1}$ is conjugate to $p_k$. By (B.8) and (B.6), we get

$$p_k^\top A p_k = r_k^\top A p_k + \beta_{k-1} p_{k-1}^\top A p_k = r_k^\top A p_k,$$

which with (B.5) and (B.7) implies,

$$r_{k+1}^\top r_k = r_k^\top r_k - \alpha_k p_k^\top A r_k = 0.$$

Therefore, by (B.5), we have

$$r_{k+1}^\top r_{k+1} = r_{k+1}^\top r_k - \alpha_k r_{k+1}^\top A p_k = -\alpha_k r_{k+1}^\top A p_k,$$

thus the formula for $\beta$ becomes

$$\beta_k = -\frac{r_{k+1}^\top A p_k}{p_k^\top A p_k} = \frac{1}{\alpha_k} \frac{r_{k+1}^\top r_{k+1}}{p_k^\top A p_k} = \frac{r_{k+1}^\top r_{k+1}}{r_k^\top r_k}.$$

Now we summarize the formula for the conjugate gradient method:

$$p_0 = r_0 = b - Ax_0, \tag{B.9a}$$

$$\alpha_k = \frac{\|r_k\|^2}{p_k^\top A p_k}, \tag{B.9b}$$

$$x_{k+1} = x_k + \alpha_k p_k, \tag{B.9c}$$

$$r_{k+1} = r_k - \alpha_k A p_k, \tag{B.9d}$$

$$\beta_k = \frac{\|r_{k+1}\|^2}{\|r_k\|^2} \tag{B.9e}$$

$$p_{k+1} = r_{k+1} + \beta_k p_k. \tag{B.9f}$$

The following important property of conjugate gradient method can be shown by induction:

**Theorem B.6.** *For the conjugate gradient method defined above, the search direction $p_k$ and the residue $r_k$ satisfies:*

$$r_k^\top r_j = p_k^\top A p_j = 0, \quad \forall k \neq j.$$

An implication of this property is that the conjugate gradient method in theory converges in at most $n$ steps if the matrix $A$ has size $n \times n$, because $r_j$ belongs to $\mathbb{R}^n$, in which there are at most $n$ orthogonal vectors. On the other hand, the iteration (B.9) is unstable subject to round-off errors. Nonetheless, the CG method (B.9) is still a good choice for finding an approximate solution in many applications.

**Remark B.5.** *Notice that we have assumed $A$ is real symmetric positive definite. For solving $Ax = b$ with a real symmetric positive semi-definite matrix $A$, e.g., the Poisson equation with purely Neumann b.c., then* (B.9) *might diverge if $Ax = b$ does not have a solution. Notice that $f(x) = \frac{1}{2}x^\top Ax - b^\top x$ always has minimizers for $A \geq 0$, even if $Ax = b$ is an inconsistent linear system. In other words, if $Ax = b$ is an inconsistent linear system, then the iteration* (B.9) *may diverge when the residue $r_k$ reaches the magnitude around* $\min_x \|Ax - b\|$.

In practice, usually it is not affordable to have $n$ iteration steps thus it is important to analyze the convergence rate, for which we have

**Theorem B.7.** *If $0 < \lambda_1 \leq \lambda_2 \leq \cdots \leq \lambda_n$ are eigenvalues of the positive definite matrix $A$, then the error in the conjugate gradient method satisfies*

$$\|e_k\|_A \leq 2\left(\frac{\sqrt{\lambda_n} - \sqrt{\lambda_1}}{\sqrt{\lambda_n} + \sqrt{\lambda_1}}\right)^k \|e_0\|_A.$$

The proofs of these two theorems can be found in [Strikwerda (2004)]. A weaker estimate can be given by the condition number $\kappa$ of $A$:

$$\|e_k\|_A \leq 2\left(\frac{\sqrt{\kappa} - 1}{\sqrt{\kappa} + 1}\right)^k \|e_0\|_A.$$

The dominating operations in each iteration of either steepest descent or conjugate gradient are matrix-vector multiplications. For many problems, $A$ is sparse and one matrix-vector multiplication requires $\mathcal{O}(m)$ operations where $m$ is the number of nonzero entries in $A$.

Suppose we wish to reduce the norm of the error by a factor of $\varepsilon$, which is to achieve $\|e_k\|_A \leq \varepsilon\|e_0\|_A$. Then by the convergence rate, the maximum number of iterations required in steepest descent is

$$k \leq \left\lceil \frac{1}{2}\kappa \ln\left(\frac{1}{\varepsilon}\right)\right\rceil,$$

and the maximum number of iterations in CG is

$$k \leq \left\lceil \frac{1}{2}\sqrt{\kappa} \ln\left(\frac{2}{\varepsilon}\right)\right\rceil.$$

So the time complexity of steepest descent is $\mathcal{O}(m\kappa)$ and the time complexity of CG is $\mathcal{O}(m\sqrt{\kappa})$. Both have the space complexity $\mathcal{O}(m)$.

For the second order finite difference scheme solving the Poisson equation with Dirichlet boundary conditions on $[0, 1]$ with $n$ grid points in 1D (with $n = N \times N$ grid points on $[0, 1] \times [0, 1]$ in 2D; with $n = N \times N \times N$ grid points on $[0,1]\times[0,1]\times[0,1]$ in 3D), the condition number of the coefficient $A$ is $\kappa=\frac{1-\cos(N\theta)}{1-\cos\theta}$ where $\theta=\pi\frac{1}{N+1}$. We have

$$\kappa=\frac{1-\cos(N\theta)}{1-\cos\theta}=\frac{1+\cos(\theta)}{1-\cos\theta}\approx\frac{1+1+\frac{1}{2}\theta^2}{1-1+\frac{1}{2}\theta^2}=\mathcal{O}(N^2)=\mathcal{O}(n^{\frac{2}{d}}),$$

where $d$ is the dimension. Therefore, steepest descent has complexity $\mathcal{O}(n^2)$ and CG has complexity $\mathcal{O}(n^{\frac{3}{2}})$ for 2D problems, and steepest descent has complexity $\mathcal{O}(n^{\frac{5}{3}})$ and CG has complexity $\mathcal{O}(n^{\frac{4}{3}})$ for 3D problems. Recall that the direct inversion method via eigenvalue decomposition in Chapter 2 has the same complexity $\mathcal{O}(n^{\frac{d+1}{d}})$, but it is only available for the discrete Laplacian of second order finite difference on a rectangular domain. The CG method can be applied to the discrete Laplacian of any domain, e.g., the L shape in Problem B.2.

In contrast, the multigrid method has the optimal complexity $\mathcal{O}(n)$ in any dimension, but we do not consider more complicated methods like the multigrid method in this book.

**Remark B.6.** *A complexity like $\mathcal{O}(n)$ means $\mathcal{O}(n)=Cn$ with a hidden constant $C$ independent of $n$. Due to this hidden constant, which might be huge, for solving a problem with fixed $n$, a method with optimal complexity such as the multigrid method may not necessarily be faster than CG method or the direct method in Chapter 2. The optimal complexity is advantageous for large enough $n$. For example, for inverting $K\otimes I+I\otimes K$ on a grid like $200\times 200$, very likely the method in Chapter 2 is faster in CPU time (depending on your code, which software, and your hardware) than the multigrid method. For inverting $K\otimes I\otimes I+I\otimes K\otimes I+I\otimes I\otimes K$ on a $1000^3$ grid in MATLAB on one modern GPU card (Nvidia A100 80G, see [Liu et al. (2024b); Hao* et al. *(2025)]), the method in Section 2.9 costs less than one second, which is quite difficult to beat.*

**Problem B.3.** *Verify the condition number for the discrete Laplacian matrices: $K$ in 1D, $K\otimes I+I\otimes K$ in 2D, and $K\otimes I\otimes I+I\otimes K\otimes I+I\otimes I\otimes K$ in 3D, where $I$ and $K$ are matrices of size $N\times N$. See Chapter 2 for eigenvalues of $K$.*

## B.4 The preconditioned conjugate gradient method

For solving $Ax=b$ we can consider solving instead an equivalent system

$$PAP^Ty=Pb$$

where $P$ is a *preconditioner* matrix. Recall that the performance of CG method solving $Ax = b$ depends on the condition number $\kappa(A)$. If we can find a nonsingular matrix $P$ such that $\kappa(PAP^T) \ll \kappa(A)$, then we can more efficiently solve the system $PAP^Ty = Pb$ with CG then find $x$ by $x = P^\top y$.

For instance, if $A = LL^\top$ is the Cholesky factorization ($L$ is lower triangular), then $P = L^{-1}$ is the most ideal preconditioner, because $PAP^T = I$ thus $\kappa(P^1AP^T) = 1$. The full Cholesky factorization costs $\mathcal{O}(n^3)$ which is as expensive as solving $Ax = b$ by Gaussian elimination. On the other hand, we can use a cheaper approximate Cholesky factorization to construct a preconditioner, i.e., if $A \approx \tilde{L}\tilde{L}^\top$ ($\tilde{L}$ is lower triangular) then use $P = \tilde{L}^{-1}$. For a sparse coefficient matrix $A = LL^\top$, the matrix $L$ may have nonzero fill-in for the zero entries of $A$. The incomplete Cholesky factorization will have zero fill-in, returning an approximation $A \approx \tilde{L}\tilde{L}^\top$.

By applying the conjugate gradient method to $PAP^Ty = Pb$, we get the following algorithm:

$$\begin{aligned}
\bar{p}_0 &= \bar{r}_0 = Pb - PAP^\top y_0,\\
\bar{\alpha}_k &= \frac{\|\bar{r}_k\|^2}{\bar{p}_k^\top PAP^\top \bar{p}_k},\\
y_{k+1} &= y_k + \bar{\alpha}_k \bar{p}_k,\\
\bar{r}_{k+1} &= \bar{r}_k - \bar{\alpha}_k PAP^\top \bar{p}_k,\\
\bar{\beta}_k &= \frac{\|\bar{r}_{k+1}\|^2}{\|\bar{r}_k\|^2}\\
\bar{p}_{k+1} &= \bar{r}_{k+1} + \bar{\beta}_k \bar{p}_k.
\end{aligned}$$

With the change of variables $x = P^\top y$ and $p = P^\top \bar{p}$, multiplying certain rows by either $P$ or $P^\top$, we get

$$\begin{aligned}
p_0 &= P^\top \bar{r}_0 = P^\top Pb - P^\top PAx_0,\\
\bar{\alpha}_k &= \frac{\|\bar{r}_k\|^2}{p_k^\top Ap_k},\\
x_{k+1} &= x_k + P^\top \bar{\alpha}_k P^{-T} p_k = x_k + \bar{\alpha}_k p_k,\\
\bar{r}_{k+1} &= \bar{r}_k - \bar{\alpha}_k PAp_k = \bar{r}_k - P\bar{\alpha}_k Ap_k,\\
\bar{\beta}_k &= \frac{\|\bar{r}_{k+1}\|^2}{\|\bar{r}_k\|^2},\\
p_{k+1} &= P^\top \bar{r}_{k+1} + P^\top \bar{\beta}_k P^{-T} p_k = P^\top \bar{r}_{k+1} + \bar{\beta}_k p_k.
\end{aligned}$$

Now introduce new variables $r_0 = b - Ax_0$ and $r = P^{-1}\bar{r}$, then we get

$$\begin{aligned}
r_0 &= b - Ax_0, \quad p_0 = P^\top P r_0 \\
\bar{\alpha}_k &= \frac{r_k^\top P^\top P r_k}{p_k^\top A p_k}, \\
x_{k+1} &= x_k + \bar{\alpha}_k p_k, \\
r_{k+1} &= r_k - \bar{\alpha}_k A p_k, \\
\bar{\beta}_k &= \frac{r_{k+1}^\top P^\top P r_{k+1}}{r_k^\top P^\top P r_k}, \\
p_{k+1} &= P^\top P r_{k+1} + \bar{\beta}_k p_k,
\end{aligned}$$

Finally, let $M = P^{-1}[P^{-1}]^\top$ and $z = M^{-1}r = P^\top P r$. Set $\bar{\alpha} = \alpha$ and $\bar{\beta} = \beta$, then we get the following implementation:

$$r_0 = b - Ax_0, \quad z_0 = M^{-1}r_0, \quad p_0 = z_0 \tag{B.10a}$$

$$\alpha_k = \frac{r_k^\top z_k}{p_k^\top A p_k}, \tag{B.10b}$$

$$x_{k+1} = x_k + \alpha_k p_k, \tag{B.10c}$$

$$r_{k+1} = r_k - \alpha_k A p_k, \tag{B.10d}$$

$$z_{k+1} = M^{-1} r_{k+1} \tag{B.10e}$$

$$\beta_k = \frac{z_{k+1}^\top r_{k+1}}{z_k r_k}, \tag{B.10f}$$

$$p_{k+1} = z_{k+1} + \beta_k p_k. \tag{B.10g}$$

In practice, we can simply use any matrix $M^{-1} \approx A^{-1}$ in the implementation (B.10) without deriving the matrix $P$. But here both $M$ and $A$ must be symmetric positive definite. Otherwise the iteration above may not converge. Also, $M$ must be a matrix that can be efficiently inverted ($Mz = r$ must be efficiently solved) for the algorithm to make any sense.

**Example B.1.** *Consider solving a 1D variable coefficient problem as discussed in Section 2.11:*

$$-(a(x)u'(x))' = f(x), \quad x \in [0, 1],$$

*with homogeneous Dirichlet boundary conditions. A conservative discretization must be used:*

$$-\frac{1}{\Delta x^2}[-a_{j-\frac{1}{2}}u_{j-1} + (a_{j-\frac{1}{2}} + a_{j+\frac{1}{2}})u_j - a_{j+\frac{1}{2}}u_{j+1}] = f_j,$$

*where* $a_{j-\frac{1}{2}} = a(x_j - \frac{1}{2}\Delta x)$. *The matrix vector form of this scheme is* $Au = f$ *where* $A$ *is a real symmetric tridiagonal matrix:*

$$A = -\frac{1}{\Delta x^2}\begin{pmatrix} a_{\frac{1}{2}} + a_{\frac{3}{2}} & -a_{\frac{3}{2}} & & \\ -a_{\frac{3}{2}} & a_{\frac{3}{2}} + a_{\frac{5}{2}} & -a_{\frac{5}{2}} & \\ & \ddots & \ddots & \ddots \end{pmatrix}.$$

*Notice that* $A$ *reduces to tridiagonal* $K$ *matrix if* $a(x) \equiv 1$. *Thus we can use the discrete Laplacian as the preconditioner* $M$ *in Preconditioned CG* (B.10) *even for solving a 2D variable coefficient problem* $-\nabla \cdot (a(x, y)\nabla u) = f$, *since* $K2D$ *matrix can be efficiently inverted by the eigenvector method as we have discussed in Chapter 2.*

**Problem B.4.** *For the same Problem B.2, also implement CG and PCG methods. For the PCG method, use CG to solve* $PAP^T y = Pb$, *and use the incomplete Cholesky factorization as the preconditioner, i.e., if* $A \approx \tilde{L}\tilde{L}^\top$ *then set* $P = \tilde{L}^{-1}$. *Use the function ichol in MATLAB for the incomplete Cholesky factorization. You can check how good the approximation* $\tilde{L}\tilde{L}^\top$ *is and the sparsity pattern of* $\tilde{L}$ *as follows:*

```
Lt=ichol(A);
spy(Lt);
norm(A-Lt*Lt','fro')/norm(A)
```

*Store cost function value* $f(x_k)$ *where* $f(x) = \frac{1}{2}x^\top Ax - x^\top b$ *for each iteration. Stop the iteration if the relative residue is smaller than some tolerance parameter, e.g.,* $10^{-10}$ *or* $10^{-15}$.

# Appendix C

# Splitting Methods in Convex Minimization

The content in this appendix chapter serves as a quick reference to the splitting methods mentioned in this book. Interested readers may also read [Ryu and Yin (2022)] for more on related concepts and methods.

## C.1 Indicator function and closed convex proper functions

We first explain why the indicator function of a set $\Omega \subset \mathbb{R}^n$ should be defined as

$$\iota_\Omega(\mathbf{x}) = \begin{cases} 0, & \mathbf{x} \in \Omega, \\ +\infty, & \mathbf{x} \notin \Omega. \end{cases} \tag{C.1}$$

If not using $+\infty$, consider a function

$$I_\Omega(\mathbf{x}) = \begin{cases} 0, & \mathbf{x} \in \Omega, \\ M, & \mathbf{x} \notin \Omega, \end{cases}$$

for a very large number $M$. Then it may seem that $I_\Omega(\mathbf{x})$ can serve the same purpose numerically as (C.1). But a well defined function on the whole space $\mathbb{R}^n$ must be a continuous function if it is convex [Rockafellar (1970), Corollary 10.1.1], thus the function $I_\Omega(\mathbf{x})$ cannot be convex. To this end, one must consider a closed convex proper function, which will be defined as follows.

The indicator function (C.1) is an *extended* function

$$f : \mathbb{R}^n \longrightarrow \mathbb{R} \cup \{\pm\infty\}.$$

The *epigraph* of an extended function $f$ is defined as

$$\text{epi} f = \{(\mathbf{x}, a) \in \mathbb{R}^n \times \mathbb{R} : f(\mathbf{x}) \leq a\}.$$

An extended function is called *closed* if its epigraph is a closed set in $\mathbb{R}^{n+1}$. The *domain* of an extended function is denoted as

$$\text{dom} f = \{\mathbf{x} \in \mathbb{R}^n : f(\mathbf{x}) \in \mathbb{R}\}.$$

The indicator function (C.1) is a closed extended function if and only if $\Omega$ is a closed set, see [Beck (2017)].

A *convex* extended function is defined as an extended function satisfying:

$$f(\lambda\mathbf{x} + (1-\lambda)\mathbf{y}) \leq \lambda f(\mathbf{x}) + (1-\lambda)f(\mathbf{y}), \quad \forall \mathbf{x}, \mathbf{y} \in \text{dom} f, \ \forall \lambda \in (0,1).$$

An extended function is called *proper* if it never maps to negative infinity. For instance, the indicator function (C.1) is a proper function. Thus if $\Omega$ is a closed convex set, then the indicator function (C.1) is a closed convex proper function, to which many results about convex functions defined on $\mathbb{R}^n$ can be extended.

A closed extended function can be proven to be lower semi-continuous [Beck (2017), Theorem 2.6]. Thus the indicator function (C.1) for a closed convex set $\Omega$ is also a lower semi-continuous convex proper function.

## C.2 Proximal operator and Douglas-Rachford splitting

Consider a general composite minimization

$$\min_{\mathbf{x}} \ f(\mathbf{x}) + g(\mathbf{x}), \tag{C.2}$$

where the functions $f(\mathbf{x})$ and $g(\mathbf{x})$ are closed convex proper functions. For a closed convex proper function $f(\mathbf{x})$, its subdifferential $\partial f(\mathbf{x}_0)$ at a point $\mathbf{x}_0$ is defined as a set of slopes of subtangent lines:

$$\partial f(\mathbf{x}_0) = \{\mathbf{v} \in \mathbb{R}^n : f(\mathbf{x}) \geq f(\mathbf{x}_0) + \langle \mathbf{v}, \mathbf{x} - \mathbf{x}_0 \rangle, \quad \forall \mathbf{x} \in \text{dom}(f)\}.$$

An element in the subdifferential set $\partial f$ is called a subgradient. If $f(\mathbf{x})$ is differentiable at $\mathbf{x}_0$, then the subgradients coincide with the gradient, i.e., $\partial f(\mathbf{x}_0) = \{\nabla f(\mathbf{x}_0)\}$. For example, for $f(x) = |x|$, the subdifferential set is

$$\partial f(x) = \begin{cases} \{1\}, & x > 0 \\ \{-1\}, & x < 0 \\ [-1,1], & x = 0 \end{cases}.$$

Let $\mathbb{I}$ be the identity operator. The proximal operator with a given parameter $\gamma > 0$ for the function $f$ is defined as

$$\text{Prox}_f^\gamma(\mathbf{x}) := (\mathbb{I} + \gamma\partial f)^{-1}(\mathbf{x}) = \text{argmin}_{\mathbf{z}} f(\mathbf{z}) + \frac{1}{2\gamma}\|\mathbf{z} - \mathbf{x}\|_2^2, \quad \gamma > 0. \tag{C.3}$$

Since $f(\mathbf{z}) + \frac{1}{2\gamma}\|\mathbf{z}-\mathbf{x}\|_2^2$ is a closed strongly convex proper function, it has a unique minimizer [Beck (2017), Theorem 5.25], thus $\mathrm{Prox}_f^\gamma(\mathbf{x})$ is well defined.

For the indicator function $\iota_\Omega$ of a closed convex set, its proximal operator is the Euclidean projection onto the set $\Omega$.

Assume the proximal operators have explicit formulae or can be efficiently approximated. If $g(x)$ is differentiable but $f(x)$ is not differentiable, then the simplest method for (C.2) is the subgradient method:

$$\mathbf{x}_{k+1} = \mathbf{x}_k - \gamma_k(\nabla g(\mathbf{x}_k) + \mathbf{v}_k), \quad \mathbf{v}_k \in \partial f(\mathbf{x}_k),$$

where $\gamma_k$ is a step size and any subgradient $\mathbf{v}_k$ can be chosen. However, the subgradient method converges very slowly and a method using the proximal operator $\mathrm{Prox}_f^\gamma(\mathbf{x})$ is usually faster. The simplest splitting method to use the proximal operator is the forward-backward splitting:

$$\mathbf{x}_{k+1} = (I+\gamma\partial f)^{-1}(I-\gamma\nabla g)(\mathbf{x}_k) = \mathrm{Prox}_f^\gamma(\mathbf{x}_k - \gamma\nabla g(\mathbf{x}_k)). \tag{C.4}$$

The splitting method (C.4) is also referred to as the *proximal gradient method*, or *projected gradient method* when $\mathrm{Prox}_f^\gamma$ is a projection operator, e.g., when $f$ is an indicator function.

Next, we describe the splitting methods with two proximal operators. Define reflection operators as

$$\mathrm{R}_f^\gamma = 2\,\mathrm{Prox}_f^\gamma - \mathbb{I},$$

then the general version of Douglas-Rachford splitting (DRS) [Lions and Mercier (1979)] can be given as

$$\begin{cases} \mathbf{y}_{k+1} &= [(1-\lambda)\mathbb{I} + \lambda\frac{\mathbb{I}+\mathrm{R}_f^\eta\,\mathrm{R}_g^\eta}{2}](\mathbf{y}_k), \quad \lambda \in (0,2) \\ &= \mathbf{y}_k - \lambda\mathbf{x}_k + \lambda\,\mathrm{Prox}_f^\eta(2\mathbf{x}_k - \mathbf{y}_k) \\ \mathbf{x}_k &= \mathrm{Prox}_g^\eta(\mathbf{y}_k) \end{cases} \tag{C.5}$$

which converges for any step size $\eta > 0$ and any relaxation parameter $\lambda \in (0,2)$.

If $\lambda = 2$, (C.5) reduces to

$$\text{Peaceman-Rachford}: \begin{cases} \mathbf{y}_{k+1} &= \mathrm{R}_f^\eta\,\mathrm{R}_g^\eta(\mathbf{y}_k) \\ \mathbf{x}_k &= \mathrm{Prox}_g^\eta(\mathbf{y}_k) \end{cases},$$

which converges only if either $f$ or $g$ is strongly convex. In this book, we only use the case $\lambda = 1$:

$$\text{Douglas-Rachford}: \begin{cases} \mathbf{y}_{k+1} &= \frac{\mathbb{I}+\mathrm{R}_f^\eta\,\mathrm{R}_g^\eta}{2}(\mathbf{y}_k) = \mathbf{y}_k - \mathbf{x}_k + \mathrm{Prox}_f^\eta(2\mathbf{x}_k - \mathbf{y}_k) \\ \mathbf{x}_k &= \mathrm{Prox}_g^\eta(\mathbf{y}_k) \end{cases}. \tag{C.6}$$

These splittings above converge for convex problems with any step size $\eta > 0$: $\mathbf{x}_k$ will converge to the minimizer $\mathbf{x}_*$ and $\mathbf{y}_k$ is only an auxiliary variable converging to a fixed point $\mathbf{y}_*$ to the iteration operator. On the other hand, the error $\|\mathbf{y}_k - \mathbf{y}_*\|$ decreases monotonically w.r.t. $k$ but $\|\mathbf{x}_k - \mathbf{x}_*\|$ usually does not decrease monotonically.

Its extension to three functions

$$\min_{\mathbf{x}} f(\mathbf{x}) + g(\mathbf{x}) + h(\mathbf{x}),$$

is called the Davis-Yin splitting [Davis and Yin (2017)] given as

$$\begin{cases} \mathbf{x}_{k+\frac{1}{2}} &= \mathrm{Prox}_g^\eta(\mathbf{z}_k) \\ \mathbf{x}_{k+1} &= \mathrm{Prox}_f^\eta(2\mathbf{x}_{k+\frac{1}{2}} - \mathbf{z}_k - \eta\nabla h(\mathbf{x}_{k+\frac{1}{2}})) \\ \mathbf{z}_{k+1} &= \mathbf{z}_k + \mathbf{x}_{k+1} - \mathbf{x}_{k+\frac{1}{2}} \end{cases},$$

which can be proven convergent for convex functions $f, g, h$ with $\nabla h$ being Lipschitz continuous with Lipschitz constant $L$, and $\eta < \frac{2}{L}$.

When $h(\mathbf{x}) \equiv 0$, Davis-Yin splitting reduces to Douglas-Rachford splitting. When $g(\mathbf{x}) \equiv 0$, Davis-Yin splitting reduces to the forward-backward splitting. Thus Davis-Yin splitting is also called Forward-Douglas-Rachford splitting [Raguet (2019)].

## C.3 Convex conjugate and Fenchel duality

The *convex conjugate* is also called Legendre Transform, Fenchel Transform and Fenchel dual. For an extended function $f : \mathbb{R}^n \longrightarrow [-\infty, +\infty]$, its convex conjugate is defined as

$$f^*(\mathbf{x}) = \max_{\mathbf{y}\in\mathbb{R}^n} \langle \mathbf{x}, \mathbf{y}\rangle - f(\mathbf{y}).$$

We only list some useful facts. For further details on duality, we refer the interested reader to the classical reference [Rockafellar (1970)].

**Theorem C.8.** *For any proper function $f : \mathbb{R} \longrightarrow (-\infty, +\infty]$, $f^*(\mathbf{x})$ is a closed convex function on its domain* $\mathrm{dom}(f^*) = \{\mathbf{x} : f^*(\mathbf{x}) < +\infty\}$ *even if $f(\mathbf{x})$ is not convex.*

**Theorem C.9.** *For any convex proper function $f : \mathbb{R} \longrightarrow (-\infty, +\infty]$, $f^*(\mathbf{x})$ is a closed convex proper function.*

**Theorem C.10.** *The convex conjugate satisfies:*

- *For any $f : \mathbb{R}^n \longrightarrow (-\infty, +\infty]$, $f(\mathbf{x}) + f^*(\mathbf{y}) \geq \langle \mathbf{x}, \mathbf{y}\rangle, \quad \forall \mathbf{x}, \mathbf{y}$.*

- *For any* $f : \mathbb{R}^n \longrightarrow [-\infty, +\infty]$, $f(\mathbf{x}) \geq f^{**}(\mathbf{x}), \quad \forall \mathbf{x}$.
- *For any closed convex proper function* $f$, $f(\mathbf{x}) = f^{**}(\mathbf{x}), \quad \forall \mathbf{x}$.

**Example C.2.** *Below are two quick examples:*

$$f(x) = e^x \Rightarrow f^*(x) = \begin{cases} x \log x - x, & x > 0 \\ 0, & x = 0 \\ +\infty, & x < 0 \end{cases}.$$

$$f(x) = ax + b \Rightarrow f^*(x) = \begin{cases} -b, & x = a \\ +\infty, & x \neq a \end{cases}.$$

**Example C.3.** *If* $f(\mathbf{x}) = \|\mathbf{x}\|$ *for some norm, then the convex conjugate is the indicator function of the dual norm unit ball*

$$f^*(\mathbf{x}) = \begin{cases} 0, & \|\mathbf{x}\|_* \leq 1 \\ +\infty, & \|\mathbf{x}\|_* > 1 \end{cases}.$$

(1) *For the vector 2-norm* $\|\mathbf{x}\|_2$, $\|\mathbf{x}\|_* = \|\mathbf{x}\|_2$.
(2) *For the vector 1-norm* $\|\mathbf{x}\|_1$, $\|\mathbf{x}\|_* = \|\mathbf{x}\|_\infty = \max_i |x_i|$.
(3) *For the vector* $\infty$*-norm* $\|\mathbf{x}\|_\infty$, $\|\mathbf{x}\|_* = \|\mathbf{x}\|_1 = \sum_i |x_i|$.
(4) *For the vector p-norm* $\|\mathbf{x}\|_p$, $\|\mathbf{x}\|_* = \|\mathbf{x}\|_q$, *where*

$$\|\mathbf{x}\|_p = \left[\sum_i |x_i|^p\right]^{\frac{1}{p}}, \quad \frac{1}{p} + \frac{1}{q} = 1.$$

**Theorem C.11** (Moreau-Decomposition). *For any closed convex proper function* $f : \mathbb{R} \longrightarrow (-\infty, +\infty]$, *for any* $\mathbf{x}$ *and* $\eta > 0$,

$$\mathrm{Prox}_f^\eta(\mathbf{x}) + \eta\, \mathrm{Prox}_{f^*}^{\frac{1}{\eta}}(\frac{\mathbf{x}}{\eta}) = \mathbf{x}.$$

Thus by Moreau-Decomposition, we have an expression for $\mathrm{Prox}_{f^*}$ whenever we have it for $\mathrm{Prox}_f$.

**Theorem C.12** (Fenchel Duality Theorem). *Let* $f$ *and* $g$ *be two closed convex proper functions. Assume the intersection of relative interior of* $\mathrm{dom}(f)$ *and relative interior of* $\mathrm{dom}(g)$ *is not empty, then*

$$\min_{\mathbf{x}}[f(\mathbf{x}) + g(\mathbf{x})] = -\min_{\mathbf{y}}[f^*(\mathbf{y}) + g^*(-\mathbf{y})]$$

**Theorem C.13.** *For a closed convex proper function* $f$, *the following are equivalent:*

*(1)* $f(\mathbf{x}) + f^*(\mathbf{y}) = \langle \mathbf{x}, \mathbf{y} \rangle$.
*(2)* $\mathbf{y} \in \partial f(\mathbf{x})$.
*(3)* $\mathbf{x} \in \partial f^*(\mathbf{y})$.

Therefore, for two closed convex proper functions, we have the following primal dual relation:

$$\mathbf{y}^* = \text{argmin}_{\mathbf{y}} f^*(\mathbf{y}) - \langle \mathbf{x}^*, \mathbf{y} \rangle \Leftrightarrow \mathbf{x}^* \in \partial f^*(\mathbf{y}^*)$$

$$\mathbf{x}^* = \text{argmin}_{\mathbf{x}} \langle \mathbf{x}, \mathbf{y}^* \rangle + g(\mathbf{x}) \Leftrightarrow -\mathbf{y}^* \in \partial g(\mathbf{x}^*).$$

## C.4 Douglas-Rachford splitting, ADMM and their equivalence

**How many different Douglas-Rachford splittings?** Now consider solving

$$\min_{\mathbf{x}} [f(\mathbf{x}) + g(\mathbf{x})] = -\min_{\mathbf{y}} [f^*(\mathbf{y}) + g^*(-\mathbf{y})],$$

for which there are at least four choices to apply the Douglas-Rachford splitting fix point iteration $\mathbf{y}_{k+1} = T(\mathbf{y}_k)$:

(1) $T = \frac{1}{2}[\mathbb{I} + \mathrm{R}_{f(\mathbf{x})} \mathrm{R}_{g(\mathbf{x})}]$.
(2) $T = \frac{1}{2}[\mathbb{I} + \mathrm{R}_{g(\mathbf{x})} \mathrm{R}_{f(\mathbf{x})}]$.
(3) $T = \frac{1}{2}[\mathbb{I} + \mathrm{R}_{f^*(\mathbf{y})} \mathrm{R}_{g^*(-\mathbf{y})}]$.
(4) $T = \frac{1}{2}[\mathbb{I} + \mathrm{R}_{g^*(-\mathbf{y})} \mathrm{R}_{f^*(\mathbf{y})}]$.

For the first two choices, for the Peaceman-Rachford splitting, it can be easily proven that they are the same if using a special initial guess:

**Theorem C.14.** *The sequence produced by*

$$\begin{cases} \mathbf{z}_{k+1} & = \mathrm{R}_g^\eta \mathrm{R}_f^\eta(\mathbf{z}_k) \\ \mathbf{x}_k & = \mathrm{R}_f^\eta(\mathbf{z}_k) \end{cases}, \quad \mathbf{z}_0 = \mathrm{R}_g^\eta(\mathbf{y}_0),$$

*is the same as the sequence produced by*

$$\begin{cases} \mathbf{y}_{k+1} & = \mathrm{R}_f^\eta \mathrm{R}_g^\eta(\mathbf{y}_k) \\ \mathbf{x}_k & = \mathrm{R}_g^\eta(\mathbf{y}_k) \end{cases}, \quad \forall \mathbf{y}_0.$$

**Remark C.7.** *For general Douglas-Rachford splitting, though the same result cannot be shown, in practice the difference in numerical performance between two different versions caused by switching $f$ and $g$ is marginal and minimal.*

Now the question is, does it make a difference if using Douglas-Rachford splitting on the Fenchel's dual problem? It turns out that there is still no difference.

For solving $\min_{\mathbf{x}}[F(\mathbf{x})+G(\mathbf{x})]$, with step size $\eta > 0$, the general version of Douglas-Rachford splitting can be written as

$$\mathbf{v}_{k+1} = [(1-\lambda)\mathbb{I} + \lambda\frac{\mathbb{I} + \mathrm{R}_F^\eta \mathrm{R}_G^\eta}{2}](\mathbf{v}_k), \quad \lambda \in (0,2).$$

For the primal problem $\min_{\mathbf{x}}[f(\mathbf{x}) + g(\mathbf{x})]$, we take $G(\mathbf{x}) = f(\mathbf{x})$ and $F(\mathbf{x}) = g(\mathbf{x})$, then

$$\text{DR on (P)}: \begin{cases} \mathbf{v}_{k+1} & = [(1-\lambda)\mathbb{I} + \lambda\frac{\mathbb{I}+\mathrm{R}_g^\eta \mathrm{R}_f^\eta}{2}](\mathbf{v}_k), \quad \lambda \in (0,2) \\ & = \mathbf{v}_k - \lambda\mathbf{x}_k + \lambda\,\mathrm{Prox}_f^\eta(2\mathbf{x}_k - \mathbf{v}_k) \\ \mathbf{x}_k & = \mathrm{Prox}_f^\eta(\mathbf{v}_k) \end{cases}. \quad \text{(C.7)}$$

For the dual problem $\min_{\mathbf{y}}[f^*(\mathbf{y}) + g^*(-\mathbf{y})]$, we take $F(\mathbf{y}) = g^*(-\mathbf{y})$ and $G(\mathbf{y}) = f^*(\mathbf{y})$, then

$$\mathrm{Prox}_F^\eta(\mathbf{u}) = -\mathrm{Prox}_{g^*}^\eta(-\mathbf{u}).$$

Using step size $\tau > 0$, we have

$$\text{DR on (D)}: \begin{cases} \mathbf{u}_{k+1} & = [(1-\lambda)\mathbb{I} + \lambda\frac{\mathbb{I}+\mathrm{R}_F^\tau \mathrm{R}_G^\tau}{2}](\mathbf{u}_k), \quad \lambda \in (0,2) \\ & = \mathbf{u}_k - \lambda\mathbf{y}_k - \lambda\,\mathrm{Prox}_{g^*}^\tau(-2\mathbf{y}_k + \mathbf{u}_k) \\ \mathbf{y}_k & = \mathrm{Prox}_{f^*}^\tau(\mathbf{u}_k) \end{cases}. \quad \text{(C.8)}$$

**Theorem C.15.** *The general Douglas-Rachford splitting on the primal problem* (C.7) *is exactly the same as general Douglas-Rachford splitting on the dual problem* (C.8) *if* $\eta = \frac{1}{\tau}$. *In particular,* $\mathbf{x}_k \to \mathbf{x}_*, \mathbf{y}_k \to \mathbf{y}_*$ *and*

$$\mathbf{u}_k = \frac{\mathbf{v}_k}{\eta}, \quad \mathbf{y}_k = \frac{\mathbf{v}_k - \mathbf{x}_k}{\eta}.$$

**Problem C.5.** *Prove the two theorems in this section. If using* (C.8), *how to recover the physical variable* $\mathbf{x}$ *from its iterate* $\mathbf{u}_k$ *or* $\mathbf{y}_k$*?*

**Primal Dual Hybrid Gradient (PDHG) method.** For a given composite problem in Theorem C.12, we have the following three equivalent formulation:

$$\text{Primal Problem (P)}: \quad \min_{\mathbf{x}}[f(\mathbf{x}) + g(\mathbf{x})]$$

$$\text{Dual Problem (D)}: -\min_{\mathbf{y}}[f^*(\mathbf{y}) + g^*(-\mathbf{y})]$$

$$\text{Primal Dual (PD)}: \min_{\mathbf{x}}\max_{\mathbf{y}}[\langle \mathbf{x}, \mathbf{y}\rangle - f^*(\mathbf{y}) + g(\mathbf{x})]$$

$$\text{Primal Dual relation}: \mathbf{x}_* \in \partial f^*(\mathbf{y}_*), \quad \mathbf{y}_* \in -\partial g^*(\mathbf{x}_*).$$

In (PD), the cost function is

$$L(\mathbf{x}, \mathbf{y}) = \langle \mathbf{x}, \mathbf{y} \rangle - f^*(\mathbf{y}) + g(\mathbf{x}),$$

and for finding the saddle point $\min_{\mathbf{x}} \max_{\mathbf{y}} L(\mathbf{x}, \mathbf{y})$, a simple method is to use implicit gradient descent/ascent:

$$\frac{\mathbf{x}_{k+1} - \mathbf{x}_k}{\eta} = -\frac{\partial L(\mathbf{x}_{k+1}, \mathbf{y}_k)}{\partial \mathbf{x}} = -\mathbf{y}_k - \partial g(\mathbf{x}_{k+1})$$
$$\frac{\mathbf{y}_{k+1} - \mathbf{y}_k}{\eta} = \frac{\partial L(\mathbf{x}_{k+1}, \mathbf{y}_{k+1})}{\partial \mathbf{y}} = \mathbf{x}_{k+1} - \partial f^*(\mathbf{y}_{k+1})$$

which gives the Arrow-Hurwitz method (1958):

$$\text{Arrow-Hurwitz}: \quad \begin{cases} \mathbf{x}_{k+1} & = \text{Prox}_g^\eta[\mathbf{x}_k - \eta \mathbf{y}_k] \\ \mathbf{y}_{k+1} & = \text{Prox}_{f^*}^\tau[\mathbf{y}_k + \tau \mathbf{x}_{k+1}] \end{cases}, \quad \eta > 0, \tau > 0.$$

For the Arrow-Hurwitz method to converge, the step sizes must be small enough. A better method is the Primal Dual Hybrid Gradient (PDHG) method introduced around 2010:

$$\text{PDHG}: \quad \begin{cases} \mathbf{x}_{k+1} & = \text{Prox}_g^\eta[\mathbf{x}_k - \eta \mathbf{y}_k] \\ \mathbf{y}_{k+1} & = \text{Prox}_{f^*}^\tau[\mathbf{y}_k + \tau(2\mathbf{x}_{k+1} - \mathbf{x}_k)] \end{cases}, \quad \eta > 0, \tau > 0, \eta\tau \le 1.$$

**Theorem C.16.** *The PDHG method with $\tau = \frac{1}{\eta}$ is equivalent to the Douglas-Rachford splitting $\frac{\mathbb{I} + \mathrm{R}_f \mathrm{R}_g}{2}$. Thus the PDHG method with $\tau = \frac{1}{\eta}$ converges for any $\eta > 0$ if $f$ and $g$ are two closed convex proper functions satisfying assumptions in the Fenchel's duality Theorem.*

*Proof.* Define $\mathbf{v}_k = \mathbf{x}_k - \eta \mathbf{y}_k$, then the PDHG method above with $\tau = \frac{1}{\eta}$ becomes

$$\begin{cases} \mathbf{x}_{k+1} & = \text{Prox}_g^\eta[\mathbf{v}_k] \\ \mathbf{v}_{k+1} & = \mathbf{v}_k - \mathbf{x}_{k+1} + \text{Prox}_f^\eta[(2\mathbf{x}_{k+1} - \mathbf{v}_k)] \end{cases}, \quad \eta > 0.$$

□

**A simple version of ADMM.** The Alternating Direction Method of Multipliers (ADMM) [Glowinski and Marroco (1975); Gabay and Mercier (1976)] is a widely used popular method. We first consider its simplest version. For solving $\min_{\mathbf{x}} f(\mathbf{x}) + g(\mathbf{x})$, we rewrite it as

$$\min f(\mathbf{w}) + g(\mathbf{z}), \quad \mathbf{w} = \mathbf{z}.$$

For such a constrained minimization, the *Lagrangian* is defined as

$$L(\mathbf{w}, \mathbf{z}, \mathbf{y}) = f(\mathbf{w}) + g(\mathbf{z}) - \langle \mathbf{y}, \mathbf{w} - \mathbf{z} \rangle,$$

where $\mathbf{y}$ is the Lagrangian multiplier. For finding a saddle point to the Lagrangian, the Augmented Lagrangian is given as

$$\mathcal{L}(\mathbf{w}, \mathbf{z}, \mathbf{y}) = f(\mathbf{w}) + g(\mathbf{z}) - \langle \mathbf{y}, \mathbf{w} - \mathbf{z} \rangle + \frac{\tau}{2}\|\mathbf{w} - \mathbf{z}\|^2.$$

The ADMM method with step sizes $\tau > 0$ and $\sigma > 0$ is given as

$$\begin{aligned}
\mathbf{z}_{k+1} &= \operatorname{argmin}_{\mathbf{z}} \mathcal{L}(\mathbf{w}_k, \mathbf{z}, \mathbf{y}_k) \\
\mathbf{w}_{k+1} &= \operatorname{argmin}_{\mathbf{w}} \mathcal{L}(\mathbf{w}, \mathbf{z}_{k+1}, \mathbf{y}_k) \\
\mathbf{y}_{k+1} &= \mathbf{y}_k + \sigma \frac{\partial \mathcal{L}}{\partial y}(\mathbf{w}_{k+1}, \mathbf{z}_{k+1}, \mathbf{y}_k)
\end{aligned}$$

which is equivalent to

$$\text{(ADMM)}: \quad \begin{aligned}
\mathbf{z}_{k+1} &= \operatorname{argmin}_{\mathbf{z}} g(\mathbf{z}) - \langle \mathbf{y}_k, \mathbf{w}_k - \mathbf{z} \rangle + \frac{\tau}{2}\|\mathbf{w}_k - \mathbf{z}\|^2 \\
\mathbf{w}_{k+1} &= \operatorname{argmin}_{\mathbf{w}} f(\mathbf{w}) - \langle \mathbf{y}_k, \mathbf{w} - \mathbf{z}_{k+1} \rangle + \frac{\tau}{2}\|\mathbf{w} - \mathbf{z}_{k+1}\|^2 \\
\mathbf{y}_{k+1} &= \mathbf{y}_k - \sigma(\mathbf{w}_{k+1} - \mathbf{z}_{k+1})
\end{aligned}$$

**Theorem C.17.** *The ADMM method with $\sigma = \tau = \eta$ is equivalent to the Douglas-Rachford splitting $\frac{\mathbb{I}+\mathrm{R}_F^\eta \mathrm{R}_G^\eta}{2}$ on the dual problem with $F = g^*(-\mathbf{y})$ and $G = f^*(\mathbf{y})$. Thus the ADMM method converges for any two closed convex proper functions if using step size $\sigma = \tau > 0$.*

*Proof.* For the DR splitting $\mathbf{v}_{k+1} = \frac{\mathbb{I}+\mathrm{R}_F^\eta \mathrm{R}_G^\eta}{2}(\mathbf{v}_k)$, $\mathbf{y}_k = \mathrm{Prox}_G(\mathbf{v}_k)$, define $\frac{\mathbf{v}_{k+1}-\mathbf{y}_k}{\eta} = \mathbf{z}_{k+1}$ and $\frac{\mathbf{v}_k - \mathbf{y}_k}{\eta} = \mathbf{w}_k$, then it can be verified. □

**Problem C.6.** *Finish the proof above.*

**Problem C.7.** *Use the definition of the proximal operator to show that ADMM method with $\sigma = \tau$ is equivalent to*

$$\begin{cases}
\mathbf{z}_{k+1} & = \mathrm{Prox}_g^{\frac{1}{\tau}}[\mathbf{w}_k - \mathbf{y}_k/\tau] \\
\mathbf{w}_{k+1} & = \mathrm{Prox}_f^{\frac{1}{\tau}}[\mathbf{z}_{k+1} + \mathbf{y}_k/\tau] \\
\mathbf{y}_{k+1} & = \mathbf{y}_k - \tau(\mathbf{w}_{k+1} - \mathbf{z}_{k+1})
\end{cases} \tag{C.9}$$

**Problem C.8.** *Start with the general DR splitting on the dual problem $\mathbf{v}_{k+1} = [(1-\lambda)\mathbb{I} + \lambda \frac{\mathbb{I}+\mathrm{R}_F^\eta \mathrm{R}_G^\eta}{2}(\mathbf{v}_k)]$, $\mathbf{y}_k = \mathrm{Prox}_G(\mathbf{v}_k)$ to derive a general ADMM method with a relaxation parameter $\lambda \in (0, 2)$.*

**Equivalence of popular algorithms.** Now consider the discussed algorithms for solving

$$\text{Primal Problem (P)}: \quad \min_{\mathbf{x}}[f(\mathbf{x})+g(\mathbf{x})]$$
$$\text{Dual Problem (D)}: -\min_{\mathbf{y}}[f^*(\mathbf{y})+g^*(-\mathbf{y})]$$
$$\text{Primal Dual (PD)}: \min_{\mathbf{x}}\max_{\mathbf{y}}[\langle \mathbf{x},\mathbf{y}\rangle - f^*(\mathbf{y})+g(\mathbf{x})],$$

the following are exactly equivalent:

(1) PDHG on (PD) with step size choice $\tau=\frac{1}{\eta}$.
(2) Douglas-Rachford splitting with $\lambda=1$ on (P) with step size $\eta>0$.
(3) ADMM on (D) with step size $\tau=\sigma=\eta$.
(4) Douglas-Rachford splitting with $\lambda=1$ on (D) with step size $\frac{1}{\eta}>0$.
(5) ADMM on (P) with step size $\tau=\sigma=\frac{1}{\eta}$.

These algorithms are popular choices for many applications, and in practice their performance is often similar to one another, which is not a surprise once we know about the equivalence above. For making a choice of which algorithm to use, the equivalence above is also useful since they are the same algorithm if using proper parameters.

**ADMM and DRS on more general problems.**

The TV norm minimization in Section 4.2 can be denoted as $\min_{\mathbf{x}}[f(K\mathbf{x})+g(\mathbf{x})]$ and rewritten as

$$\min_{\mathbf{v},\mathbf{x}}[f(\mathbf{v})+g(\mathbf{x})], \quad \mathbf{v}-K\mathbf{x}=0.$$

It can be casted into a more general problem

$$\min_{\mathbf{x},\mathbf{y}} f(\mathbf{x})+g(\mathbf{y}), \quad A\mathbf{x}+B\mathbf{y}=C,$$

for which the Lagrangian is given as

$$L(\mathbf{x},\mathbf{y},\mathbf{z})=f(\mathbf{x})+g(\mathbf{y})+\langle \mathbf{z}, A\mathbf{x}+B\mathbf{y}-C\rangle.$$

The augmented Lagrangian with a parameter $\sigma>0$ is given as

$$\mathcal{L}_\sigma(\mathbf{x},\mathbf{y},\mathbf{z})=f(\mathbf{x})+g(\mathbf{y})+\langle \mathbf{z}, A\mathbf{x}+B\mathbf{y}-C\rangle+\frac{\sigma}{2}\|A\mathbf{x}+B\mathbf{y}-C\|^2.$$

The ADMM method with step sizes $\tau>0$ and $\sigma>0$ is given as

$$\mathbf{x}_{k+1}=\operatorname{argmin}_{\mathbf{x}}\mathcal{L}_\sigma(\mathbf{x},\mathbf{y}_k,\mathbf{z}_k)$$
$$\mathbf{y}_{k+1}=\operatorname{argmin}_{\mathbf{y}}\mathcal{L}_\sigma(\mathbf{x}_{k+1},\mathbf{y},\mathbf{z}_k)$$
$$\mathbf{z}_{k+1}=\mathbf{z}_k+\tau\frac{\partial\mathcal{L}_\sigma}{\partial z}(\mathbf{x}_{k+1},\mathbf{y}_{k+1},\mathbf{z}_k)$$

which is equivalent to

$$
\text{(ADMM)}: \quad
\begin{aligned}
\mathbf{x}_{k+1} &= \operatorname{argmin}_{\mathbf{x}} f(\mathbf{x}) + \langle \mathbf{z}_k, A\mathbf{x} + B\mathbf{y}_k - C\rangle \\
&\quad + \frac{\sigma}{2}\|A\mathbf{x} + B\mathbf{y}_k - C\|^2 \\
\mathbf{y}_{k+1} &= \operatorname{argmin}_{\mathbf{y}} g(\mathbf{y}) + \langle \mathbf{z}_k, A\mathbf{x}_{k+1} + B\mathbf{y} - C\rangle \\
&\quad + \frac{\sigma}{2}\|A\mathbf{x}_{k+1} + B\mathbf{y} - C\|^2 \\
\mathbf{z}_{k+1} &= \mathbf{z}_k + \tau(A\mathbf{x}_{k+1} + B\mathbf{y}_{k+1} - C).
\end{aligned}
$$

From the Lagrangian, we can derive the dual problem:

$$
\begin{aligned}
\text{(P)}: &\quad \min_{\mathbf{x},\mathbf{y}} f(\mathbf{x}) + g(\mathbf{y}), \quad A\mathbf{x} + B\mathbf{y} = C \\
\text{(D)}: &\quad -\min_{\mathbf{z}}[f^*(-A^\top \mathbf{z}) + g^*(-B^\top \mathbf{z}) + \langle \mathbf{z}, C\rangle],
\end{aligned}
$$

**Theorem C.18.** *Assume some technical conditions for matrices $A, B$ and convex functions $f, g$ so that (P) $\Leftrightarrow$ (D) and $F(\mathbf{z}) = f^*(-A^\top \mathbf{z})$ and $G(\mathbf{z}) = g^*(-B^\top \mathbf{z}) + \langle \mathbf{z}, C\rangle$ are well defined, then the ADMM method with $\sigma = \tau = \eta$ is equivalent to the Douglas-Rachford splitting $\frac{\mathbb{I}+\mathrm{R}_F^\eta \mathrm{R}_G^\eta}{2}$ on the dual problem. Thus the ADMM method converges for any two closed convex proper functions if using step size $\sigma = \tau > 0$.*

See also [Gil Torres *et al.* (2025)] and references therein for the exact equivalent relation.

# Bibliography


Absil, P.-A., Mahony, R., and Sepulchre, R. (2008). *Optimization Algorithms on Matrix Manifolds* (Princeton University Press).

Alliez, P., Meyer, M., and Desbrun., M. (2002). Interactive geometry remeshing, *Proceeding of ACM SIGGRAPH* .

Beck, A. (2017). *First-Order Methods in Optimization* (SIAM).

Belkin, M. and Niyogi, P. (2002). Laplacian eigenmaps for dimensionality reduction and data representation, *Neural Computation* **15**, pp. 1373–1396.

Belkin, M., Sun, J., and Wang, Y. (2009). Constructing Laplace operator from point clouds in $\mathbb{R}^d$, in *Proceedings of the Twentieth Annual ACM-SIAM Symposium on Discrete Algorithms* (Philadelphia, PA, USA), pp. 1031–1040.

Benamou, J.-D. and Brenier, Y. (2000). A computational fluid mechanics solution to the Monge-Kantorovich mass transfer problem, *Numerische Mathematik* **84**, 3, pp. 375–393.

Bérard, P., Besson, G., and Gallot, S. (1994). Embedding Riemannian manifolds by their heat kernel, *Geometric and Functional Analysis* **4**, 4, pp. 373–398.

Berger, B., Kleinberg, J., and Leighton, T. (1999). Reconstructing a three-dimensional model with arbitrary errors, *Journal of the ACM (JACM)* **46**, 2, pp. 212–235.

Biswas, P., Liang, T.-C., Toh, K.-C., Ye, Y., and Wang, T.-C. (2006). Semidefinite programming approaches for sensor network localization with noisy distance measurements, *IEEE transactions on automation science and engineering* **3**, 4, p. 360.

Bonito, A., Demlow, A., and Nochetto, R. H. (2020). Finite element methods for the Laplace–Beltrami operator, in *Handbook of Numerical Analysis*, Vol. 21 (Elsevier), pp. 1–103.

Bonneel, N., Rabin, J., Peyré, G., and Pfister, H. (2015). Sliced and radon Wasserstein barycenters of measures, *Journal of Mathematical Imaging and Vision* **51**, 1, pp. 22–45.

Botsch, M., Kobbelt, L., Pauly, M., Alliez, P., and Lévy, B. (2010). *Polygon mesh processing* (AK Peters/CRC Press).

Bronstein, A. M., Bronstein, M. M., Kimmel, R., Mahmoudi, M., and Sapiro, G. (2010). A Gromov-Hausdorff framework with diffusion geometry for topologically-robust non-rigid shape matching, *International Journal of Computer Vision* **89**, 2-3, pp. 266–286.

Bronstein, M. M. and Kokkinos, I. (2010). Scale-invariant heat kernel signatures for non-rigid shape recognition, *IEEE Conference on Computer Vision and Pattern Recognition (CVPR)*, pp. 1704–1711.

Brunton, S. L. and Kutz, J. N. (2019). *Data-Driven Science and Engineering: Machine Learning, Dynamical Systems, and Control* (Cambridge University Press).

Cancès, E., Chakir, R., and Maday, Y. (2010). Numerical analysis of nonlinear eigenvalue problems, *Journal of Scientific Computing* **45**, 1-3, pp. 90–117.

Candès, E. and Recht, B. (2008). Exact matrix completion via convex optimization. *Found. of Comput. Math.* **9**, pp. 717–772.

Chan, T. F., Esedoglu, S., and Nikolova, M. (2006). Algorithms for finding global minimizers of image segmentation and denoising models, *SIAM Journal on Applied Mathematics* **66**, pp. 1632–1648.

Chan, T. F. and Vese, L. (2001). Active contours without edges, *IEEE Transactions on Image Processing* **10**, pp. 266–277.

Chavel, I. (1984). *Eigenvalues in Riemannian geometry*, Vol. 115 (Academic press).

Chavel, I. (2006). *Riemannian Geometry: A Modern Introduction, Cambridge Studies in Advanced Mathematics*, Vol. 108, 2nd edn. (Cambridge University Press).

Chen, Z., Lu, J., Lu, Y., and Zhang, X. (2025). Fully discretized Sobolev gradient flow for the Gross-Pitaevskii eigenvalue problem, *Mathematics of Computation* **94**, 356, pp. 2723–2760.

Chung, F. R. K. (1997). *Spectral Graph Theory* (American Mathematical Society).

Ciarlet, P. G. (2002). *The Finite Element Method for Elliptic Problems* (SIAM).

Coifman, R. and Lafon, S. (2006). Diffusion maps, *Applied and Computational Harmonic Analysis* **21**, pp. 6–30.

Collatz, L. (1960). *The numerical treatment of differential equations* (Springer Berlin, Heidelberg).

Crane, K. (2018). Discrete differential geometry: An applied introduction, SIGGRAPH Course Notes.

Crippen, G. M. and Havel, T. F. (1988). *Distance geometry and molecular conformation*, Vol. 74 (Research Studies Press Taunton, UK).

Cross, L. J. and Zhang, X. (2024a). On the monotonicity of $Q^2$ spectral element method for Laplacian on quasi-uniform rectangular meshes, *Communications in Computational Physics* **35**, 1, pp. 160–180.

Cross, L. J. and Zhang, X. (2024b). On the monotonicity of $Q^3$ spectral element method for Laplacian, *Annals of Applied Mathematics* **40**, 2, pp. 161–190.

Davis, D. and Yin, W. (2017). A three-operator splitting scheme and its optimization applications, *Set-valued and variational analysis* **25**, pp. 829–858.

Desbrun, M., Meyer, M., Schröder, P., and Barr, A. H. (1999). Implicit fairing of irregular meshes using diffusion and curvature flow, *Proceedings of SIGGRAPH* , pp. 317–324.

Dijkstra, E. W. (1959). A note on two problems in connexion with graphs, *Numerische Mathematik* **1**, 1, pp. 269–271.

do Carmo, M. P. (1976). *Differential Geometry of Curves and Surfaces* (Prentice Hall).

Dziuk, G. and Elliott, C. M. (2013). Finite element methods for surface PDEs, *Acta Numerica* **22**, pp. 289–396.

Eck, M., DeRose, T., Duchamp, T., Hoppe, H., Lounsbery, M., and Stuetzle, W. (1995). Multiresolution analysis of arbitrary meshes, *Proceeding of ACM SIGGRAPH* .

Evans, L. C. (2012). *An introduction to stochastic differential equations*, Vol. 82 (American Mathematical Soc.).

Fang, X. and Toh, K.-C. (2013). Using a distributed SDP approach to solve simulated protein molecular conformation problems, in *Distance Geometry* (Springer), pp. 351–376.

Feng, K. (1965). Difference schemes based on variational principle, *Journal of Applied and Computational Mathematics* **2**, 4, pp. 238–262.

Ferreira, J. A. and Grigorieff, R. D. (1998). On the supraconvergence of elliptic finite difference schemes, *Applied Numerical Mathematics* **28**, 2, pp. 275–292.

Fox, L. (1947). Some improvements in the use of relaxation methods for the solution of ordinary and partial differential equations, *Proceedings of the Royal Society of London. Series A. Mathematical and Physical Sciences* **190**, 1020, pp. 31–59.

Gabay, D. and Mercier, B. (1976). A dual algorithm for the solution of nonlinear variational problems via finite element approximation, *Computers & Mathematics with Applications* **2**, 1, pp. 17–40.

Gil Torres, E., Jacobs, M., and Zhang, X. (2025). Asymptotic linear convergence of ADMM for isotropic TV norm compressed sensing, *arXiv preprint arXiv:2505.01240* .

Glowinski, R. and Marroco, A. (1975). Sur l'approximation, par éléments finis d'ordre un, et la résolution, par pénalisation-dualité d'une classe de problèmes de dirichlet non linéaires, *Revue Française d'Automatique, Informatique et Recherche Opérationnelle Série Rouge. Analyse Numérique* **9**, R2, pp. 41–76.

Goldstein, T. and Osher, S. (2009). The split Bregman method for L1-regularized problems, *SIAM Journal on Imaging Sciences* **2**, 2, pp. 323–343.

Gu, X., Wang, Y., Chan, T. F., Thompson, P., and Yau, S. T. (2004). Genus zero surface conformal mapping and its application to brain surface mapping, *IEEE Transaction on Medical Imaging* **23**, pp. 949–958.

Gu, X. and Yau., S. (2002). Computing conformal structures of surfaces, *Communications in Information and Systems* **2**, 2, pp. 121–146.

Gu, X. and Yau, S. T. (2003). Global conformal surface parameterization, *Symposium on Geometry Processing* , pp. 127–137.

Hao, W., Lee, S., and Zhang, X. (2025). An efficient quasi-Newton method with tensor product implementation for solving quasi-linear elliptic equations and systems, *Journal of Scientific Computing* **103**, 3, p. 89.

Henning, P. and Jarlebring, E. (2025). The Gross–Pitaevskii equation and eigenvector nonlinearities: numerical methods and algorithms, *SIAM Review* **67**, 2, pp. 256–317.

Hu, J. and Zhang, X. (2023). Positivity-preserving and energy-dissipative finite difference schemes for the Fokker-Planck and Keller-Segel equations, *IMA Journal of Numerical Analysis* **43**, 3, pp. 1450–1484.

Ji, X. and Zha, H. (2004). Sensor positioning in wireless ad-hoc sensor networks using multidimensional scaling, in *INFOCOM 2004. Twenty-third Annual Joint Conference of the IEEE Computer and Communications Societies*, Vol. 4 (IEEE), pp. 2652–2661.

Jost, J. (2008). *Riemannian geometry and geometric analysis* (Springer Science & Business Media).

Kanai, T., Suzuki, H., and Kimura., F. (1998). Three-dimensional geometric metamorphosis based on harmonic maps, *The Visual Computer* **14**, 4, pp. 166–176.

Kantorovich, L. and Krylov, V. (1960). Approximate methods of higher analysis, *Bull. Amer. Math. Soc* **66**, 3, pp. 146–147.

Kimmel, R. and Sethian, J. A. (1998). Computing geodesic paths on manifolds, *Proceedings of the National Academy of Sciences* **95**, 15, pp. 8431–8435.

Lai, R. and Chan, T. F. (2011). A framework for intrinsic image processing on surfaces, *Computer Vision and Image Understanding* **115**, 12, pp. 1647–1661.

Lai, R. and Li, J. (2018). Solving partial differential equations on manifold from incomplete inter-point distance, *SIAM Journal on Scientific Computing* **39**, 5, pp. 2231–2256.

Lai, R., Liang, J., and Zhao, H. (2013). A local mesh method for solving PDEs on point clouds, *Inverse Prob. and Imaging* **7**, 3, pp. 737–755.

Lai, R. and Lu, J. (2018). Point cloud discretization of Fokker–Planck operators for committor functions, *Multiscale Modeling & Simulation* **16**, 2, pp. 710 726.

Lai, R., Shi, Y., Scheibel, K., Fears, S., Woods, R., Toga, A. W., and Chan, T. F. (2010). Metric-induced optimal embedding for intrinsic 3D shape analysis, *Computer Vision and Pattern Recognition (CVPR)*, pp. 2871–2878.

Lai, R., Wen, Z., Yin, W., Gu, X., and Lui, L.M. (2014). Folding-free global conformal mapping for genus-0 surfaces by harmonic energy minimization. *Journal of Scientific Computing*, **58**, 3, pp. 705–725.

Lai, R. and Zhao, H. (2017). Multiscale nonrigid point cloud registration using robust sliced-Wasserstein distance via Laplace-Beltrami eigenmap, *SIAM Journal on Imaging Sciences* **10**, 2, pp. 449–483.

Lavenant, H., Claici, S., Chien, E., and Solomon, J. (2018). Dynamical optimal transport on discrete surfaces, *ACM Transactions on Graphics (TOG)* **37**, 6, pp. 1–16.

Lee, J. M. (2013). *Introduction to Smooth Manifolds*, 2nd edn. (Springer).

Lee, J. M. (2018). *Introduction to Riemannian Manifolds*, 2nd edn. (Springer).

Leung, N.-H. Z. and Toh, K.-C. (2009). An SDP-based divide-and-conquer algorithm for large-scale noisy anchor-free graph realization, *SIAM Journal on Scientific Computing* **31**, 6, pp. 4351–4372.

LeVeque, R. J. (2007). *Finite difference methods for ordinary and partial differential equations: steady-state and time-dependent problems* (SIAM).

Levina, E. and Bickel, P. J. (2004). Maximum likelihood estimation of intrinsic dimension, in *Advances in neural information processing systems*, pp. 777–784.

Levy, B. (2006). Laplace-Beltrami eigenfunctions: Towards an algorithm that understands geometry, in *IEEE International Conference on Shape Modeling and Applications, invited talk.*

Levy, B., Petitjean, S., Ray, N., and Maillot, J. (2002). Least squares conformal maps for automatic texture atlas generation, *Proceeding of ACM SIGGRAPH* .

Li, H., Appelo, D., and Zhang, X. (2022). Accuracy of Spectral Element Method for Wave, Parabolic, and Schrödinger Equations, *SIAM Journal on Numerical Analysis* **60**, 1, pp. 339–363.

Li, H., Xie, S., and Zhang, X. (2018). A high order accurate bound-preserving compact finite difference scheme for scalar convection diffusion equations, *SIAM Journal on Numerical Analysis* **56**, 6, pp. 3308–3345.

Li, H. and Zhang, X. (2020a). On the monotonicity and discrete maximum principle of the finite difference implementation of $C^0$-$Q^2$ finite element method, *Numerische Mathematik* **145**, 2, pp. 437–472.

Li, H. and Zhang, X. (2020b). Superconvergence of $C^0$-$Q^k$ finite element method for elliptic equations with approximated coefficients, *Journal of Scientific Computing* **82**, 1, p. 1.

Li, H. and Zhang, X. (2020c). Superconvergence of high order finite difference schemes based on variational formulation for elliptic equations, *Journal of Scientific Computing* **82**, 2, p. 36.

Li, H. and Zhang, X. (2024). A high order accurate bound-preserving compact finite difference scheme for two-dimensional incompressible flow, *Commun. Appl. Math. Comput.* **6**, 1, pp. 113–141, doi:10.1007/s42967-022-00227-9.

Li, H. and Zhang, X. (2025). A monotone $Q^1$ finite element method for anisotropic elliptic equations, *Beijing Journal of Pure and Applied Mathematics* **2**, 1, pp. 183–217.

Liang, J., Lai, R., Wong, T., and Zhao, H. (2012). Geometric understanding of point clouds using Laplace-Beltrami operator, *Computer Vision and Pattern Recognition (CVPR)* , pp. 214–221.

Liang, J. and Zhao, H. (2013). Solving partial differential equations on point clouds, *SIAM Journal on Scientific Computing* **35**, 3, pp. A1461–A1486.

Lieb, E. H., Seiringer, R., and Yngvason, J. (2001). Bosons in a trap: A rigorous derivation of the Gross-Pitaevskii energy functional, *The Stability of Matter: From Atoms to Stars: Selecta of Elliott H. Lieb* , pp. 685–697.

Lin, T. and Zha, H. (2008). Riemannian manifold learning, *Pattern Analysis and Machine Intelligence, IEEE Transactions on* **30**, 5, pp. 796–809.

Lions, P.-L. and Mercier, B. (1979). Splitting algorithms for the sum of two nonlinear operators, *SIAM Journal on Numerical Analysis* **16**, 6, pp. 964–979.

Little, A. V., Jung, Y.-M., and Maggioni, M. (2009a). Multiscale estimation of intrinsic dimensionality of data sets. in *AAAI fall symposium: manifold learning and its applications*, Vol. 9, p. 04.

Little, A. V., Lee, J., Jung, Y.-M., and Maggioni, M. (2009b). Estimation of intrinsic dimensionality of samples from noisy low-dimensional manifolds in high dimensions with multiscale SVD, in *Statistical Signal Processing, 2009. SSP'09. IEEE/SP 15th Workshop on* (IEEE), pp. 85–88.

Liu, C., Gao, Y., and Zhang, X. (2024a). Structure preserving schemes for Fokker-Planck equations of irreversible processes, *Journal of Scientific Computing* **98**, article number 4.

Liu, C. and Zhang, X. (2023). A positivity-preserving implicit-explicit scheme with high order polynomial basis for compressible Navier-Stokes equations, *Journal of Computational Physics* **493**, p. 112496.

Liu, X., Shen, J., and Zhang, X. (2024b). A simple GPU implementation of spectral-element methods for solving 3D Poisson type equations on rectangular domains and its applications, *Communications in Computational Physics* **36**, 5, p. 1157–1185.

Lui, L., Gu, X., Chan, T., and Yau, S.-T. (2008). Variational method on Riemann surfaces using conformal parameterization and its applications to image processing, *Methods and Applications of Analysis* **15(4)**, pp. 513–538.

Mémoli, F. and Sapiro, G. (2005). Distance functions and geodesics on submanifolds of $\mathbb{R}^d$ and point clouds, *SIAM Journal on Applied Mathematics* **65**, 4, pp. 1227–1260.

Meyer, M., Desbrun, M., Schröder, P., and Barr, A. H. (2003). Discrete differential-geometry operators for triangulated 2-manifolds, in H.-C. Hege and K. Polthier (eds.), *Visualization and Mathematics III* (Springer), pp. 35–57.

Millman, R. S. and Parker, G. D. (1977). *Elements of Differential Geometry* (Prentice Hall).

Milnor, J. W. (1963). *Morse theory*, 51 (Princeton University Press).

Niyogi, P., Smale, S., and Weinberger, S. (2008). Finding the homology of submanifolds with high confidence from random samples, *Discrete & Computational Geometry* **39**, 1-3, pp. 419–441.

Osher, S. and Fedkiw, R. (2003). *Level Set Methods and Dynamic Implicit Surfaces* (Springer).

Osher, S. and Sethian, J. A. (1988). Fronts propagating with curvature-dependent speed: Algorithms based on Hamilton–Jacobi formulations, *Journal of computational physics* **79**, 1, pp. 12–49.

Ovsjanikov, M., Ben-Chen, M., Solomon, J., Butscher, A., and Guibas, L. (2012). Functional maps: a flexible representation of maps between shapes, *ACM Transactions on Graphics (TOG)* **31**, 4, p. 30.

Peherstorfer, B., Gugercin, S., and Willcox, K. (2016). Data-driven operator inference for nonintrusive model reduction, *Computer Methods in Applied Mechanics and Engineering* **306**, pp. 196–215.

Pinkall, U. and Polthier, K. (1993). Computing discrete minimal surfaces and their conjugates, in *Experimental Mathematics*, pp. 15–36.

Plemmons, R. J. (1977). M-matrix characterizations. I—nonsingular M-matrices, *Linear Algebra and its applications* **18**, 2, pp. 175–188.

Qian, J., Zhang, Y.-T., and Zhao, H.-K. (2007). Fast sweeping methods for Eikonal equations on triangular meshes, *SIAM Journal on Numerical Analysis* **45**, 1, pp. 83–107.

Raguet, H. (2019). A note on the forward-Douglas–Rachford splitting for monotone inclusion and convex optimization, *Optimization Letters* **13**, 4, pp. 717–740.

Raviv, D., Bronstein, A. M., Bronstein, M. M., Kimmel, R., and Sochen, N. (2011). Affine-invariant diffusion geometry for the analysis of deformable 3D shapes, *CVPR* , pp. 2361–2367.

Recht, B., Fazel, M., and Parrilo, P. (2010). Guaranteed minimum-rank solutions of linear matrix equations via nuclear norm minimization, *SIAM Review* **52**, 3, pp. 471–501.

Reuter, M. (2010). Hierarchical shape segmentation and registration via topological features of Laplace-Beltrami eigenfunctions, *International Journal of Computer Vision* **89**, pp. 287–308.

Reuter, M., Wolter, F.-E., and Peinecke, N. (2006). Discrete Laplace–Beltrami operators for shape analysis and segmentation, *Applied Mathematics and Computation* **4**, 1, pp. 353–361.

Rockafellar, R. T. (1970). *Convex analysis* (Princeton University Press).

Rudin, L. I., Osher, S., and Fatemi, E. (1992). Nonlinear total variation based noise removal algorithms, *Physica D: nonlinear phenomena* **60**, 1-4, pp. 259–268.

Rustamov, R. M. (2007). Laplace-Beltrami eigenfunctions for deformation invariant shape representation, *Eurographics Symposium on Geometry Processing*.

Ryu, E. K. and Yin, W. (2022). *Large-scale convex optimization: Algorithms & analyses via monotone operators* (Cambridge University Press).

Sauer, K. and Bouman, C. (1992). Bayesian estimation of transmission tomograms using segmentation based optimization, *IEEE Transactions on Nuclear Science* **39**, 4, pp. 1144–1152.

Schoen, R. and Yau, S.-T. (1994). *Lectures on Harmonic Maps* (International Press).

Schonsheck, S. C., Bronstein, M. M., and Lai, R. (2021). Nonisometric surface registration via conformal Laplace–Beltrami basis pursuit, *Journal of Scientific Computing* **86**, 3, p. 30.

Sethian, J. A. (1996). A fast marching level set method for monotonically advancing fronts, *Proceedings of the National Academy of Sciences* **93**, 4, pp. 1591–1595.

Sethian, J. A. (1999). *Level Set Methods and Fast Marching Methods* (Cambridge University Press).

Shen, J. and Zhang, X. (2022). Discrete maximum principle of a high order finite difference scheme for a generalized Allen-Cahn equation, *Communications in Mathematical Sciences* **20**, 5, pp. 1409–1436.

Shi, Y., Lai, R., Krishna, S., Sicotte, N., Dinov, I., and Toga, A. W. (2008). Anisotropic Laplace-Beltrami eigenmaps: bridging Reeb graphs and skeletons, in *Computer Vision and Pattern Recognition Workshops*, pp. 1–7.

Shi, Y., Lai, R., and Toga, A. (2012). Unified geometry and topology correction for cortical surface reconstruction with intrinsic Reeb analysis, *Medical Image Computing and Computer-Assisted Intervention (MICCAI)* , pp. 601–608.

Spivak, M. (1965). *Calculus on manifolds* (W. A. Benjamin, Inc., New York-Amsterdam).

Springborn, B., Schröder, P., and Pinkall, U. (2008). Conformal equivalence of triangle meshes, *ACM Transactions on Graphics (TOG) - Proceedings of ACM SIGGRAPH 2008* **27**, 3.

Strikwerda, J. C. (2004). *Finite difference schemes and partial differential equations* (SIAM).

Sun, J., Ovsjanikov, M., and Guibas, L. (2009). A concise and provably informative multi-scale signature based on heat diffusion, in *Computer graphics forum*, Vol. 28 (Wiley Online Library), pp. 1383–1392.

Tasissa, A. and Lai, R. (2018). Exact reconstruction of Euclidean distance geometry problem using low-rank matrix completion, *IEEE Transactions on Information Theory* **65**, 5, pp. 3124–3144.

Uhlenbeck, K. (1976). Generic properties of eigenfunctions, *Amer. J. of Math.* **98**, 4, pp. 1059–1078.

Vallet, B. and Lévy, B. (2008). Spectral geometry processing with manifold harmonics, *Computer Graphics Forum (Proceedings Eurographics)* .

Wardetzky, M., Mathur, S., Kälberer, F., and Grinspun, E. (2007). Discrete Laplace operators: No free lunch, in *Eurographics Symposium on Geometry Processing*, pp. 33–37.

Wen, Z. and Yin, W. (2013). A feasible method for optimization with orthogonality constraints, *Mathematical Programming* **142**, 1-2, pp. 397–434.

Xu, J. and Zikatanov, L. (1999). A monotone finite element scheme for convection-diffusion equations, *Mathematics of Computation* **68**, 228, pp. 1429–1446.

Yu, J., Lai, R., Li, W., and Osher, S. (2023). Computational mean-field games on manifolds, *Journal of Computational Physics* **484**, p. 112070.

Yu, J., Lai, R., Li, W., and Osher, S. (2024). A fast proximal gradient method and convergence analysis for dynamic mean field planning, *Mathematics of Computation* **93**, 346, pp. 603–642.

Zhang, X. (2025). Recent progress on $Q^k$ spectral element method: Accuracy, monotonicity and applications, in S. Chun, J.-H. Jung, E.-J. Park, and J. Shen (eds.), *Spectral and High-Order Methods for Partial Differential Equations ICOSAHOM 2023, Lecture Notes in Computational Science and Engineering*, Vol. 142 (Springer, Cham), pp. 113–137.

Zhao, H. (2005). A fast sweeping method for Eikonal equations, *Mathematics of computation* **74**, 250, pp. 603–628.

# Index